Physical Properties of Tissue

A Comprehensive Reference Book

To remember my mother

Physical Properties of Tissue

A Comprehensive Reference Book

Francis A. Duck
Royal United Hospital and University of Bath,
Bath, England

First published in hardback by Academic Press Limited
1990

All rights having reverted to the author,
this paperback edition is published by the
Institute of Physics and Engineering in Medicine
2012

Fairmount House, 230 Tadcaster Road
YorkYO24 1ES
ISBN 978 1 903613 50 4

Prepared for printing by the Charlesworth Group, Huddersfield, UK
www.charlesworth-group.com

Printed on demand by Lightning Source UK, Milton Keynes, United Kingdom

Preface and Acknowledgements

During the process of compiling this book I have drawn on a wide literature base both within medical science and outside it. I hope that those specialists upon whose work I have drawn will feel that I have summarised their specialism with sufficient care and condensed it without trivialisation. I believe that in some areas I have been able to build usefully upon early surveys of tissue properties and that the revised compilations of data in this book have current value. It is to be hoped that the specialist in ultrasound or dielectric properties of tissue may perhaps find the information on thermal or x-ray properties valuable, and that the expert on skin or blood may learn from parallel knowledge of other body tissues. For those bioengineers, physicists or physiologists dealing with a wider range of problems the book is intended to provide a broad-ranging reference allowing immediate access to a literature base which may not be known in detail. For the clinical diagnosticians, particularly radiologists, I have compiled data on normal body tissue which forms the underlying basis of their investigations. The understanding of some aspects of physiotherapy, hyperthermia and surgery using microwaves, diathermy, ultrasound or lasers depends in part on a thorough knowledge of tissue properties. Lastly, those concerned with environmental hazards, whether these be from ultraviolet radiation, gamma radiation or microwaves, require knowledge of the relevant physical property of tissue in order that some sensible estimate of hazard may be made. It is my hope that parts of this compilation of data may prove to be a valuable resource for some or all of these groups of scientists, either as a simple data bank or as a guide into a literature.

Whilst every effort has been made in transcribing the data from the original sources, constructing the tables and cross-checking the values to validate their accuracy, it is perhaps inevitable that minor errors may still be included. The responsibility for such errors is, of course, entirely mine. The serious user of the data included in this book is strongly recommended to consult the original reference before making a final decision about its use. All measurements are fully referenced and the sources should be accessible through any academic library service.

I extend my thanks to many friends and colleagues who have given valuable guidance and help during the preparation for this book, and without whose advice the task would have been very much more laborious. The original suggestion for a book bearing this title was suggested by James Drake who was most helpful in its early stages of preparation. Sally Clift, Brian Diffey, Lindsay Grant, Dick Lerski, Francis Ring, Hazel Starritt, Ian Swain and Peter Wells have all, at times, made helpful comments and their input has been greatly appreciated. Martin Schonhoff helped with translation from some German texts. Lastly, I wish to express my particular thanks to Stephen Lillicrap for his continual support, and his reassurance that the effort involved was worthwhile.

Francis Duck

Preface to the 2012 Paperback Edition

When this book was first published by Academic Press in 1990, it was well received and much in demand, but when it went out of print, the original publishers chose not to reprint and copyright reverted to me.

Over the years, particularly with the development of e-commerce, a healthy market has developed in pre-owned copies of books. From the many requests that I have received over the years, I know that very few such copies of the present book have appeared.

As technology has changed, my professional body, the Institute of Physics and Engineering in Medicine, has taken advantage of new publishing and printing technologies with its long-established, peer-reviewed books series. IPEM's book publishing division has had, as its primary business, new or revised and updated works of particular interest to the various specialties of medical physics and biomedical engineering, commissioned by IPEM's Science Board and written mainly by IPEM members, but occasionally it has published works jointly with or on behalf of other organisations in related fields.

I am pleased that, on this occasion, IPEM has agreed to re-publish *Physical Properties of Tissue* as a paperback edition, unchanged except for this additional preface and minor errata. It will be printed digitally, on demand. European and international sales will be on-line from amazon.co.uk or amazon.com but IPEM members can continue to enjoy member discounts by ordering from IPEM's office in York and, from late summer 2012, from IPEM's on-line book store.

It is now over twenty years since the tables of data presented in this book were compiled. Since that time, some of the gaps have been filled by scientists working in laboratories throughout the world. Some new data have confirmed previous measurements, serving primarily to improve knowledge of normal ranges of tissue properties, Other measurements have explored, for example, a particular property over a wider frequency range or over an extended range of tissue age. Such new data is absent from this reprint. Nevertheless, data from well-performed experiments, whenever they were performed, remain for ever valid, and can continue to provide a basic resource as new observations go on being added.

It is my sincere hope that this republished work will be as well received and be of as much value to a new generation of medical science professionals, be they bioengineers, physicists or physiologists, or clinicians, especially radiologists, and those working in environmental protection, especially in the areas of ionising and non-ionising radiations, as it has been to a previous generation.

Francis Duck

Errata

Page 107 Table 4.15 The reference for "Vertebra, human" is McKelvie & Palmer 1987.
Page 111 Table 4.18. Column 4 heading should read dB cm^{-1} K^{-1}.
Pages 118,119 Tables 4.21 and 4.22. Titles should read Scattering Coefficient per Unit Volume.

Contents

Chapter 1

Introduction

In 1929 a book was published under the title *The Physics of X-Ray Therapy*. Its author, then only 25 years old, remarked when discussing the problems facing the radiotherapist planning a radiation treatment that, 'for sufficiently obvious reasons, it is not easy to obtain data for the complex mass of tissues constituting the body of the patient'. W.V.Mayneord was at the time physicist to the radiotherapeutic department of the Cancer Hospital (Free) in London, and within a little over ten years would be appointed to a personal chair in Physics Applied to Medicine in the University of London. Since that time the requirement for data on the 'complex mass of tissue' has extended far beyond that which Professor Mayneord was discussing for ionising radiation dosimetry. Other therapies in use at that time, using ultraviolet radiation, heat, light and electricity have since evolved. New diagnostic tools using ultrasound and nuclear magnetic resonance have emerged. Laser therapy is now in widespread use. In all these areas of medicine the need for knowledge of tissue properties, particularly physical properties, remains essential so that each therapeutic and diagnostic tool may be used with maximum understanding and to greatest advantage.

This book, then, describes quantitatively the physical properties of mammalian tissues. The classical areas of physics, namely heat, light, sound, properties of matter, electricity and magnetism, are considered in turn. In each case, the properties which are used to describe the tissue are reviewed, with emphasis being placed on a thorough documentation of reported measurements of these properties. Ultrasonic rather than sonic properties are reported, and whilst information on the magnetic properties of tissue is minimal, a substantial review of some nuclear magnetic resonance properties is included. 'Light' is broadened to include both ultraviolet and infrared radiation. In addition the properties of mammalian tissue in relation to ionising radiation, x-radiation and gamma radiation, and charged and uncharged particles are also reviewed. Lastly, some relevant data concerning the composition of human tissue are included.

Throughout the book it not the microscopic or biochemical components of tissue which are described. Rather it is, first and foremost, the large-scale properties an organ, or the tissue comprising part of an organ, which are considered. Whilst there is considerable value in the study and description of, for instance, the dielectric or ultrasonic properties of biochemical solutions or cell suspensions, it is not primarily within the scope of this book to evaluate and explain the physical properties of tissue through the investigation of its component parts. One exception to this rule is for blood, where values are reported for the whole range of haematocrit from pure plasma to packed red cells.

One constituent of tissue which has been singled out for special attention is water. About three-quarters of the total mass of soft tissue is water, and some of the physical properties of tissue are dominated by its presence. Ultrasound propagates through soft tissue with a velocity close to that of water. X-ray photons are attenuated similarly in water and tissue, and the thermal capacity and conductivity of soft tissue and water are similar. On the other hand there are some obvious situations where soft tissue properties differ substantially from those of water. Light is strongly scattered by tissue; ultrasound is absorbed strongly. The magnetic relaxation of hydrogen protons is substantially faster in soft tissue than in pure water. However, whether tissue properties are close to those of water or very different from them, water is a valuable reference point against which to compare the physical property of any tissue. In addition the properties of pure water have been widely studied and reported. Throughout this book, relevant values for water of the physical quantity under discussion have been included to allow comparison to be made easily.

Each chapter is organised in a similar way. Firstly, basic definitions are given of the quantities to be used. No values are tabulated without an associated definition. Standard units are given, together with acceptable alternatives. Simple statements of the theoretical basis of the properties under discussion are also given. Sufficient discussion is included to allow a correct interpretation of the data included in the tables, but without being tutorial in nature. Fuller texts are referenced for those whose knowledge is limited. Whilst for many physical quantities simple expressions can be used to describe the behaviour of homogeneous, linear, isotropic media, these expressions cannot usually be applied to mammalian tissue without reservation because it is an inhomogenous mixture, both physically and chemically, which may well be, in addition, non-linear and anisotropic. Considerable efforts have been made over the years to develop sound theoretical bases for the behaviour of all of the physical properties of tissue. Much progress has been made, particularly in the area of ionising radiation. For this reason only the data presented in Chapter 7, dealing with ionising radiation, is dominated by calculated rather than experimental values, although the latter are included for comparison. For other properties there is much less agreement about the theoretical basis for the prediction of tissue properties. Therefore it is largely the results of experimental investigations rather than predictions based on theory which constitute the bulk of the data included in the tables. For each quantity the limitations associated with a simple theory are identified and reference is made to relevant sources which discuss more thorough analyses.

There are few simple mathematical models which can be used to describe the physical behaviour of tissue. The simplest model assumes a linear relationship between two quantities and a property of the material is defined as the ratio between the two. Obvious examples are electrical resistivity and Young's modulus of elasticity. For many properties it has been usual to assume a linear relationship until proved otherwise, and this assumption has from time to time caused problems. For instance, Young's modulus is relevant only if Hooke's Law applies: equal stress causes equal strain. In fact the extension or contraction of tissues, and in particular of soft tissue, is highly non-linear, and linear assumptions can be applied only approximately and over a narrow band of stress. Another area where linear assumptions have often been applied incorrectly is in the analysis of the temperature dependence of a physical quantity. The temperature coefficient is rarely constant over a wide temperature range and assumptions of linearity can be justified only over narrow bands of temperature.

Other properties may best be quantified on the basis of exponential laws. Attenuation of wave transmission can be quantified simply in this way as can the magnetic relaxation of nuclei. A single value of attenuation coefficient, or of relaxation time, fully describes the property of the material, provided that the assumption of an single exponential process is in fact true. In reality, relaxation, for instance, may best be described by a multi-exponential function. Again, the attenuation loss may not be exponential, as is the case with broad spectrum x-rays, and care is needed in the use of simple attenuation coefficients. It is worth adding that as scattering is usually anisotropic and may well involve multiple scattering processes, its analysis and measurement remains a substantial problem for all radiations.

Several other empirical and semi-empirical mathematical models have been used to quantify the observed physical behaviour of tissue. Relaxation processes, which may be observed in the alteration with frequency of quantities such as ultrasonic attenuation, dielectric constant or nuclear magnetic relaxation time, have been analysed in this way. Values from such analyses have been included in the tables of data. A commonly used, simple expression is that of a power law of the form $a=bf^c$, $f_1<f<f_2$, which may be used to describe the variation of quantity, a, with frequency, f, in terms of coefficients, b and c, in the frequency range from f_1 to f_2. This model has been used to characterise the variation with frequency of, for instance, ultrasonic attenuation and NMR relaxation over a narrow frequency band. However, a complete analysis over a wide range of frequencies requires a more complex model and expressions of the Cole-Cole type have been used, for instance, in the analysis of dielectric properties over several orders of magnitude of frequency.

Following the description of theory and definitions of quantities, a brief outline giving a historical perspective of the development of the present state of knowledge is given, identifying valuable reviews and surveys which have already been published. Provided the experimental methods have not subsequently been shown to be unsound, measurements on tissues made by early workers are usually valuable reference points in a developing science, and data from these studies have sometimes been retained in the tables. On the other hand, more recent experimental methods have in some cases rendered earlier measurements redundant, and they are referenced only for their historical interest.

The bulk of each chapter consists of tables of values drawn from the literature together with some explanation and discussion. There has been some inevitable selection from the mass of available data for inclusion in the tables, although the surveys have been kept as wide as possible. When selection has been necessary, preference has generally been given to values obtained from human tissue at body temperature *in vivo*, provided that the measurement methods used were not substantially prejudiced by the problems of *in-vivo* measurement. In addition, other general preference hierarchies have been applied. Values for human tissue have been selected in preference to those from other animals, although some animal values have often been retained for comparison. Measurements at body temperature have always been preferred to those at room temperature, and values of temperature coefficients have always been included where available. More recent data have usually superseded older data. No independent value judgements have been made as to the quality of any measurement obtained from the literature: it has been assumed that, at the time of publication, the reported values were the best available. Only in the light of subsequent discussion and criticism in later publications has earlier material been set aside.

There are two other main categories of data which may be absent. The first consists of those measurements which have been published in the open literature, but which have not come to my attention. The English language literature has been searched quite thoroughly in order to compile the tables included here, but inevitably some material will have been missed. To the authors of these papers I give my sincere apologies. Of equal importance is the scientific material in journals and books published in languages other than English. Although material from these sources is cited occasionally, mostly *via* other references, the majority of the literature outside English language journals has not been reviewed.

The second main category of omission, for which I make no apology, consists of data contained within useful studies which have remained unpublished in individual laboratories, or at best have been included only in university theses. Such theses are cited intermittently in this book when referenced by others, but only when the material appears to be unique. Careful experimental measurements of the properties of tissue are not sufficiently plentiful to warrant leaving them only in local libraries.

The main quantitative information is presented in the tables. In addition, graphs have been included where it seemed valuable to present a qualitative and generalised indication of the variation of a particular quantity. This has usually been done to show, for instance, the variation of a tissue property with temperature or with frequency, or its overall variability. These graphs are not intended as primary reference material and should not be relied upon to provide quantitative data. They are included only to give the reader a simple overview of a particular, generally non-linear dependence, and the tables of data should always be used for detailed reference.

The tabulated tissue values should be used with some care, and attention should be given to the discussion associated with these tables. The values presented do not have the status of physical constants; indeed they are highly variable in some cases. This variability has a variety of causes. Firstly, the tissue within any particular organ may itself be variable in the physical property under discussion. The label 'liver', 'spleen' or 'bone' does not describe a simple standard material but refers generally to a range of materials even within one particular organ. In some cases, tissues within an organ may be more clearly specified and a narrower variability observed. For instance, cortical bone differs from trabecular bone; the properties of kidney vary between the cortex and the medulla and those of the brain between grey and white matter. Even so, natural biological variability results in considerable ranges for all the physical properties listed. When a measurement has been made on a single tissue sample, or no range of values has been reported, only the single value is included in the table. If two measurements were reported, both values are included. For three or more measurements the overall range is listed. If sufficient measurements have been made, a standard deviation may be given. Great care should be taken when using values from the tables to recognise that this natural variability is real and to include note of the variation when using the data.

It is usual for experimentalists to report estimates of errors as an assessment of the precision of the experiment. Judgements about absolute accuracy are often not made when reporting the results of a particular set of measurements. This knowledge comes more often from the comparison of measurements of the same quantity using different techniques. Some quantities, such as density or ultrasonic velocity, should be capable of measurement with both high precision and a high degree of accuracy. Other measurements may have poor precision or be inaccurate for a variety of unrecognised reasons, or have causes whose effects cannot easily be quantified. NMR relaxation time measurement *in vivo* may well fall into this category at present. True variability of tissue can be assessed properly only if the measurements are in the first category, in which the overall percentage error in measurement is much less than the percentage variation in tissue. For quantities in the second category, incorrect conclusions about tissue variability can arise. It is recommended that those using particular values from the tables for serious research refer to the original references for a full discussion of measurement errors.

In addition to the natural variability within any organ, other factors can result in an overall spread of the data. Particular factors are discussed alongside each of the physical properties listed, and their importance judged and quantified as far as possible. Ageing of the tissue causes alterations in some properties, and where values are available the question of ageing is addressed. Fetal or immature tissue may also differ from adult tissue, in particular in the case of the fetus because of its relatively high water content. Of importance too is the relevance to human tissue of measurements carried out on animal tissues. Generally many more measurements have been made on animal tissues and where possible comparisons are made. Sex differences may or may not occur, depending usually on the importance of differences in tissue composition between the sexes.

For many measurements simple empirical equations have been reported which have been suggested for use as a quantitative summary of the observed variation of a particular tissue property with some other factor. For instance, both ultrasonic velocity and thermal conductivity have been related quantitatively to the partial constituent parts of tissue, water, lipid and protein. Similarly, empirical relationships relating the properties of blood to its percentage red blood cell concentration have also been reported. When such empirical equations have been reported they have been included in the text and reference may be made to the tabulations giving percentage constituents of tissues, to enable the prediction of properties for selected tissues (Chapter 9). As noted earlier, water dominates many tissue properties.

In addition the presence of fat also has a strong influence, significantly lowering the density, thermal conductivity, ultrasonic velocity and linear x-ray attenuation, and modifying nuclear magnetic relaxation. The amount of protein, especially collagen, also affects the tissue and this is seen especially in its mechanical properties.

Most of the measurements included are for normal tissues. For some physical properties, particularly those associated with clinical diagnosis, a large quantity of experimental data has also been reported in the literature relating pathology to these quantities. Examples are linear X-ray attenuation coefficient, ultrasonic attenuation and NMR relaxation time. This material has not been included in detail. However, some values have been tabulated for pathological tissues which are intended primarily for illustrative rather than reference purposes.

A final factor which may cause the physical properties of tissue to alter is death. It is clearly much easier to make careful experimental measurements on a prepared excised tissue sample *in vitro* in the laboratory than to make the same measurement on a living animal or human subject. However, it may be that these measurements bear little relation to the living state if important alterations in the physical property to be measured occur between the times of death and measurement. Furthermore, fixation of the tissue may preserve it biologically but substantially alter it physically. The question of the value and relevance of *in-vitro* measurements to the *in-vivo* state is discussed in each chapter.

In addition to biological factors, changes in some physical factors result in alterations in physical properties. The frequency and temperature dependence of many properties has already been mentioned. The geometric direction of measurement may also be important for some properties if the tissue is anisotropic. Many tissues, such as cortical bone, tendon and muscle have an organised structure which results in anisotropy of some physical properties. Examples are the strength of bone, the resistivity of muscle and the NMR relaxation of tendon. If such anisotropy exists but is not recognised, the measured variability of a property will increase. Lastly, the range of linearity for any property needs to be known and the implications for measurement outside this range recognised. In some cases, in the propagation of ultrasound through tissue, for instance, the non-linear property of the tissue may be quantified using a suitable parameter.

For many physical quantities, materials have been investigated whose properties make them suitable as tissue substitutes. Terms such as 'phantom materials' or 'tissue mimics' have also been used to describe these media. For completeness some information is included in each chapter about such tissue substitute materials, their constituents and some relevant physical properties. Properties of agar or agarose gels have also been added since these gels form the basis of many tissue substitute materials.

Some conventions used in the tables

The organs and tissues have been listed alphabetically. In some cases fluid, soft tissues and hard tissues have been listed in separate tables. The terms 'fat', 'adipose tissue', and 'fatty tissue' have been loosely used in the literature and unless specifically indicated otherwise (for instance renal fat, orbital fat) all values in these categories are included under the label 'fatty tissue'. Mammary gland is listed under breast. Heart muscle is listed under muscle:cardiac. Blood components, plasma, serum and packed erythrocytes are listed under blood. Usually any particular entry for tissue and property starts with values for human tissue. Data for animal tissue follow and are listed in no particular order.

Entries in the form 'A±B' indicate an average value A and a standard deviation B. Entries in the form 'C–D' or 'C to D' indicate an overall range. A hyphen (–) has been used throughout as a ditto mark to indicate a repeated word or phrase from the line above; commas are used as separators. Thus '–,–,longitudinal' below 'bone,cortical,transverse' means 'bone,cortical,longitudinal'. A variety of symbols used in the tables relate to the footnotes.

Chapter 2

Thermal Properties of Tissue

Heat transport in materials may occur by conductive, convective or radiative processes. In this chapter, thermal conduction through tissue is described, together with its heat capacity. Convective heat transport is discussed only when relating thermal conductivity to tissue perfusion. The properties of tissues related to radiant heat are dealt with in Chapter 3.

2.1 Thermal conductivity, thermal diffusivity and thermal inertia

2.1.1 *Terminology and definitions*

The **thermal conductivity**, k, of a substance is defined as the quantity of heat, Q, transmitted due to a unit temperature gradient, in a direction normal to a surface of unit area in unit time under steady-state conditions and where heat transfer is dependent only on the temperature gradient.

$$\frac{Q}{A} = - k\frac{\partial T}{\partial x} \qquad (2.1)$$

where $\partial T/\partial x$ is the temperature gradient in the direction of heat flow, and A the cross-sectional area. The negative sign indicates that the heat flux is in the direction of decreasing temperature. The SI unit for thermal conductivity is watt per metre kelvin ($W\ m^{-1}\ K^{-1}$). Other units, and conversion factors are given in Table 2.1.

Where steady-state conditions are not relevant, the quantity **thermal diffusivity**, $\alpha = k/\rho C$, is used, where C is the specific heat capacity and ρ is the density. α is related to the spatial and temporal variation of temperature, T, in the medium by

Table 2.1 Conversion factors for thermal units

	To convert from	to	multiply by
k	$W\ m^{-1}\ K^{-1}$	$W\ cm^{-1}\ K^{-1}$	10^{-2}
		$cal\ s^{-1}\ cm^{-1}\ {}^{o}C^{-1}$	$0.2388x10^{-2}$
		$BTU\ hr^{-1}\ ft^{-1}\ {}^{o}F^{-1}$	0.5778
C	$J\ g^{-1}\ K^{-1}$	$J\ kg^{-1}\ K^{-1}$	10^{3}
		$cal\ g^{-1}\ {}^{o}C^{-1}$	0.2388
		$BTU\ lb^{-1}\ F^{-1}$	0.2388
α	$cm^{2}\ s^{-1}$	$m^{2}\ s^{-1}$	10^{-4}
		$ft^{2}\ hr^{-1}$	3.875
$k\rho C$	$W^{2}\ s\ cm^{-4}\ K^{-2}$	$W^{2}\ s\ m^{-4}\ K^{-2}$	10^{8}
		$cal^{2}\ s^{-1}\ cm^{-4}\ {}^{o}C^{-2}$	$5.712x10^{-2}$
		$BTU^{2}\ hr^{-1}\ ft^{-4}\ {}^{o}F^{-2}$	861.6

$$\frac{\partial T}{\partial t} = \alpha \nabla^2 T \qquad (2.2)$$

The numeric value of α determines the relative time rate of temperature change, and is thus a measure of the ability of a thermally perturbed system to relax back to steady–state conditions. The SI unit for α is metre^2 per second ($m^2\ s^{-1}$). Other units, and conversion factors are given in Table 2.1.

Living tissue is perfused, and the passage of blood modifies the heat transfer process. Furthermore, metabolic processes generate heat within the tissue. A formulation taking these factors into account has been given by Pennes (1948) and Perl (1962) as follows:

$$\rho C\frac{\partial T}{\partial t} = \nabla(k\nabla T) - m_b C_b (T - T_b) + Q_m \qquad (2.3)$$

where m_b, C_b and T_b are the mass flow rate, specific heat and temperature of the perfusing blood respectively, and Q_m is the rate of metabolic heat production. Equation 2.3 is the so–called bio–heat equation. The measurement of thermal conductivity *in-vivo* using thermal clearance methods has been widely used as a technique to estimate tissue blood flow both of the skin and of deeper tissue. A practical quantity, the **effective thermal conductivity**, k_{eff}, is used under these circumstances to give an equation with a parameter which is flow dependent.

$$\rho C \frac{\partial T}{\partial t} = \nabla . (k_{eff}\ \nabla T) \qquad (2.4)$$

k_{eff} is found to vary proportionally with the square root of the perfusion rate (Jain *et al* 1979).

An equivalent **effective thermal diffusivity**, $\alpha_{eff} = k_{eff}/\rho C$, can also be defined.

Another thermal quantity, that of thermal inertia, was introduced by Buettner (1951) in considering the thermal response of skin during exposure to radiation. The temperature rise, ΔT, at the surface of an opaque semi-infinite solid following irradiation during time t is:

$$\Delta T = \frac{2AH\sqrt{t}}{\sqrt{(\pi . k\rho C)}} \tag{2.5}$$

where H is the irradiating intensity and A the surface absorptance. The product $k\rho C$ is termed the **thermal inertia** and has SI units of watt^2 second per metre^4 kelvin^2 (W^2 s m^{-4} K^{-2}). Other units and conversion factors are given in Table 2.1.

2.1.2 *Measurement of thermal conductivity*

An extensive review of methods which may be used to measure the thermal properties of materials is given by Touloukian (1964), and a useful summary of methods relevant for biological materials given in a general survey by Bowman *et al* (1975). The techniques used may be categorised as invasive or non-invasive, and in each case may enable steady-state or non-steady-state measurements to be made. Historically, many measurements of k were made using a guarded hot plate (e.g. Poppendiek *et al*, 1966). The experimental geometry is intended to ensure that heat flow between two conductive plates is entirely in the axial direction of the specimen. Equation 2.1 is applied directly to calculate k. The method has been applied exclusively to *in-vitro* samples of tissue. More recently, invasive probes have been developed which involve the implantation within the specimen of heat sources (or sinks) which may also serve as temperature sensors. Invasive thermal diffusion methods use a heated thermocouple (e.g. Grayson *et al*, 1971) or a thermistor probe (Chen *et al*, 1981; Balasubramaniam and Bowman, 1977; Valvano *et al*, 1984). The thermistor probe, typically 0.5 mm diameter, is inserted into the tissue through a hypodermic needle, which is then removed to minimise the heat conduction along the metal. The thermistor is first used passively to measure the ambient temperature, and then heated. The electrical power required to maintain the thermistor at a steady temperature is related to the effective thermal conductivity of the surrounding tissue (Valvano *et al*, 1984). Alternatively, the time change of temperature following a transient pulse of energy to the probe is analysed (Chen *et al*, 1981). In either case the method depends upon adequate modelling of the thermal characteristics of the probe and surrounding tissue. In spite of the possible criticism that some tissue damage may occur during probe insertion and cause some modification in thermal behaviour, reliable values have been obtained using these probes in both perfused and non-perfused tissues.

A set of 'semi-invasive' techniques has been investigated in which temperatures have been measured using cutaneous and subcutaneous thermocouples with surface heat fluxes provided by various non–invasive sources. Henriques (1947) and Hensel and Bender (1956) describe the use of contact heating. Other techniqes have used electromagnetic irradiation: for instance infrared (Dersken *et al*, 1957), microwave (Cook, 1952) or optical radiation (Kraning, 1973).

Completely non–invasive methods depend upon the fact that when two homogeneous solids initially at different temperatures are brought into contact the interface temperature takes an intermediate value dependent upon the thermal inertias of the two contacting bodies. Tanasawa and Katsuda (1972) have used a technique based upon that described by Vendrik and Vos (1957). Totally non–contact methods use external radiation to heat tissue, and observe the subsequent time–course of skin temperature with a radiometer. Results of measurements of $k\rho C$ using this method of measurement have been reported by Lipkin and Hardy (1954), Dersken *et al* (1957) and Kraning (1973) and have been used to give estimates of k using the assumption that ρC for tissue has the same value as that for water.

2.1.3 *Historical background*

Earliest measurements of the conduction of heat through skin were reported by Klug in 1874. Bordier (1898) showed that beef muscle conducted heat almost twice as well as fat. A value of 5×10^{-4} cal cm^{-1} s^{-1} $°C^{-1}$ was given by Lefevre (1901) for the average conductivity of peripheral tissues in man, based on the assumption that the thermal gradient extended 20 mm inwards from the skin surface. Subsequent work has been tabulated by Chato (1969) and with additional material by Bowman *et al* (1975). Values for skin have been surveyed by Cohen (1977) and for bone by Lundskog (1972). The difficulties of making accurate measurements in bone have been discussed by Nelson *et al* (1986). Data on foodstuffs, and in particular measurements made on thermal properties of various meats, are extensively reported in various texts (Mohsenin, 1980; Rha, 1975; Morley, 1972). A useful tabulation of the thermal properties of meat containing values for both k and α has been assembled by Morley (1972), and a substantial bibliography for thermal properties of food compiled by Mellor (1980).

2.1.4 *Values of thermal conductivity for tissue*

Measured values for the thermal conductivity of normal, unperfused tissues are given in Table 2.2. Thermal diffusivity values are given in Table 2.3, and for thermal inertia in Table 2.4. Where measurements have been reported for human tissues from any particular organ these values have been entered first. The majority of the measurements are from non–human tissue. A small amount of selection from the available literature has been necessary, primarily excluding older or duplicate results. A number of factors of importance when using these data are discussed afterwards.

Table 2.2 Thermal conductivity, k, of normal unperfused tissues

Tissue	Temp °C	Conductivity $W\ m^{-1}\ K^{-1}$	Reference
Adrenal gland,human	37	0.363–0.485	Bowman 1981
Artery,human,aorta	35	0.476±0.041	Valvano & Chitsabesan 1987
-	90	0.612±0.012	
–,–,–	25–80	0.444	Van Gemert *et al* 1986
–,dog,femoral	25–80	0.404	-
–,–,carotid	25–80	0.449	-
Blood:whole,human, 43% Hct	24–38	0.530	Poppendiek *et al* 1966
-	36.6–39.6	0.507–0.513	Spells 1960
–,44% Hct	37	0.484–0.491	Bowman 1981
–,haemolysed	21	0.492±0.009	Balasubramaniam & Bowman 1977
–,dog,44% Hct	37	0.54	Singh & Blackshear 1967
–,rat	35.9–37.2	0.519–0.532 (av 0.528)	-
also	see Equations 2.13 and 2.14		
Blood:packed cells human	~37	0.482	Spells 1960
Blood:plasma,human	24–38	0.572	Poppendiek *et al* 1966
–,–	37	0.582±0.03	Bowman 1981
–,dog	37	0.60±0.02	Singh & Blackshear 1967
Bone,human,rib	37	0.373–0.496	Bowman 1981
–,–,sternum	*in-vivo*	0.36 (av)	Graf & Stein 1957
–,–,–,age < 60 yr	*in-vivo*	0.29–0.47	-
–,–,–,age > 60 yr	*in-vivo*	0.23–0.37	-
Bone:trabecular,cow	25	0.27–0.33	Clattenburg *et al* 1975
Bone:marrow,cow	24–38	0.22	Poppendiek *et al* 1966
Brain,human, 78% water	5–20	0.528	Cooper & Trezek 1972
–,–,white,71% water	5–20	0.503	-
–,–,grey,83% water	5–20	0.565	-
–,–	37	0.503–0.576	Bowman 1981
–,–,cortex	37	0.515	Valvano *et al* 1985
Brain,cow	24–38	0.497	Poppendiek *et al* 1966
Breast,human	37	0.499±0.004	Bowman 1981
–,animal,lactating		0.45±0.05	Linzell 1953
Colon,human	37	0.556±0.009	Bowman 1981
Eye:lens,rabbit	*in-vivo*	0.40±0.1	Lagendijk 1982
Eye:vitreous,rabbit	*in-vivo*	0.60	-
–,human	24–38	0.594	Poppendiek *et al* 1966
Eye:aqueous,human	24–38	0.578	-

cont.

Table 2.2 cont. Thermal conductivity

Tissue	T,°C	k, W m^{-1} K^{-1}	Reference
Fat,human, subcutaneous	*in-vivo*	0.23–0.27	Hensel & Doerr 1959
–	37	0.200–0.246	Bowman 1981
–,subcutaneous, 4.8 mm deep	37	0.268	–
6.4 mm deep	37	0.248	–
9.8 mm deep	37	0.219	–
–,–,splenic	37	0.334	Valvano *et al* 1985
–,cow,2.0% water	20	0.093	Lapshin 1954
–,–,15.2% water	20	0.180	–
–,–,29.5% water	20	0.345	–
–,–,7% water	0	0.204	Cherneeva 1956
–,–,–	30	0.237	–
–,–,perirenal	37	0.188	Morley 1966
–,pig,subcutaneous	30–48	0.15–0.17	Henriques & Moritz 1947
–,rabbit,perirenal	*in-vivo*	0.17	Chen *et al* 1981
Gastric juice,human	24–38	0.445	Poppendiek *et al* 1966
Kidney,human	37	0.513–0.564	Bowman 1981
–,–,cortex	37	0.547	Valvano *et al* 1985
–,–,medulla	37	0.540	–
–,–,pelvis	37	0.551	–
–,dog,cortex	37	0.538	–
–,–,medulla	37	0.555	–
–,–,pelvis	37	0.532	–
–,cow	24–38	0.525	Poppendiek *et al* 1966
–,pig,cortex	37	0.540	Valvano *et al* 1985
–,sheep	37.8	0.488	Bowman *et al* 1974
Liver,human	37	0.467–0.527	Bowman 1981
–,–	37	0.512	Valvano *et al* 1985
–,dog	*in-vivo*	0.510	Chen *et al* 1981
–,–	*in-vivo*	0.56,0.59	Cooper & Trezek 1972
–,–	21	0.550±0.010	Balasubramaniam & Bowman 1977
–,cow	24–38	0.488	Poppendiek *et al* 1966
–,pig	37	0.528	Valvano *et al* 1985
–,rabbit	37	0.563	–
–,rat,rabbit	*in-vivo*	0.494	Grayson 1952
Lung,human	37	0.302–0.550	Bowman 1981
–,–	37	0.451	Valvano *et al* 1985
–,cow	24–38	0.282	Poppendiek *et al* 1966
–,pig	37	0.316	Valvano *et al* 1985
Milk,cow	24–38	0.531	Poppendiek *et al* 1966

cont.

Table 2.2 cont. Thermal conductivity

Tissue	T,°C	k, W m^{-1} K^{-1}	Reference
Muscle:cardiac,			
–,human	37	0.492–0.562	Bowman 1981
–,–	37	0.537	Valvano *et al* 1985
–,dog	*in-vivo*	0.49	Chen *et al* 1981
–,–	37	0.536	Valvano *et al* 1985
–,pig	37	0.533	–
–,rat	37	0.52,0.53	Valvano *et al* 1984
Muscle:skeletal, human	37	0.449–0.546	Bowman 1981
–,–	*in-situ*	0.48–0.53	Hensel & Doerr 1959
–,cow	24–38	0.528	Poppendiek *et al* 1966
–,calf	24–38	0.546	–
–,cow,along fibres	32	0.434	Hill *et al* 1967
–,–,across fibres	36	0.467	–
–,pig	30–48	0.43–0.51	Henriques & Moritz 1947
–,–,along fibres	42.9	0.485	Hill *et al* 1967
–,–,across fibres	44.8	0.530	–
–,rat	37	0.51–0.53	Valvano *et al* 1984
Pancreas,human	37	0.294–0.588	Bowman 1981
–,–	37	0.542	Valvano *et al* 1985
–,dog	37	0.510	–
–,pig	37	0.477	–
Skin,human	*in-vivo*	0.293±0.016	Hensel & Bender 1956
–,–	*in-vivo*	0.385–0.393	Van der Staak *et al* 1968
–,–	37	0.266±0.007	Bowman 1981
–,–,1.6 mm deep	37	0.498±0.001	–
Skin:dermis,pig	30–48	0.36–0.38	Henriques & Moritz 1947
Skin:epidermis,pig	30–48	0.16–0.25	–
Spleen,human	37	0.539	Valvano *et al* 1985
–,pig	37	0.533	–
Stomach,human	37	0.489–0.565	Bowman 1981
Tooth:dentine,root	29	0.45	Soyenkoff & Okun 1958
–,–,crown	29	0.40	–
–	50	0.56(0.47–0.69)	Craig & Peyton 1961
Tooth:enamel	26–29	0.65	Soyenkoff & Okun 1958
–	50	0.93(0.88–1.07)	Craig & Peyton 1961
Thyroid,human	37	0.526–0.533	–
Urine,human	24–38	0.561	Poppendiek *et al* 1966

Table 2.3 Thermal diffusivity, α, of normal, unperfused tissues

Tissue	Temp °C	Diffusivity $cm^2\ s^{-1}x10^3$	Reference
Artery,human,aorta	35	1.27±0.07	Valvano & Chitsabesan 1987
–,–,–	90	1.56±0.05	
–,–,–	25–80	1.22	Van Gemert *et al* 1986
–,dog,femoral	25–80	1.08	–
–,–,carotid	25–80	1.17	–
Blood:haemolised human	21	1.19±0.05	Balasubramaniam & Bowman 1977
Blood:plasma,human	21	1.21±0.05	
Brain,human	5–20	1.38±0.11	Cooper & Trezek 1972
–,–,white matter	5–20	1.34±0.10	–
–,–,grey matter	5–20	1.43±0.09	–
–,–,cerebral cortex	37	1.47	Valvano *et al* 1985
Kidney,human	5–20	1.32±0.12	Cooper & Trezek 1972
–,–,cortex	37	1.47	Valvano *et al* 1985
–,–,medulla	37	1.48	–
–,–,pelvis	37	1.37	–
–,dog	21	1.55–1.83	Bowman *et al* 1974
–,sheep	37.8	1.68	–
–,pig,cortex	37	1.43	Valvano *et al* 1985
–,rabbit	37	1.41	–
Liver,human	5–20	1.50±0.10	Cooper & Trezek 1972
–,–	37	1.41	Valvano *et al* 1985
–,dog	21	1.63±0.2	Balasubramaniam & Bowman 1977
–,–	*in-vivo*	1.50,1.57	Cooper & Trezek 1972
–,sheep	21	1.63±0.2	Balasubramaniam & Bowman 1977
–,pig	37	1.44	Valvano *et al* 1985
–,rabbit	37	2.03	–
–,rat	37	1.30–1.41	Valvano *et al* 1984
Lung,human	37	1.31	Valvano *et al* 1985
–,pig	37	0.99	–
–,rabbit	37	1.37	Valvano *et al* 1985
Muscle:cardiac,human	5–20	1.48±0.09	Cooper & Trezek 1972
–,–	37	1.47	Valvano *et al* 1985
–,dog	37	1.51	–
–,–	21	1.47–1.55	Bowman *et al* 1974
–,rat	37	1.52–1.65	Valvano *et al* 1984
Muscle:skeletal,dog		1.83	Bowman *et al* 1974
–,cow	30	1.25	Cherneeva 1956
–,pig	30	1.25	–
–,sheep	21	1.59±0.08	Balasubramaniam & Bowman 1977

cont.

Table 2.3 cont. Thermal diffusivity

Tissue	T,°C	α,cm² s⁻¹x10³	Reference
Muscle,rat	37	1.40–1.53	Valvano *et al* 1984
Spleen,human	37	1.44	Valvano *et al* 1985
–,dog	5–20	1.38±0.15	Cooper & Trezek 1972
–,pig	37	1.41	Valvano *et al* 1985
–,sheep	19.8	1.55	Bowman *et al* 1974
Fat,human,splenic	37	1.31	Valvano *et al* 1985
–,cow,7% water	30	0.78	Cherneeva 1956
–,–,7% water	0	0.53	–
–,–,2% water	20	0.553	Lapshin 1954
–,–,15% water	20	0.633	–
–,–,29.5% water	20	0.706	–
Frozen Tissue			
Muscle,cow,74.5% water			
high fat	–20	4.31	Cherneeva 1956
low fat	–20	4.72	–
–,pig,76.8% water	–20	3.89	–
Fat,cow,7% water	–20	1.36	–

Table 2.4 Thermal inertia, kρC, of tissues

Tissue	kρC W² s cm⁻⁴ K⁻² x10³	Reference
Bone,human	8.75	Dersken *et al* 1957
Fat,human	4.55±0.53	Dersken *et al* 1957
–,pig	3.35	Tanasawa & Katsuda 1972
Kidney,sheep,*in–vivo*	14.18	Bowman *et al* 1975
Liver,pig	10.08	Tanasawa & Katsuda 1972
Muscle,human,dry	19.78	Lipkin & Hardy 1954
–,pig	10.22	Tanasawa & Katsuda 1972
Skin,human,dry	9.62	Lipkin & Hardy 1954
–,–,moist	13.1	–
–,–,*in–vivo*	15.8	–
–,–,forearm	15.1	Dersken *et al* 1957
–,–,blackened	25.4	Kraning 1973
–,–,heel,blackened	12.3	–
–,–	18.9±1.4	Hendler *et al* 1958
–,–,forearm,blackened, irradiance 100–400 mcal cm⁻² s⁻¹	13.1–31.7	Stoll & Greene 1959

Table 2.5 Temperature coefficients for thermal conductivity and thermal diffusivity of tissue: 3–45°C

Tissue	Temperature coefficients	
	Conductivity % °C^{-1}	Diffusivity % °C^{-1}
Kidney,human,cortex	0.26	0.43
–,–,medulla	0.22	0.43
–,–,pelvis	0.40	0.08
–,pig,cortex	0.24	0.30
–,rabbit	0.27	0.21
Liver,human	0.25	0.28
–,pig	0.16	0.43
–,rabbit	0.56	1.30
Lung,pig	0.95	1.15
–,rabbit	0.78	0.77
Muscle:cardiac,human	0.24	0.39
–,pig	0.28	0.40
Spleen,human	0.26	0.37
–,pig	0.26	0.33

Source: Valvano *et al* (1985).

2.1.5 *Factors affecting thermal conductivity*

2.1.5.1 Temperature

The thermal conductivity of tissues at temperatures above freezing may show a very slight positive temperature coefficient. Hill *et al* (1967) reported the temperature dependence over a range –25°C to 60°C for a variety of meats and showed a greatest variation of about 0.2% °C^{-1} for unfrozen pig muscle. Van Gemert *et al* (1986) reported a slight increase in thermal conductivity for carotid and femoral arteries over the temperature range 25°C to 85°C. Detailed measurements of the thermal properties of a wide range of animal and human tissues *in-vitro* in the temperature range 3 °C to 45 °C are given by Valvano *et al* (1985). Temperature coefficients derived from this report are given in Table 2.5. By combining data from all tissues studied, except fat, lung and tumour, Valvano *et al* derived the following empirical expressions:

$$k = 0.4882 + 0.001265T \qquad r = 0.642 \qquad (2.6)$$

$$\alpha = 1.304 + 0.00519T \qquad r = 0.510 \qquad (2.7)$$

k is in W m^{-1} K^{-1} and α is in cm^2 s^{-1}; T is in °C.

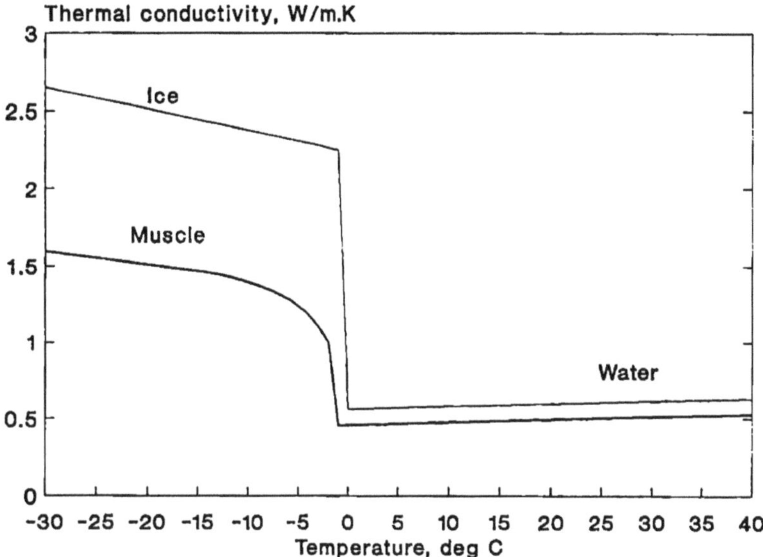

Figure 2.1 The variation of thermal conductivity with temperature, for muscle and water/ice.

The temperature dependence of k for blood has been given by Singh and Blackshear (1967) as about 0.4% $^{o}C^{-1}$ (Equation 2.14).

Thermal conductivity increases abruptly on freezing and continues to increase with decreasing temperature (Figure 2.1). Frozen muscle typically has a value of k about three times greater than that of unfrozen muscle (Table 2.6).

2.1.5.2 Tissue perfusion

Blood perfusion through tissues increases the effective values of both thermal conductivity and diffusivity. An increase in k_{eff} with the square root of the perfusion rate was reported by Jain *et al* (1979). Investigating thermal transport in an isolated rat liver, Valvano *et al* (1984) measured an increase in k_{eff} from 0.53 W m^{-1} K^{-1} when unperfused to 0.69 and 0.80 W m^{-1} K^{-1} at perfusion rates of 0.05 and 0.1 g cm^{-3} s^{-1} respectively. The wide range of values of k_{eff} for bone *in-vivo* tabulated by Bowman *et al* (1975), 0.158 to 3.08 W m^{-1} K^{-1}, reflects the strong dependence of k_{eff} on perfusion for this tissue also. All the values included in the tables are for unperfused tissues. Mean values for normal perfusion rates for several adult human tissues are included in Appendix A, and may be used to help to predict values of k_{eff} under resting conditions.

Table 2.6 Thermal conductivity of frozen tissues

Tissue	Temp °C	Conductivity W m^{-1} K^{-1}	Reference
Blood:plasma	−10	2.03	Rinfret 1969
−	−100	3.19	−
Blood:whole	−10	1.64	−
−	−100	2.66	−
Blood:packed cells	−10	1.24	−
−	−100	2.26	−
Bone:femur,cow	−19	0.33	Morley 1966
Fat,cow,7% water	−20	0.244	Cherneeva 1956
−,−	−19	0.298	Morley 1966
Muscle,cow,along fibres	−20	1.565	Lentz 1961
−,−,across fibres	−20	1.168	−
−,along fibres	−11.5	1.395	Hill *et al* 1967
−,−,across fibres	−13	1.331	−
−,pig,along fibres	−23.5	1.59	Lentz 1961
−,−,across fibres	−23.0	1.37	−
−,−,along fibres	−13.2	1.424	Hill *et al* 1967
−,−,across fibres	−14.3	1.298	−
−,lamb	−19	1.77	Morley 1966

2.1.5.3 Changes following death

It is generally considered that the thermal conductivity does not alter during the period immediately following death. Grayson (1952) could demonstrate no change for rat and rabbit liver during 2 hours following death and other authors also report no observed alterations. Cooper and Trezek (1972) observed a slight reduction in conductivity at 24 hours, but suggested that the change resulted from a temperature change in the tissue. However, changes do ultimately occur. Hatfield and Pugh (1951) noted that whilst fat preparations kept a day or two at room temperature showed only a slight change in conductivity, large changes in muscle values were seen, usually increasing with time following death. Some high values of conductivity reported for bone have been criticised partly because of the procedures of drying and rehydration which were used in the preparation of the specimens (Nelson *et al*, 1986). Bowman *et al* (1974) have reported a drop of about 20% in the conductivity of sheep kidney by 4 hours post-mortem. However, it is possible this could have resulted from a slight drying of the specimen.

2.1.5.4 Tissue composition

Many authors (for example Spells, 1960; Poppendiek *et al*, 1966) have recognised that there is a close to linear relationship between thermal conductivity of many tissues and water content, particularly for water content greater than 50%. Equations 2.8 and 2.9 give predicting expressions derived by Sweat (1975) following a regression analysis of published data on a range of meat products:

$$k = 0.0798 + 0.00517W \qquad (0 \text{ to } 60^{\circ}C) \qquad (2.8)$$

$$k = 0.284 + 0.0194W - 0.00923T \quad (-40 \text{ to } -50^{\circ}C) \qquad (2.9)$$

where W is the percentage water content, T the temperature in °C and k is in SI units. Equation 2.8 applies to tissues with water contents in the range 60% to 80%, and 2.9 to frozen tissue with water of 65% to 85%. However, deviations from these simple relationships do occur, and these equations must be used with caution. Taking a wider range of body fluids with a wider range of water content, the data of Spells (1960) have been analysed by Cooper and Trezek (1971) who give a linear fit to the data as:

$$k = 0.0502 + 0.00577W \qquad (2.10)$$

Poppendiek *et al* (1966) have suggested that tissues may be considered more accurately for thermal analysis as being composed of water, protein and fat. Thermal conductivity may then be expressed as

$$k = \rho \sum_{n=1}^{3} k_n \omega_n / \rho_n \qquad (2.11)$$

where k_n, ρ_n and ω_n are thermal conductivity, density and mass fraction of the nth component respectively, and ρ the density of the composite material. Bowman (1981) suggests the following values; k(water) = 0.603 W m^{-1} K^{-1}, k(fat) = 0.22 W m^{-1} K^{-1}, and k(protein) = 0.195 W m^{-1} K^{-1}. Cooper and Trezek (1971) used the slightly lower values for k(fat) and k(protein) of 0.19 and 0.18 W m^{-1} K^{-1} respectively, accounted for in part by the lower temperature used in their study. Poppendiek took the value of k for cellulose nitrate as being typical of all proteins. Calculations using Equation 2.11, using literature values for composition have been shown to give good fits for thermal conductivity in the majority of cases, depending on the choice of thermal conductivity values for the constituents.

An expression for the variation of thermal diffusivity of tissue with water content has been given by Riedel (1969) as:

$$\alpha_t = 8.8\times10^{-4} + (\alpha_w - 8.8\times10^{-4})x_w \qquad (2.12)$$

where α_w is the thermal diffusivity of water, and x_w the mass fraction of water. α_t and α_w are in $cm^2\ s^{-1}$.

2.1.5.5 Anisotropy

Slight anisotropy of thermal conductivity is observed in muscle from a variety of animals (Hill *et al*, 1967; Lentz, 1961). At temperatures above freezing, different values of conductivity were reported depending on whether the measurement was made parallel or perpendicular to the muscle fibres. However, the reports are not consistent in which is the greater, and it is likely that different handling procedures, such as prior freezing times and rates, could have been a primary factor contributing to the differences reported. However, when frozen, the difference in conductivities is consistent, with the conductivity parallel to the fibres always greater than that perpendicular to the fibres (Table 2.6).

For perfused tissue *in vivo* the effective thermal conductivity, k_{eff}, will show anisotropy depending on the pattern of local tissue perfusion.

2.1.5.6 Tissue pathology

In an extensive series of measurements on a wide range of human tissues *in vitro*, Bowman (1981) reports comparative conductivity values of tumour and adjacent normal tissue. It is noted that all except scirrhous carcinoma of the breast and metastatic colonic carcinoma of the liver showed an increase relative to the surrounding host tissue (see Table 2.7). Lung tissue appeared to vary with age, the lowest conductivity of 0.302 $W\ m^{-1}\ K^{-1}$ being from a young male and the highest (0.550 $W\ m^{-1}\ K^{-1}$) from older diseased lungs. It was suggested that the latter may have been more collapsed and infiltrated with fluid.

Grayson (1967) observed a consistent pattern of reduction in the conductivity of canine cardiac muscle following infarction. Measurements were made *in vitro* at 23°C with the tissue embedded in 10% gelatin gels. The mean conductivity for infarcted myocardium was 0.427±0.017 $W\ m^{-1}\ K^{-1}$ compared with 0.506±0.008 $W\ m^{-1}\ K^{-1}$ for normal myocardium. It was argued on evidence of density measurements that the change had a more complex basis than a simple alteration in water content.

In a study of human and canine arteries, Valvano and Chitsabesan (1987) and van Gemert *et al* (1986) report no significant difference between the thermal properties of normal human aorta and those for fatty or fibrous plaque. For calcified plaque conductivity and diffusivity were slightly higher.

Table 2.7 Thermal conductivity of some pathological tissues

Tissue	Temp °C	Conductivity W m^{-1} K^{-1}	Reference
Aorta,			
calcified plaque	35	0.502±0.059	Valvano & Chitsabesan 1987
–,–	90	0.619±0.085	–
–,fibrous plaque	25–80	0.476	Van Gemert *et al* 1986
–,fatty plaque	25–80	0.442	–
Tumour			
rat,Walker 256	*in-vivo*	0.32±0.09	Jain *et al* 1979
Breast,human,normal	37	0.499±0.004	Bowman 1981
scirrhous		0.397±0.004	–
muscious (colloid)		0.527±0.041	–
adenocarcinoma	37	0.564	Valvano *et al* 1985
Colon,human,normal	37	0.556±0.009	Bowman 1981
metastatic colonic		0.556±0.012	–
colon	19	0.564	Valvano *et al* 1985
Liver,human,normal	37	0.572±0.009	Bowman 1981
metastatic colonic		0.520±0.008	–
normal	37	0.508±0.011	–
metastatic pancreatic		0.562±0.021	–
Lung,human,normal	37	0.518±0.021	–
squamous cell		0.666±0.018	–
Pancreas,human,normal		0.354±0.005	–
metastatic		0.478±0.039	–
normal	37	0.468±0.006	–
metastatic gastric		0.492±0.054	–

2.1.6 *Thermal conductivity of blood*

Poppendiek (1964–66) gives an expression for the dependence of thermal conductivity of blood with per cent haematocrit, H, as

$$k = 0.571 - 0.00121H \tag{2.13}$$

A similar expression which takes account of the temperature dependence of thermal conductivity has been given by Singh and Blackshear (1967):

$$k = 0.6 - 0.0012H - 0.00226(37 - T) \tag{2.14}$$

where T is the temperature in °C. These authors also report that k was independent of shear rate in flowing blood up to a shear rate of 4150 s^{-1}.

Table 2.8 Thermal conductivity and diffusivity of ice and water

Temperature °C	Conductivity $W\ m^{-1}\ K^{-1}$	Diffusivity $cm^2\ s^{-1} x10^3$
−150	5.70±15%	
−100	3.95±10%	
−50	2.85±10%	
0	2.25±5%	
0	0.560	1.31
10	0.581	1.38
20	0.600	1.44
40	0.628	1.52
60	0.653	1.59
80	0.669	1.65

Sources: Powell (1958), Ratcliffe (1962), Dickerson (1968).

2.1.7 *Thermal conductivity of some materials other than tissue*

2.1.7.1 Water and ice

Values of thermal conductivity and thermal diffusivity for water and thermal conductivity for ice are given in Table 2.8. The thermal conductivity values for water, k_w, are taken from a review by Powell (1958). They are marginally higher than those predicted by the expression given in the International Critical Tables:

$$k_w = 0.587[1 + 0.00281(T - 20)] \quad 0<T<80°C \qquad (2.15)$$

Expressions for the thermal conductivity and also thermal diffusivity of water, α_w, are given by Valvano *et al* (1985). For thermal diffusivity,

$$\alpha_w = 1.339 + 0.00473T \qquad 0<T<45°C \qquad (2.16)$$

where α_w is in $cm^2\ s^{-1}$.

The conductivity of ice, k_i, has been reviewed by Ratcliffe (1962) who gives the following expression for the variation of k_i with temperature:

$$k_i = \frac{780}{\theta} - 0.615 \qquad 120<\theta<273\ K \qquad (2.17)$$

2.1.7.2 Aqueous solutions

The conductivity of a simple aqueous solution may in some instances be represented by the expression:

$$k = k_w[1 - \alpha(1 - x_w)] \tag{2.18}$$

where α is a constant for a particular solute and temperature, and x_w the mass fraction of water. However, this simple relationship is empirical and does not hold universally. For instance it does not apply for NaCl solutions. Experimental values for seawater are given by Castelli *et al* (1974). Omitting a minor pressure, the dependence of k on temperature T (°C) is given as:

$$k = 0.5529 + 1.836\times10^{-3}T - 3.306\times10^{-7}T^3 \tag{2.19}$$

For binary liquid mixtures, International Critical Tables give the equation:

$$k.\sinh(100\mu) = k_1.\sinh(P_1\mu) + k_2.\sinh(P_2\mu) \tag{2.20}$$

where P_1 and P_2 are the percentages by mass of the constituents, and μ depends upon the constituents and the temperature.

Table 2.9 Thermal conductivity and diffusivity of gels

Gel	T °C	Conductivity W m^{-1} K^{-1}	Diffusivity cm^2 s^{-1}x10^3	Reference
	Frozen			
Gelatin 6%	−30	2.28		Lentz 1961
	−10	2.09		-
Gelatin 12%	−30	2.08		-
	−10	1.89		-
Gelatin 20%	−30	1.56		-
	−10	1.57		-
	Not frozen			
Agar 1.5%	21	0.609±0.005	1.42±0.05	Balasubramaniam & Bowman 1977
Agar 1.5%		0.597±0.007	1.56±0.006	Bowman *et al* 1974
Gelatin 2%	20	0.619		Morley 1966
Carrageenan 2%	10	0.592±0.031	1.38−1.43	Kent *et al* 1984
	25	0.609±0.018	1.32−1.73	-
	40	0.645±0.022	1.41−1.60	-
19%	37	0.496		Grayson 1952
28%	37	0.464		-

2.1.7.3 Gels

The thermal conductivity and diffusivity of some gels are given in Table 2.9.

2.1.7.4 Tissue substitute materials

The thermal properties of some tissue substitute materials which have been developed for use in microwave and x-ray applications are included in Table 2.10. The compositions and microwave properties of these materials are given in Table 6.24, and x-ray data are given in Tables 7.12, 7.13 and 7.14. These thermal properties have been reported by Leonard *et al* (1984), Ho *et al* (1971), Durney *et al* (1980) and Guy (1971) for the microwave tissue substitutes, and by Domen (1980) for the A150 plastic material for x-radiation.

2.1.7.5 Gases

The thermal conductivity of air at 37.8°C is 0.0267 W m^{-1} K^{-1}. The thermal conductivity of water vapour at saturation at 37.8°C is 0.0187 W m^{-1} K^{-1}.

Table 2.10 Thermal properties of tissue substitute materials

Material		k W m^{-1} K^{-1}	C J g^{-1} K^{-1}	ρ g cm^{-3}	α cm^2 s^{-1}
Brain,2.45GHz	a	0.478±0.015	3.41	0.98±0.03	0.00143
-	c		3.48	0.96	
Muscle,2.45GHz	a	0.535±0.003	3.70	1.00±0.02	0.00145
-,27MHz	a	0.657±0.028	3.58	1.10±0.01	0.00167
-	b	0.544	3.60	1.0	0.0015
-	c		3.52	0.97	
Fat/Bone	b	0.190	1.00	1.3	0.0015
-	c		1.21–1.55	1.29–1.38	
A150 plastic	d	5.3x10^{-3}±1.4%	1.72±1.3%	1.126	2.72x10^{-3} ±0.4%

Sources: a Leonard *et al* (1984); b Ho *et al* (1971) and Guy (1971); c Durney *et al* (1980); d Domen (1980).

2.2 Specific heat capacity and latent heat

2.2.1 *Terminology and definitions*

The **specific heat capacity**, C, of a substance is the quantity of heat Q required to raise the temperature of a unit mass of the substance one degree.

$$C = \frac{Q\Delta T}{M} \qquad (2.21)$$

where M is the mass and ΔT the change in temperature. The SI unit of specific heat capacity is joule per kilogram kelvin (J kg^{-1} K^{-1}), with a common unit being J g^{-1} K^{-1}.

The freezing of tissue is not characterised by the discontinuity in enthalpy which exists for many pure substances. The first-order transition from ice to water shows this discontinuity and the heat released is termed the **latent heat of fusion**. However, since the fluid water in tissue does not all freeze at the same temperature the concept of latent heat is not relevant, and therefore latent heats of fusion for tissues cannot be given. It has also been suggested that the concept of specific heat is also inappropriate for frozen tissue since there is no way to separate the heats associated with freezing from those associatd with change in temperature. For this reason **enthalpy** or total heat content is often reported for food industry calculations (Riedel, 1957b; Fleming, 1969), referred to an arbitrary zero reference temperature, commonly −40°C. In the tables included here only values of specific heat for tissues are given, accepting the argument outlined above. For tabulations and diagrams of enthalpy values, reference may be made to food science texts (Dickerson, 1968; Rha, 1975).

2.2.2 *Values of specific heat of tissue*

Values of specific heats of various tissues are given in Table 2.11 for normal temperatures and 2.12 for frozen tissue. Details are discussed below.

2.2.3 *Factors affecting specific heat*

2.2.3.1 Temperature

At temperatures below freezing the specific heat varies markedly with temperature in a manner depending strongly on the tissue water content (Figure 2.2). Because the water within the tissue does not all form ice at the same temperature the specific heat of ice cannot be used to characterise tissue. Riedel (1957a) suggested that all free water can be considered as

Table 2.11 Specific heat capacity, C, of normal tissues

Tissue	Specific heat $J\ g^{-1}\ K^{-1}$	Reference
Blood:whole,human	3.84	Atzler & Richter 1920
–,–,42.4→30.0 Hct.	3.60→3.72	Mendlowitz 1948
–,dog	3.60	Jakovlev 1940
	also see Equations 2.25, 2.26	
Blood:plasma,human	3.93	Mendlowitz 1948
Bone:cortical,human	1.3	Henschel 1943
–,cow	1.33	Morley 1986
–,pig	1.36±0.10	Lehmann & Johnson 1958
Bone:trabecular,cow	1.15–1.73	Clattenburg *et al* 1975
–,pig	2.11±0.06	Lehmann & Johnson 1958
Bone marrow,pig	2.74±0.03	–
Brain:white,human	3.60	Cooper & Trezek 1972
Brain:grey,human	3.68	–
Eye:lens,rabbit	3.0±0.5	Lagendijk 1982
Eye:vitreous,rabbit	4.2	–
Fat,dog	2.30–2.68	Jakovlev 1940
–,pig	2.25,2.40	Henriques & Moritz 1947
–,–	2.60±0.06	Lehmann & Johnson 1958
–,–	2.68–3.92	Robinson 1972
–,cow	3.59	–
Kidney,human	3.89	Cooper & Trezek 1972
–	3.60	Ordinanz 1946
Liver,human	3.60	Cooper & Trezek 1972
–,cow	3.37	Robinson 1972
–,dog	3.51	Cooper & Trezek 1972
Muscle:cardiac,human	3.72	–
Muscle:skeletal,cow	3.43	Ordinanz 1946
–,–	3.81	Cherneeva 1956
–,pig	3.72,3.87	Henriques & Moritz 1947
–,–	3.14±0.07	Lehmann & Johnson 1958
–,–	3.06	Ordinanz 1946
–,sheep,90% water	3.89	–
–,rabbit	3.60	Jakovlev 1940
Skin:dermis,pig	3.15,3.28	Henriques & Moritz
Skin:epidermis,pig	3.53,3.71	1947
–,rabbit	3.47	Jakovlev 1940
Spleen,human	3.72	Cooper & Trezek 1972
Tooth:dentine	1.17	Barker *et al* 1972
–	1.26	Henschel 1943
–	1.59±0.04	Brown *et al* 1970
Tooth:enamel	0.75	Barker *et al* 1972
–	0.712±0.04	Brown *et al* 1970

Table 2.12 Specific heat capacity of frozen tissue, J g^{-1} K^{-1}

Temp °C	-40	-30	-20	-10	-5	-1	0	10
Beef,74.5% water (Riedel, 1957a)	1.9	2.1	2.5	4.5	12.1	146	3.6	3.6
Beef,78.5% water (Cherneeva, 1956)		2.89	3.31	4.40	5.07		3.81	

Temp °C	-40	-23	-12	-4.4	-1.1	0	10
Kidney,79.8% water	1.63	2.14	4.61	20.1	209	4.10	3.77
Lamb,64.9% water	1.30	1.93	3.60	13.0	96.3	3.39	3.39
(Fleming, 1969)							

Temp °C	-169	-87	-57	-34	-22
Beef (Moline *et al*, 1961)	1.05	1.63	1.88	2.43	3.68

Temp °C	-160	-120	-80	-40
Fat,beef	0.85	1.00	1.55	2.30
-,pig	0.88	1.03	1.21	1.60
(Moline *et al*, 1961)				

Temp °C	-30	-20	-10	-5	0	30
Fat,beef 7% water (Cherneeva, 1956)	1.72	1.97	2.43	3.01	4.10	3.35

frozen only at temperatures below -40°C, and that there remains about 10% of 'bound' water, in muscle, which never freezes (Figure 2.3). Moline *et al* (1961) have investigated the specific heat of a number of frozen substances including gelatins, fats, and beef over the temperature range -30°C to -160°C. They report a linear increase in specific heat capacity with temperature of about 0.8% °C^{-1} in the range -160°C to -50°C. Above -50 °C the specific heat rises, increasing more rapidly as the freezing point is approached (Figure 2.2).

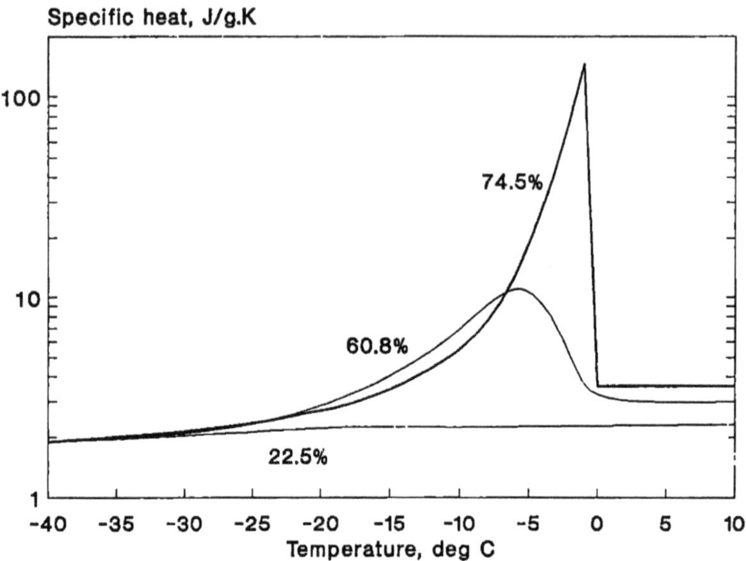

Figure 2.2 The variation of specific heat with temperature and with water content of muscle. From Riedel (1957a).

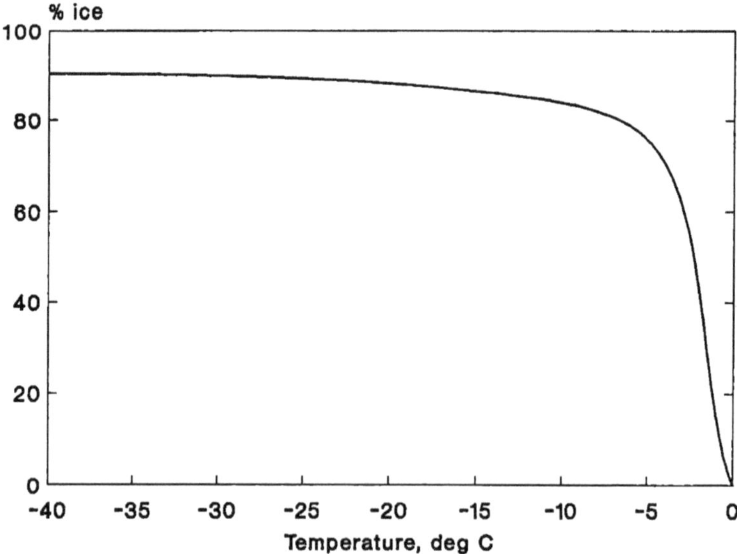

Figure 2.3 The freezable mass fraction of muscle. Riedel (1957a), Plank (1960).

2.2.3.2 Tissue composition

In composite materials C may be estimated as the sum of heat capacities of each fractional component. Reporting an examination of the specific heat of foodstuffs, Siebel (1892) reasoned that it should be possible to calculate their specific heat above the freezing point by summing the fractional contributions from two components, water and solid matter. The specific heat of the latter was taken as 0.2 cal g^{-1} $°C^{-1}$ (~ 0.84 J g^{-1} K^{-1}). Following a similar argument, Cooper and Trezek (1971) have proposed the following mixture equation for the calculation of thermal capacities:

$$C = \sum_{n=1}^{3} \omega_n C_n \tag{2.22}$$

where ω_n is the mass fraction of the nth component and C_n its specific heat. The three components are water, fat and protein. Values for C for each component may be found in the tables. The specific heat of gelatin (Table 2.14; Hampton and Mennie, 1934) has been often used for protein. Alternatively the specific heat for dehydrated collagen given by Kanagy (1955) of 1.56±0.04 J g^{-1} K^{-1} may be used. Kanagy reports that hydrated collagen at 17.03% water content has a specific heat of 1.79±0.04 J g^{-1} K^{-1}.

A common assumption is that the proportions of fat and protein are equal. Under these circumstances it is possible to reduce Equation 2.22 to one involving only percentage water content. Riedel (1956b, 1957a) gives such an expression using measurements from a variety of meat and fish tissues. Expressed in SI units this equation is:

$$C = 1670 + 25.1W \tag{2.23}$$

where W is the percentage water content. This equation may be used to give good estimates of specific heat capacity if the tissue water content is known and no other experimental data is available. A similar equation derived empirically is quoted by Cooper and Trezek (1971). These authors also compared experimental data from tissues and gelatin gels, with predictions using Equation 2.17 and found, anomalously, that values for the gel are in closer agreement with values for a water/fat system than a water/protein system. Values for soft tissues, with the exception of that for brain grey matter, lay between the upper and lower limits of Equation 2.22. An extensive survey of other predicting equations for aqueous unfrozen foods with possible relevence to tissues is given by Miles *et al* (1983).

The specific heat of fat varies in a very complex way both with fatty acid composition and with temperature. Lipids may exist in either of two phases, gel or liquid crystal, and the transition between these phases is discontinuous only under conditions of extreme purity. Commonly both phases co–exist and a broad phase transition occurs. For aqueous lipid dispersions the transition can lie somewhere in the range –20°C to 70°C (Lee, 1977).

The transitions, moreover, commonly show hysteresis, exhibiting different thermal properties on heating from those on cooling. For beef fat, 40–50% of the composition consists of the unsaturated oleic acid, with palmitic and stearic acids making up the majority of the saturated acid contribution (Swern, 1979). The proportions differ for fats from other species. For instance, the oleic acid content of pig fat may reach 58.7% (Williams, 1966).

Figure 2.4 shows the variation of C with temperature, over the range -20°C to 60°C for both pig and beef fat (Riedel, 1956a). Two melting transitions are seen at about 4°C and 26°C for pig fat, and at higher temperatures for beef. Morley (1986) has demonstrated that the presence and magnitude of these peaks are dependent upon whether the measurements are made during cooling or during heating. In addition the same sample of fat can show a different response depending on the length of time it was held at 0°C before reheating. Comparable data for fat derived from human tissue appear to be unavailable.

In assessing heat transport through heat-damaged tissue the specific heat of carbonised tissue may be needed. In the absence of any experimental value, the specific heat of carbon, 0.670 J g^{-1} K^{-1} has been used.

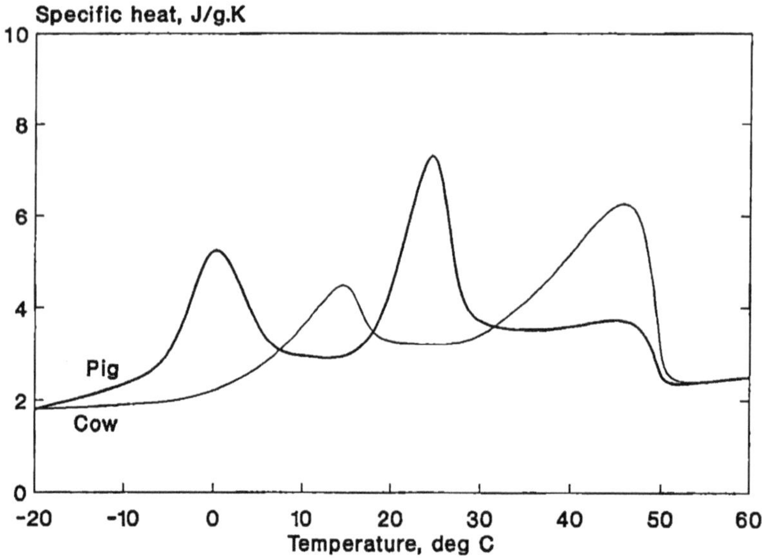

Figure 2.4 Variation of specific heat with temperature for pig and beef fat; Riedel (1956a) and Miles *et al* (1983).

Table 2.12 Numeric values for estimating specific heat for beef and pig fats from Equation 2.24

	Beef	Pig
n	2	2
$A, J\ kg^{-1}\ {}^{\circ}C^{-1}$	1.884×10^3	1.420×10^3
$B, J\ kg^{-1}\ {}^{\circ}C^{-2}$	7.008	3.67
$A_1, J\ kg^{-1}\ {}^{\circ}C^{-1}$	2.24×10^3	3.73×10^3
$B_1, J\ kg^{-1}\ {}^{\circ}C^{-3}$	20	50
$T_1, {}^{\circ}C$	13.8	1.18
$A_2, J\ kg^{-1}\ {}^{\circ}C^{-1}$	4.28×10^3	4.5×10^3
$B_2, J\ kg^{-1}\ {}^{\circ}C^{-3}$	100	150
$T_2, {}^{\circ}C$	47.85	26.85

Source: Miles *et al* (1983).

An empirical equation describing the specific heat of fat has been proposed by Latyshev and Ozerova (1976):

$$C = A + BT + \sum_{i=1}^{n} \frac{A_i}{\left[1 + \frac{B_i}{A_i}(T - T_i)^2\right]} \tag{2.24}$$

where A, B, A_i, B_i and T_i are fat-specific constants. Values for these constants for pig and beef fat are given in Table 2.12. Equation 2.24 could be used to fit the specific heat of human fat if suitable data were available since it will fit curves with any number of maxima occurring at temperatures T_i. The problem of the prediction of the behaviour of fatty tissues (rather than pure fat) is complicated by the fact that phase transitions in lipids are influenced by the presence of impurities.

2.2.4 *Specific heat of blood*

The specific heat capacity for blood has been investigated by Epstein *et al* (1963). The relationship of specific heat to percentage haematocrit, and specific gravity S was determined at 37.5°C, giving the following expressions where C is in $J\ kg^{-1}\ K^{-1}$.

$$C = 3950 - 7.24H \tag{2.25}$$

$$C = 4176 - 9623(S-1) \tag{2.26}$$

It was noted that the expression using specific gravity (2.26) gave a better fit to the data than that using haematocrit.

Table 2.13 Specific heat of ice and water

Ice		Water	
Temp °C	Specific heat J g^{-1} K^{-1}	Temp °C	Specific heat J g^{-1} K^{-1}
−160	0.97	0	4.217
−120	1.25	5	4.202
−100	1.39	10	4.192
−80	1.54	15	4.186
−60	1.68	20	4.182
−40	1.82	40	4.178
−20	1.94	60	4.184
0	2.06	80	4.196

2.2.5 *Heat of ablation*

The energy required to raise the temperature of a tissue from ambient until its structure is destroyed may be defined as its **heat of ablation.** The unit of measurement may be J cm^{-3} or J g^{-1}. It is usually measured from the slope of the relationship between tissue removed by the absorption of energy from any source and the energy invested in the ablation process. The heat of ablation obtained from the reciprocal of the slope of this surve.

The heat of ablation for water at body temperature is about 2.5 J g^{-1}. Values for tissues vary from 4.27 kJ g^{-1} for skin to 0.61 kJ g^{-1} for liver, according to Walsh and Deutsch (1989), for exposure to infrared radiation; see Table 2.14.

Table 2.14 Heats of ablation for tissues

Tissue	Heat of ablation kJ g^{-1}
Aorta	3.37
Arterial plaque, 249 nm	0.97±0.28
–, 308 nm	1.1±0.3
–, 9470 nm	2.65±0.86
Liver	0.61
Muscle:cardiac	1.51
Skin	4.27
Water	2.5

Sources: Walsh and Deutsch (1989); Furzikov *et al* (1987).

2.2.6 *Specific heat of some materials other than tissue*

2.2.6.1 Water and ice

The specific heat capacity for water and for ice is given in Table 2.13. The latent heat of fusion of the ice/water phase transition at 0°C is 333.6 J g^{-1}. The latent heat of vaporisation at the water/water vapour phase transition at 100°C is 2256.7 J g^{-1}. These values have been obtained from the standard reference tables.

2.2.6.2 Gels

The specific heat of gelatin gels of a range of concentrations has been reported by Hampton and Mennie (1934) and Horn and Mennie (1935). Values for gel concentrations in the range 9% to 100% for the temperature range 0° to 25°C are shown in Table 2.15. Values of heat capacity of the total gel and of the water in the gel were studied over the temperature range −180°C to 25°C. Moline *et al* (1961) have also reported measurements of specific heat from frozen gels. The specific heat of 100% gelatin increased from 0.63 J g^{-1} K^{-1} at −160°C to 1.30 J g^{-1} K^{-1} at −40°C, and a 26.4% gel increased from 0.88 to 1.84 J g^{-1} K^{-1} over the same temperature range.

2.2.6.3 Tissue substitute materials

The specific heat of several materials used as tissue substitutes for microwave and x-ray studies are included in Table 2.10. Source references are given in Section 2.1.5.4.

2.3 Freezing points and fat solidification temperatures

The freezing points of tissues and of body fluids are depressed slightly from that of pure water. In an ideal solution of very low percentage solute the extent of the freezing point depression depends upon the molar concentration of the solute according to the equation:

$$\Delta T_i = 1.86 m\nu \quad (2.27)$$

where m is the molality of solute, that is the number of moles of undissociated solute in 1000 g of water, and ν is the number of particles into which the solute dissociates at complete dissociation (International Critical Tables). ΔT_i is the freezing pont depression in an ideal solution in °C. Freezing point depression values are given for some biological substances in Table 2.16.

Similarly, it is to be expected that the boiling point of the fluid content of tissue will be elevated.

Table 2.15 Specific heat of gelatin gels; 0–25°C

% gel	9.0	18.7	36.0	51.5	61.8	64.0	65.6	66.2	71.2	100
C, J g^{-1} K^{-1}	4.03	3.84	3.50	3.28	2.75	2.58	2.75	2.66	2.18	1.19

Source: Hampton and Mennie (1934) and Horn and Mennie (1935).

Table 2.16 Freezing point depression for some biological substances

Tissue/fluid	Freezing point depression, °C
Bile,hepatic	0.56–0.61
Blood	0.555–0.570
Cerebrospinal fluid	0.54–0.60
Gastric juice	0.30–0.82
Pancreatic juice	0.55–0.66
Prostatic fluid	0.55–0.58
Saliva	0.07–0.70
Semen, 1 hr	0.56–0.58
16 hr after ejaculation	0.74–0.78
Sweat	0.32–0.37
Synovial fluid	0.51–0.55
Tears	0.551
Urine	0.1–2.5
Tissue	1.7–2.2

Sources: Spector (1956) and Diem and Lentner (1975), except for tissue, from Rha (1975).

Table 2.17 Melting and solidification points for fats

Species	Transition	Temperature	Reference
Human	Solidification	15°C	Harwood & Geyer 1964
Pig	Melting	42 to 45°C	Williams 1966
	Solidification	25 to 30°C	
Cow	Melting	47 to 49.5°C	Williams 1966
	Solidification	About 35°C	
Animal	Melting	44.3 to 45.4°C	Mertens 1973
		Range up to 4.2°C	

Methods for measuring the melting points of fats have been reviewed by Mertens (1973). Some values of melting and solidification points are given in Table 2.17.

2.4 Thermal expansion

Thermal expansion coefficients for tissue are included with the section on tissue density, Section 5.1.1.1, Table 5.3.

References

Atzler E. and Richter F., 1920, Über die Wärmekapazität des arteriellen und venösen Blutes, Biochem Zeit, 112, 310–312.

Balasubramaniam T.A. and Bowman H.F., 1977, Thermal conductivity and thermal diffusivity of biomaterials: a simultaneous measurement technique, J Biomech Eng, Trans ASME, 99, 148–154.

Barker R.E., Rafoth R.F. and Ward R.W., 1972, Thermally induced stresses and rapid temperature changes in teeth, J Biomed Mater Res, 6, 305–325.

Bordier H., 1898, Thermal conductivity of tissues, Arch Physiol (Paris), 5T10, 17.

Bowman H.F., 1981, Heat transfer and thermal dosimetry, J Microwave Power, 16, 121–133.

Bowman H.F., Cravalho E.G., Woods M., Dmachowski J.R. and Hansen W.P., 1974, Prog Rep 3, Nat Heart Lung Inst Grant No. 5 POIHL 14322–03: Referenced in Bowman *et al*, 1975.

Bowman H.F., Cravalho E.G. and Woods, M., 1975, Theory, measurement and application of thermal properties of biomaterials, Ann Rev Biophys Bioeng, 4, 43–80.

Brown W.S., Dewey W.A. and Jacobs H.R., 1970, Thermal properties of teeth, J Dent Res, 49, 752–755.

Buettner K., 1951, Effects of extreme heat and cold on human skin. I. Analysis of temperature changes caused by different kinds of heat application, J Appl Physiol, 3, 691–702.

Castelli V.J., Stanley E.M. and Fischer E.C., 1974, The thermal conductivity of seawater as a function of pressure and temperature, Deep–sea Res, 21, 311–319.

Challoner A.R. and Powell R.W., 1957, Thermal conductivities of liquids: new determinations for seven liquids and appraisal of existing values, Proc Roy Soc A, 238, 90–106.

Chato J.C., 1969, Heat transfer in bioengineering. In *Advanced Heat Transfer*, B.T. Chao (ed.), Univ of Illinois Press, Urbana, 395–413.

Chen M.M., Holmes K.R. and Rupinskas V., 1981, Pulse–decay method for measuring the thermal conductivity of living tissues, J Biomech Eng, Trans ASME, 103, 253–260.

Cherneeva L.I., 1956, Study of the thermal properties of foods, Report of VNIKHI, (Scientific Research Institute of the Refrigeration Industry), Gostorgisdat, Moscow.

Clattenburg R., Cohen J., Conner S. and Cook N., 1975, Thermal properties of cancellous bone, Biomed Mater Res, 9, 169–182.

Cohen M.L., 1977, Measurement of the thermal properties of human skin. A review, J Invest Dermatol, 69, 333–338.

Cook H.F., 1952, A physical investigation of heat production in human tissues when exposed to microwaves, Br J Phys, 3, 1–6.

Cooper T.E. and Trezek G.J., 1971, Correlation of thermal properties of some human tissues with water content, Aerospace Med, 42, 24–28.

Cooper T.E. and Trezek G.J., 1972, A probe technique for determing the thermal conductivity of tissue, J Heat Transfer, Trans ASME, 94, 133–140.

Craig R.G. and Peyton F.A., 1961, Thermal conductivity of tooth structure, dental cements, and amalgam, J Dent Res, 40, 411–418.

Dersken W.L., Murtha T.D. and Monahan T.I., 1957, Thermal conductivity and diathermancy of human skin for sources of intense thermal radiation employed in flash burn studies, J Appl Physiol, 11, 205–210.

Dickerson R.W., 1968, Thermal properties of foods. In *The Freezing Preservation of Foods*, 4th edition, Avi Publishing Co., Westport, Conn, vol 2, 26–51.

Diem E. and Lentner C. (eds), 1975, *Documenta Geigy, Scientific Tables*, 7th edition, Geigy Pharmaceuticals, Macclesfield.

Domen S.R., 1980, Thermal diffusivity, specific heat, and thermal conductivity of A150 plastic, Phys Med Biol, 25, 93–102.

Durney C.H., Iskander M.F., Massoudi H., Allen S.J. and Mitchell J.C., 1980, Radiofrequency radiation dosimetry handbook, 3rd edition, SAM–TR–80–32, USAF School of Aerospace Medicine.

Epstein W., Visscher M.B., Stish R. and Ballin H., 1963, Specific heat of canine blood, J Appl Physiol, 18, 843–844.

Fleming A.K., 1969, Calorimetric properties of lamb and other meats, J Food Technol, 4, 199–215.

Furzikov N.P., Karu T.I., Letokhov V.S., Beljaev A.A. and Ragimov S.E., 1987, Relative efficiency and products of atherosclerotic plaque destruction by pulsed laser radiation, Lasers Life Sci, 1, 265–274.

Graf K. and Stein E., 1957, Fortlaufende Registrierung der Knochenmarkdurchblutung des Menschen mit der Wärmeleitsonde, Zeit ges exp Medizin, 129, 1–14.

Grayson J., 1952, Internal calorimetry in the determination of thermal conductivity and blood flow, J Physiol, 118, 54–72.

Grayson J., 1967, Thermal conductivity of normal and infarcted heart muscle, Nature, 215, 767–768.

Grayson J., Coulson R.L. and Winchester B., 1971, Internal calorimetry – assessment of myocardial flow and heat production, J Appl Physiol, 30, 251–257.

Guy A.W., 1971, Analyses of electromagnetic fields induced in biological tissues by thermographic studies of equivalent phantom models, IEEE Trans Microwave Theory and Techniques, 19, 205–214.

Hampton W.F. and Mennie J.H., 1934, Heat capacity measurements on gelatin gels II, Can J Res, 10, 452–462.

Harwood H.J. and Geyer R.P. (eds), 1964, *Biology Data Book*, Fed Amer Soc Exp Biol, Washington DC, pp 380–382.

Hatfield H.S. and Pugh L.G.C., 1951, Thermal conductivity of human fat and muscle, Nature, 168, 918–919.

Hendler E., Crosbie R. and Handy J.D., 1958, Measurement of heating of the skin during exposure to infrared radiation, J Appl Physiol, 12, 177–185.

Henriques F.C. 1947, Studies of thermal injury, VIII. Automatic recording caloric applicator and skin tissue and skin surface thermocouples, Rev Sci Instrum, 18, 673–80.

Henriques F.C. and Moritz A.R., 1947, Studies of thermal injury I. the conduction of heat to and through skin and the temperatures attained therein. A theoretical and an experimental investigation, Am J Path, 23, 531–549.

Henschel C.J., 1943, Heat impact of revolving instruments on vital dentin tubules, J Dent Res, 22, 323–333.

Hensel H. and Bender F., 1956, Fortlaufaude Bestimmung der Hautdurchblutung am Menshen mit einem elektrishen Wärmeleitmesser, Pflug Archiv, 263, 603–614.

Hensel H. and Doerr F.F., 1959, Untersuchungen mit enem neuen Haut-Wärmeleitmesser, Pflug Arch ges Physiol, 270, 78 (abstract).

Hill J.E., Leitman J.D. and Sunderland J.E., 1967, Thermal conductivity of various meats, Food Technol, 21, 1143–1148.

Ho H.S., Guy A.W., Sigelmann R.A. and Lehmann J.F., 1971, Microwave heating of simulated human limbs by aperture sources, IEEE Trans Microwave Theory and Techniques, 19, 224–231.

Horn W.R. and Mennie J.H., 1935, Heat capacity measurements on gelatin gels III, Can J Res, 12, 702–706.

International Critical Tables of Numeric Data, Physics, Chemistry and Technology (1926–33), McGraw-Hill, New York.

Jain R.K., Grantham F.H. and Gullino P.M., 1979, Blood flow and heat transfer in Walker 256 mammary carcinoma, J Natl Cancer Inst, 62, 927–931.

Jakovlev V.P., 1940, Arch Sc Biol (USSR), 58, 58; cited in Stow and Schieve, 1959.

Kanagy J.R., 1955, Specific heats of collagen and leather, J Am Leather Chem Assoc, 50, 444–453.

Kent M., Christiansen K., van Haneghem I.A., *et al*, 1984, Cost 90 collaborative measurements of thermal properties of foods, J Food Eng, 3, 117–150.

Klug F., 1874, Research on the conduction of heat in the skin, Zeit für Biol, 10, 73.

Kraning K.K., 1973, Heat conduction in blackened skin accompanying pulsatile heating with a Xenon flash lamp, J Appl Physiol, 35, 281–287.

Lagendijk J.J.W., 1982, A mathematical model to calculate temperature distributions in human and rabbit eyes during hyperthermic treatment, Phys Med Biol, 27, 1301–1311.

Lapshin A., 1954, Heat and temperature conduction of raw fat material and molten fat, Myas Ind SSSR, 25, 55–56; cited in Morley, 1972.

Latyshev V.P. and Ozerova T.M., 1976, Kholod Tekn, 5, 37; cited in Miles *et al*, 1983.

Lee A.G., 1977, Lipid phase transitions and phase diagrams I. Lipid phase transitions, Biochim Biophys Acta, 472, 237–281.

Lefevre J., 1901, Studies on the thermal conductivity of skin *in-vivo* and the variations induced by changes in the surrounding temperature, J de Phys.

Lehmann J.F. and Johnson E.W., 1958, Some factors influencing the temperature distribution in thighs exposed to ultrasound, Arch Phys Med Rehab, 39, 347-356.

Lentz C.P., 1961, Thermal conductivity of meats, fats, gelatin gels, and ice, Food Technol, 15, 243-247.

Leonard J.B., Foster K.R. and Athey T.W., 1984, Thermal properties of tissue-equivalent phantom materials, IEEE Trans Biomed Eng, 31, 533-536.

Linzell J.L., 1953, Internal calorimetry in the measurement of blood flow with heated thermocouples, J Physiol, 121, 390-402.

Lipkin M. and Hardy J.D., 1954, Measurements of some thermal properties of human tissues, J Appl Physiol, 7, 212-217.

Lundskog J., 1972, Heat and bone tissue, Scand J Plast Reconstr Surg, Suppl 9.

Mellor J.D., 1980, Thermophysical properties of foodstuffs 4 - general bibliography, Bull Int Inst Refrig, 60(3), 493-515.

Mendlowitz M., 1948, The specific heat of human blood, Science, 107, 97-98.

Mertens W.G., 1973, Fat melting point determinations: a review, J Amer Oil Chem Soc, 50, 115-119.

Miles C.A., van Beek G. and Veerkamp C.H., 1983, Calculations of thermophysical properties of foods. In *Physical Properties of Foods*, R. Jowitt *et al* (eds), Applied Science, Ch.16, 269-312.

Mohsenin N.N., 1980, *Thermal Properties of Foods and Agricultural Materials*, Gordon and Breach, London.

Moline S.W., Sawdye J.A., Short A.J. and Rinfret A.P., 1961, Thermal properties of foods at low temperatures. 1. Specific heat, Food Technol, 15, 228-231.

Morley M.J., 1966, Thermal conductivity of muscles, fats and bones, J Food Technol, 1, 303-311.

Morley M.J., 1972, Thermal properties of meat: tabulated data, Special report no. 1, Meat Research Institute, Bristol.

Morley M.J., 1986, Product characteristics governing heat transfer during chilling, Meat Chilling 1986, Int Inst Refrig Commission C2, Bristol, 1986/3.

Nelson C.G., Krishnan E.C. and Neff J.R., 1986, Consideration of physical parameters to predict thermal necrosis in acrylic cement implants at the site of giant cell tumours of bone, Med Phys, 13, 462-468.

Ordinanz W.O., 1946, Specific heats of foods in cooking, Food Industries, 18, 101.

Pennes H.H., 1948, Analysis of tissue and arterial blood temperature in the resting human forearm, J Appl Physiol, 1, 93-122.

Perl W., 1962, Heat and matter distribution in body tissues and determination of tissue blood flow by local clearance methods, J Theoretical Biol, 2, 201-235.

Plank R., 1960, *Handbuch der Kältetechnik*, Springer-Verlag, Berlin, 220.

Poppendiek H.F., 1964–66, Thermal and electrical conductivity of biological fluids and tissues, DDC AD nos 608 768: 613 560: 624 897:630 303 and 630 712, Reported in Chato (1969).

Poppendiek H.F., Randall R., Breeden J.A., Chambers J.E. and Murphy J.R., 1966, Thermal conductivity measurements and predictions for biological fluids and tissues, Cryobiology, 3, 318–327.

Powell R.W., 1958, Thermal conductivities and expansion coefficients of water and ice, Adv Phys, 7, 276–297.

Ratcliffe E.H., 1962, The thermal conductivity of ice new data on the temperature coefficient, Phil Mag, 8th Ser., 7, 1197–1203.

Rha C., 1975, Thermal properties of foods. In *Theory, Determination and Control of Physical Properties of Food Materials*, C.Rha (ed.), Riedel, Dardrecht, Holland, 311–355.

Riedel L., 1956a, Enthalpie und spezifische Wärme von Fetten und Olen im Schmelz bereich, Kältetechnik, 8(3), DKV Arbeitsblatt, 8–10.

Riedel L., 1956b, Kalorimetrische Untersuchungen über das Gefrieren von Seefischen, Kältetechnik, 8, 374–377.

Riedel L., 1957a, Kalorimetrische Untersuchungen über das Gefrieren von Fleisch, Kältetechnik, 9, 38–40.

Riedel L., 1957b, Enthalpie–Konzentrations–Diagramm für mageres Rindfleisch, Kältetechnik, 9, DKV Arbeitsblatt 8–11.

Riedel L., 1969, Temperaturleitfähigkeitsmessungen an wasserreichen Lebensmitteln, Kältetechnik–Klimatisierung, 21, 315–316

Rinfret A.P., 1969, Personal communication, cited by Chato (1969).

Robinson J.E., 1972, Unpublished data reported by Morley (1972).

Siebel E., 1892, Specific heats of various products, Ice and Refrig, 2, 256.

Singh A. and Blackshear P.L., 1967, The thermal conductivity of stationary and moving blood, Proc 7th Int Conf Med Biol Eng, Stockholm, 400.

Spector W.S., 1956, *Handbook of Biological Data*, W.B.Saunders, Philadelphia and London.

Spells K.E., 1960, The thermal conductivities of some biological fluids, Phys Med Biol, 5, 139–153.

Stow R.W. and Schieve J.F., 1959, Measurement of blood flow in minute volumes of specific tissues in man, J Appl Physiol, 14, 215–224.

Soyenkoff B.C. and Okun J.H., 1958, Thermal conductivity measurements of dental tissues with the aid of thermistors, J Am Dent Assn, 57, 23–30.

Stoll A.M. and Greene L.C., 1959, Relationship between pain and tissue damage due to thermal radiation, J Appl Physiol, 14, 373–387.

Sweat V.E., 1975, Modeling the thermal conductivity of meats, Trans ASAE, 18, 564–568.

Swern D. (ed.), 1979, *Bailey's Industrial Oil and Fat Products*, 4th edition, Vol 1, Wiley–Interscience, New York, p.343.

Tanasawa I., and Katsuda T., 1972, Seisan Kenkyu, 24, 404–406, 440–443: in Japanese, cited by Bowman *et al* (1975).

Touloukian Y.S. (ed.), 1964, *Retrieval Guide to Thermophysical Properties Research Literature*, McGraw–Hill, NY.

Valvano J.W. and Chitsabesan B., 1987, Thermal conductivity and diffusivity of arterial wall and atherosclerotic plaque, Lasers Life Sci, 1, 219–229.

Valvano J.W., Allen J.T. and Bowman H.F., 1984, The simultaneous measurements of thermal conductivity, thermal diffusivity, and perfusion in small volumes of tissue, J Biomech Eng, Trans ASME, 106, 192–197.

Valvano J.W., Cochran J.R. and Diller K.R., 1985, Thermal conductivity and diffusivity of biomaterials measured with self-heating thermistors, Int J Thermophys, 6, 301–311.

Van der Staak W.J.B.M., Brakkee A.J.M. and de Rijke-Herweijer H.E., 1968, Measurements of the thermal conductivity of the skin as an indication of skin blood flow, J Invest Dermatol, 51, 149–154.

Van Gemert M.J.C., Welch A.J., Bonnier J.J.M., *et al*, 1986, Some physical concepts in laser angioplasty, Semin Interventional Radiol, 3, 27–38.

Vendrick A.J.H. and Vos J.J., 1957, A method for the measurement of the thermal conductivity of human skin, J Appl Physiol, 11, 211–215.

Walsh J.T. and Deutsch T.F., 1989, Pulsed CO_2 laser ablation of tissue: effect of mechanical properties, IEEE Trans Biomed Eng, 36, 1195–1201.

Williams K.A., 1966, *Oils Fats and Fatty Foods*, 4th edition, Churchill, London, pp.233 and 240.

Chapter 3

Optical Properties of Tissue including Ultraviolet and Infrared Radiation

This chapter concerns the optical properties of tissue in the spectrum from ultraviolet to infrared, approximately 100 nm to 1 mm. The photobiologic scheme for the division of the optical wavelengths into bands (CIE, 1970) is shown in Table 3.1. This identifies the three spectral bands assigned to the ultraviolet spectrum (100 nm to 400 nm), and the three to the infrared spectrum (760 nm to 1 mm). Light wavelengths lie between 400 nm to 760 nm. The extreme, or vacuum ultraviolet range, from 1 nm to 100 nm, is omitted and is of little practical interest for tissues.

3.1 Optical attenuation: absorption and scatter

3.1.1 *Terminology and definitions*

At distances far from sources and tissue boundaries in a homogenous semi-infinite slab, irradiated by a plane, collimated, monochromatic beam of light the space irradiance, $\varphi(z)$, at a depth z is given by

$$\varphi(z) = A \exp(-\alpha z) \tag{3.1}$$

where α is the **effective attenuation coefficient** and A is a constant depending upon the geometries of the source and tissue. Typical units are cm^{-1}. Some sources use the symbol Σ_{eff} instead of α. The reciprocal of α is termed the **penetration depth,** δ, and is the distance required for the radiant flux to reduce to 1/e of its incident value.

The forward propagation in tissue depends both on the absorption in the tissue and on the scatter. At very short ultraviolet wavelengths and for transmission in the infrared, tissue absorption may be substantially greater than scatter. Under these circumstances Equation 3.1 represents well the reduction in intensity with distance, with α representing the absorption coefficient. On the other hand for most optical wavelengths and tissues,

Table 3.1 CIE scheme for the division of the optical spectrum

Description	Wavelength
UV-C	100 nm to 280 nm
UV-B	280 nm to 315-320 nm
UV-A	315 nm to 380-400 nm
Visible	380-400 nm to 760-780 nm
IR-A	760-780 nm to 1400 nm
IR-B	1.4 μm to 3 μm
IR-C	3 μm to 1 mm

scattering makes a substantial contribution to the forward transmission. Thus the value of α depends not only upon the **absorption coefficient,** κ_a, but also upon the **scattering coefficient,** κ_s. In addition the angular distribution of the scatter is important, described by the **scattering phase function,** $S(\theta)$. A useful simplifying parameter which summarises the scatter distribution is the **mean cosine of the scatter,** $\bar{\mu}$:

$$\bar{\mu} = \int_{-1}^{1} S(\theta) \cos\theta \, d(\cos\theta) \tag{3.2}$$

$\bar{\mu} = 0$ represents isotropic scatter while $0 < \bar{\mu} \leq 1$ represents forward scatter.

The dependence of scattering upon angle, and also upon frequency, varies with the size of the scatterers relative to the wavelength. Rayleigh scattering, where the scatterers are very much smaller than the wavelength, gives weak, nearly isotropic scatter which varies with the fourth power of the wavelength. When scattering occurs from particles whose size greatly exceeds the wavelength, so-called Mie scattering, it is highly forward directed and essentially independent of frequency. Scatter from structures of intermediate size, of the order of the wavelength, shows frequency and direction dependence between these limits, and is generally characteristic of tissue.

The effective attenuation coefficient is related to the absorption coefficient by the expression:

$$\alpha = \left[\frac{\kappa_a}{D}\right]^{\frac{1}{2}} \tag{3.3}$$

D is the **diffusion constant;**

$$D = [3(1-\bar{\mu})\kappa_s + \kappa_a]^{-1} \tag{3.4}$$

assuming that the differential scattering depends only upon the scattering angle. For isotropic scatter, and where $\kappa_s > \kappa_a$, $D \simeq (3\kappa_s)^{-1}$.

The **total attenuation coefficient**, κ, is the loss in a thin slab of material without the contribution due to forward scattering:

$$\kappa = \kappa_a + \kappa_s \tag{3.5}$$

The reciprocal of κ is the **mean free path**.

A further parameter which is related to the reflectivity of the medium is the **albedo**, c. For an extended surface the albedo is defined as the ratio of the radiant flux reflected from a surface to that incident upon it, and is also called the **remittance**, or more loosely the reflectance.

Other symbols used in the literature are Σ_a for κ_a, Σ_s for κ_s, and Σ_t for κ.

An alternative approach to the analysis of light transmission through turbid materials is the 2-flux, Kulbelka-Munk model (Kubelka and Munk, 1931; Kubelka, 1948). For the case of isotropic radiance at the surface, and assuming that the radiance remains isotropic through the slab of tissue the total radiant flux is considered to result from the combined effect of only an inward flux i and an outward flux j. Two coupled equations can be derived:

$$\begin{aligned} -di/dx &= Sj - (S + K)i \\ dj/dx &= Si - (S + K)j \end{aligned} \tag{3.6}$$

where dx is the thickness of an elemental layer of tissue, and S and K are Kubelka-Munk scattering and absorption coefficients. Care needs to be exercised in the comparison of these coefficients with the diffusion coefficients in general (Ertefai and Profio, 1985).

In general the optical parameters may be related to the Kubelka-Munk coefficients through the expressions

$$\begin{aligned} \kappa_a &= \eta K \\ \kappa_s(1-\bar{\mu}) &= \chi S \end{aligned} \tag{3.7}$$

where η and χ are scaling factors. For the Kubelka-Munk model $\eta = 0.5$ when $\kappa_s > \kappa_a$, approaching 1.0 as κ_s approaches zero. χ also increases with decreasing scatter, from 1.3 to 3.3 over the whole range (van Gemert *et al*, 1989). In a study of the optical transmission of tooth enamel, Spitzer and Ten Bosch (1975) measured χ to be in the range 1.33 to 1.44.

The optical attenuation of a tissue or fluid sample is sometimes given in terms of its **optical density**, OD, which is $\log_{10}$(1/attenuation). Since optical density clearly depends on the sample thickness, attenuation values quoted in terms of OD are not included in the tables in this chapter.

3.1.2 *Measurement of optical properties*

The measurement of the effective attenuation coefficient and penetration depth are in principle straightforward and can be performed on thick samples of tissue by standard narrow-beam experiments, taking care to avoid errors associated with forward scatter and autofluorescence. Although the direct measurement of absorption and scatter coefficients is also conceptually simple, since it involves the preparation of tissue slices of thickness of the order of 0.1 mm, it can be experimentally demanding. In addition, inhomogeneities in the tissue inevitably lead to a considerable spread in measured values.

Formerly, a wideband optical source was used with the wavelength selected by a monochromator. More recently monochromatic radiation has been generated using lasers at selected wavelengths. The integrating sphere radiometer (Clark *et al*, 1953; Dersken and Monahan, 1952) has been widely used as a detector in optical experiments. Most modern measurements have used photomultiplier detectors, which are of the solar-blind type if autofluorescence interference is to be avoided. Photoconductive cells (Dersken and Monahan, 1952) or semiconductor detectors (Parker and Diffey, 1983) have also been used.

3.1.3 *Historical background*

The study of the optical properties of tissue, and in particular of the properties of skin, has a history of many decades. Hasselbalch (1911) made early studies of ultraviolet transmission through the skin. Browning and Russ (1917) noted the greater penetration in the range 296 nm to 380 nm in comparison with that at shorter wavelegths. The penetration through skin by ultraviolet radiation was studied further by Lucas (1930), and by Bachem and Reed (1931), and by the early 1930s textbooks giving good scientific data an optical transmission, absorption and fluorescence of tissue were available (Russell and Russell, 1933). Early studies on the properties of skin in the infrared range were reported by Hardy (eg Hardy and Muschenheim, 1935; Hardy, 1939), and by Pearson and Norris (1933). Spectrophotometric studies of the colour of human skin were widely reported, including its dependence on pigmentation (Brunsting and Sheard, 1929a; Edwards and Duntley, 1939), and on blood supply (Brunsting and Sheard, 1929b). Indeed the investigation of skin pigmentation and the estimation of melanin concentration has continued to attract considerable interest (Buckley and Grum, 1961; Findlay, 1966; Ballowitz and Avery, 1970; Dawson *et al*, 1980; Kollias and Baqer, 1986). Whilst the early studies were instructive in establishing the general spectral characteristics of skin, their use as quantitative sources of data must be questioned. This largely results from the use of inadequate instrumentation, failing to account accurately for the diffuse nature of the transmission of light through tissue. There are two main sources of innacuracy. Firstly, off-axis scattered radiation may be totally internally reflected within the sample, and so be lost to the measurement. The second problem is that of autofluorescence (see Section 3.4).

Several useful reviews of the optical properties of tissue have been published, generally written in a particular clinical context. Wilson and Patterson (1986) reviewed the physics of photodynamic therapy, and McKenzie and Carruth (1984) have discussed aspects of the use of lasers in medicine. The optical properties of the tissues of the skin and the eye have been discussed in the context of safety by Sliney and Wolbarsht (1980), and skin optics reviewed by Anderson and Parrish (1981, 1982), and by van Gemert *et al* (1989).

3.1.4 *Values of optical properties of tissue*

Values of the effective optical attenuation coefficient for thick slabs, α, for a variety of human and animal tissues are given in Table 3.2. Measurements have more commonly been carried out at about 630 nm, because of the interest in photodynamic therapy using haematoporphyrin derivatives. Spectral data have also been measured for some tissues from ultraviolet wavelengths, 300 nm, to the infrared, 10.6 μm.

Penetration depth δ in mm may be calculated from the values of α using the expression $\delta = 1/\alpha$.

The optical absorption coefficient, κ_a, scatter coefficient, k_s, and the mean cosine of the scatter, $\bar{\mu}$, are given for some tissues in Table 3.3, together with values of the total attenuation coefficient for thin slabs, where this has also been reported.

Many reports in the literature have given only the percentage transmission of light at a particular wavelength through a particular organ or structure. Whilst this information is not an absolute property of the tissue, some values from the literature are also included in Table 3.4 where they usefully add to the data in Table 3.2.

Table 3.2 Effective optical attenuation coefficient, α, of tissue (thick slabs)

Tissue	λ,nm	α,mm^{-1}	Reference
Artery,human,*in-vitro*	266	30.0	Oraevsky *et al* 1988
-,-,-	355	5.5	-
-,-,-	510	1.8	-
-,-,-	1064	0.5	-
Brain,*in-vivo*,cat	405-410	4.41	Doiron *et al* 1983
-,-,pig	405-410	>10	Wilson *et al* 1984,1985
-,-,pig	500	0.91	-
-,-,cat	545	3.44	Doiron *et al* 1983
-,-,pig	550	1.1	Wilson *et al* 1984,1985
-,-,cat	577	2.59	Doiron *et al* 1983
-,-,pig	600	0.77	Wilson *et al* 1984,1985
-,-,human	630	0.45,0.56	Muller & Wilson 1986

cont.

Table 3.2 cont. Effective optical attenuation coefficient

Tissue	λ,nm	α,mm^{-1}	Reference
Brain,human,*in-vivo*	630	0.2–0.34	Wilson *et al* 1986a
–,–,cat	630	0.44–0.98	Doiron *et al* 1983
–,–,pig	630	0.37–0.45	Wilson *et al* 1984,1985
Brain,human,*in-vitro*	488	1.4–2.5	Svaasand & Ellingsen 1983
–,–,–,–	514	1.7,2.5	–
–,–,–,–	660	0.59–0.83	–
–,–,–,–	1060	0.23–0.34	–
–,–,–,newborn	488	0.59,0.77	–
–,–,–,–	514	0.59,0.91	–
–,–,–,–	660	0.19,0.27	–
–,–,–,–	1060	0.11,0.14	–
Breast,human,*in-vitro*	700	0.258	Ertefai & Profio 1985
–,–,–	800	0.218	–
–,–,–	900	0.220	–
–,–,–	1000	0.258	–
–,–,–	1050	0.261	–
Fat tissue,cow,*in-vitro*	630	0.42	Bolin *et al* 1987
–,–,–	633	0.34	Preuss *et al* 1982
Kidney,human,*in-vitro*	400	1.4	Eichler *et al* 1977
–,–,pig	400	6.7	–
–,–,human	630	0.40	–
–,–,pig	630	0.48	–
–,–,cow	633	0.92	Preuss *et al* 1982
–,–,human	800	0.37	Eichler *et al* 1977
–,–,pig	800	0.37	–
–,–,human	1000	0.63	–
–,–,pig	1000	0.53	–
Liver,*in-vivo*,dog	488-516	5.0	Kiefhaber *et al* 1977
–,–,–	1060	1.25	–
–,–,–	10600	20	–
–,*in-vitro*,human	400	5.0	Eichler *et al* 1977
–,–,pig	400	5.0	–
–,–,rabbit	500	0.8–3.3	Wilson *et al* 1984,1985
–,–,rabbit	600	1.2–3.3	–
–,–,human	630	1.1	Eichler *et al* 1977
–,–,pig	630	0.50	–
–,–,cow	633	1.07	Preuss *et al* 1982
–,–,pig	633	1.30	Doiron *et al* 1983
–,–,rabbit	700	0.67–2.0	Wilson *et al* 1984,1985
–,–,human	800	1.05	Eichler *et al* 1977
–,–,pig	800	0.34	–
–,–,rabbit	800	0.56–2.0	Wilson *et al* 1984,1985
–,–,human	1000	2.0	Eichler *et al* 1977
–,–,pig	1000	0.42	–

cont.

Table 3.2 cont. **Effective optical attenuation coefficient**

Tissue	λ,nm	α, mm^{-1}	Reference
Lung,human,deflated	633	1.1	Doiron *et al* 1983
–,–,mucosal membrane	633	0.91	–
Muscle:cardiac,cow, *in–vitro*	600	2.1	Bolin *et al* 1984
–,–,–	630	1.02	–
–,–,–	700	0.69	–
–,–,–	800	0.50	–
–,–,–	900	0.48	–
–,–,–	1000	0.52	–
–,–,–	1100	0.37	–
Muscle:skeletal,rabbit, *in–vivo*	400	0.77	Wilson *et al* 1984,1985
–,–,–	500	0.50	–
–,–,–	514	0.45–0.62	–
–,–,–	514.5	0.48–0.77	Doiron *et al* 1983
–,–,–	550	0.67	Wilson *et al* 1984,1985
–,–,–	630	0.14–0.23	Doiron *et al* 1983
–,–,–	630	0.27–0.48	Wilson *et al* 1984,1985
–,–,–	800	0.37	–
–,cow,*in–vitro*	300	0.95	Bolin *et al* 1984
–,–,–	514	0.20,0.25	Doiron *et al* 1983
–,–,–	630	0.73	Bolin *et al* 1984
–,–,–	633	0.56	Doiron *et al* 1983
–,–,–	633	0.68±0.05	McKenzie & Byrne 1988
–,–,–	700	0.37	Bolin *et al* 1984
–,–,–	800	0.30	–
–,–,–	900	0.33	–
–,–,–	1000	0.39	–
–,–,–	1100	0.31	–
Nail,human,*in–vitro*	300	6.5±1.2	Parker & Diffey 1983
–,–,–	310	6.8±1.5	–
–,–,–	350	4.1±0.6	–
–,–,–	400	2.2±0.24	–
–,–,–	500	1.2±0.12	–
–,–,–	600	0.8±0.05	–
Skin,human,white,*in–vitro*	1230	1.01±0.31	Hardy *et al* 1956
–,–,–,–	2200	2.69±0.86	–
–,–,Negro,*in–vitro*	950	1.12±0.34	–
–,–,–,–	1230	0.92±0.19	–
–,–,–,–	1680	1.34±0.28	–
–,–,–,–	2200	3.26±0.47	–
Stomach,dog,*in–vivo*	488–516	2.8	Kiefhaber *et al* 1977
–,–,–	1060	0.57	–

Table 3.3 Optical absorption and scatter coefficients for normal animal tissue

Tissue	λ nm	κ_a mm^{-1}	κ_s mm^{-1}	κ mm^{-1}	$\bar{\mu}$	Reference
Bladder,human	633	0.14	8.8		0.96	Cheong *et al* 1987
–,dog	633	0.11	4.4		0.92	–
Blood,0.4% Hct	633			2.9	0.974	Flock *et al* 1987
–,45% Hct	660		167		0.995	Schmitt 1986, and
–,–	820		125		0.994	Steinke & Shepherd 1987
Brain,pig	633			68.7±9.7	0.940±0.029	Flock *et al* 1987
–,–	630			83		Crilly 1987
–,–	633	0.026	5.7	95		Wilson *et al* 1986b,87
Breast,human	630	0.02	39.4		~0.95	Andreola *et al* 1988
–,–	950	0.067				Navarro & Profio 1988
–,–	700–1200			7–40		Crilly 1987
Fatty,pig	633			37.6±6.9	0.771±0.063	Flock *et al* 1987
Liver,human	630	0.32	41.4		~0.95	Andreola *et al* 1988
Lung,human	630	0.84	35.9		~0.95	Andreola *et al* 1988
Muscle:skeletal,cow	633	0.035±0.01	4.5±1.4			McKenzie & Byrne 1988
–,cow	633			32.8±3.7	0.954±0.016	Flock *et al* 1987
–,chicken	633			34.5±4.2	0.965±0.004	–
–,pig	633	0.1	4.0	4.1	0.97	Wilksch *et al* 1984
Skin,human	630	0.18	39.4			Andreola *et al* 1988
Skin:stratum corneum, human,*in-vivo*	248			86.6		Bruls *et al* 1984
–,–,–	265			114		–
–,–,–	280			126		–
–,–,–	302			56.4		–
–,–,–	365			22.4		–
–,–,–	546			11.6		–

cont.

Table 3.3 cont. Optical absorption and scatter coefficients

Tissue	λ,nm	κ_a,mm^{-1}	κ_s,mm^{-1}	κ,mm^{-1}	$\bar{\mu}$	Reference
Skin:epidermis, human,caucasian*	270	70	15			Wan *et al* 1981a
–,–,–	400	6	5			–
–,–,–	600	2.5	2.5			–
–,–,black*	270	122	25			–
–,–,–	400	48	9.5			–
–,–,–	600	24	7.0			–
–,–,*in–vivo*	248			110		Bruls *et al* 1984
–,–,–	265			126		–
–,–,–	280			120		–
–,–,–	302			42.3		–
–,–,–	365			23.9		–
–,–,–	546			10.5		–
Skin:dermis,human, 85% water	633	0.27	18.7	19.0	0.82	Jacques *et al* 1987
–,mouse	488	0.28±0.15	23.8±6.7	24.2	0.74±0.07	Jacques & Prahl 1987
Tooth:enamel,human	220	1.19	38			Spitzer & Ten Bosch 1975
–,–	300	0.72	23			–
–,–	400	0.09	9.3			–
–,–	550	0.07	4.0			–
–,–	700	0.07	1.2			–
–,cow	220	3.3	41			–
–,–	300	2.0	29			–
–,–	400	0.14	11.6			–
–,–	550	0.07	4.0			–
–,–	700	0.07	~0			–
Uterus,human	633	0.05	40.2		~0.95	Andreola *et al* 1988

* Calculated from Kubelka–Munk theory.

Table 3.4 Optical transmittance through some tissue segments

Tissue	λ,nm	Transmittance	Reference
Eyelid,monkey	400	0.02	Crawford & Marc 1976
–,–	500	0.07	–
–,–	600	0.15	–
–,–	700	0.20	–
–,white cat	400	0.045	–
–,–	500	0.15	–
–,–	600	0.25	–
–,–	700	0.40	–
Nictating membrane, white cat	400	0.01	–
–,–	500	0.05	–
–,–	600	0.10	–
–,–	700	0.12	–
Eye:sclera	633	$(6.5–9)x10^{-2}$	Fine *et al* 1985
Abdominal wall,22–32mm			
	400	$<10^{-6}$	Wan *et al* 1981b
–	600	$(0.5–5)x10^{-5}$	–
–	800	$10^{-3}–10^{-2}$	–
Chest wall,16–22mm	400	$<10^{-6}$	–
–	600	10^{-4} max	–
–	800	$(0.5–2)x10^{-2}$	–
Skull + scalp,9–13mm	400	$10^{-4}–10^{-5}$	–
–	600	$(0.5–1)x10^{-3}$	–
–	800	$5x10^{-2}$	–
Scrotal wall,7mm	400	10^{-5}	–
–	600	$(0.5–5)x10^{-3}$	–
–	800	0.05–0.1	–

3.1.4.1 Attenuation: variation with wavelength

Broadly, two optical windows exist in tissue. The main one lies between 600 and 1300 nm and a second from 1600 to 1850 nm. The latter window is bounded by two water absorption bands. Further commentary about the variation of optical properties with wavelength are discussed is sections dealing with particular tissues, the eye, skin and blood.

3.1.4.2 Scatter

Angular scattering functions were fully reported for skin layers by Hardy *et al* (1956) at wavelengths from 550 to 2200 nm (Figure 3.1). The wavelength dependence of scatter was demonstrated for thin samples. Scatter through thick samples demonstrated a broad scatter which was largely independent of wavelength. The reflectance scattering function was shown to be substantially unaffected by sample thickness. Results from more detailed studies in the ultraviolet bands (Bruls and van der Leun, 1984) are given in Table 3.5.

Experimental determinations of the mean cosine of scatter show that all skin layers are strongly forward scattering in the range of wavelengths from 240 nm to 633 nm. A predictive equation which may be used to give estimates of the mean scatter cosine, $\bar{\mu}$, for both epidermal and dermal skin layers has been obtained by van Gemert *et al* (1989) from an analysis of published data. This expression is:

$$\bar{\mu} \simeq 0.62 + \lambda 0.29\times10^{-3} \tag{3.8}$$

The wavelength, λ, is in nm.

3.1.5 *Factors affecting the optical properties*

It should be stressed that there is a very wide spread in the observations of optical properties which have been reported from experimental studies. This results from the substantial optical inhomogeneity of tissue, particularly at a small scale. Considerable caution should be exercised when applying the data presented in these tables, and due regard be given to the fact that optical properties appear to vary quite widely for any particular tissue type.

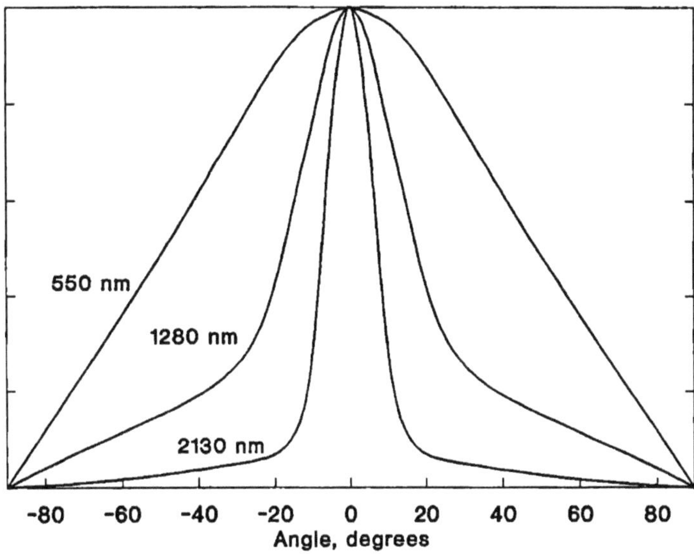

Figure 3.1 The angular scatter through a thin layer of skin, and its dependence on wavelength. From Hardy *et al* (1956).

Table 3.5 Forward scatter from human skin layers; percentage of transmitted energy within angle θ with the normal

	angle, θ	λ,nm 254	302	365	436	546	Diffuse
Stratum	2.5°	9.3	11.6	14.8	17.6	20.6	0.19
Corneum	7.5°	40.2	45.1	50.1	55.2	60.2	1.71
	12.5°	59.2	63.0	66.8	70.9	75.2	4.4
	17.5°	71.5	73.7	76.5	79.1	82.3	9.0
	22.5°	79.2	80.6	82.3	84.1	86.5	14.6
	27.5°	85.1	85.6	86.3	87.6	89.4	21.3
	42.5°	95.0	94.5	94.1	94.4	95.0	45.6
	62.5°	100	100	100	100	100	78.8
Epidermis	2.5°		1.3	1.7	2.6	4.0	0.19
	7.5°		9.4	11.9	16.1	22.5	1.71
	12.5°		20.8	24.8	30.8	39.7	4.4
	17.5°		33.2	37.5	43.6	52.5	9.0
	22.5°		45.5	49.0	54.5	62.4	14.6
	27.5°		56.9	59.8	63.9	70.2	21.3
	42.5°		83.2	84.2	85.0	87.2	45.6
	52.5°		94.1	94.3	94.3	94.9	63.0
	62.5°		100	100	100	100	78.7

Source: Bruls and van der Leun (1984).

3.1.5.1 Changes following death

In-vivo measurements are sparse. There is some evidence that the effective attenuation coefficient falls following death, in the range 400 to 800 nm (Wilson *et al*, 1984, 1985a). This change was related to blood drainage and deoxygenation since it was particularly noticeable below 600 nm where absorption by haemoglobin is important. Similarly Doiron *et al* (1983) recorded that the optical attenuation in a live cat brain was higher than that of brain tissue *in-vitro*, over a similar range of wavelengths.

3.1.5.2 Tissue composition

At infrared wavelengths the optical absorption coefficient becomes increasingly strongly dependent on the tissue water content. Walsh and Deutsch (1989) have used this fact to predict the optical attenuation of tissue, α_t, at 10.6 μm, using the simple expression $\alpha_t = \alpha_w W$. W is the percentage water content in the tissue and α_w is the absorption coefficient of water at the wavelength of interest. At 10.6 μm, $\alpha_w = 79.4\ mm^{-1}$.

The presence of fat in muscle *in-vivo* was reported by Doiron *et al* (1983) to reduce the attenuation of the muscle. For instance at 630 nm, rabbit muscle with a high fat content had $\alpha = 0.14\ mm^{-1}$, whilst the attenuation coefficient of low-fat muscle was $0.23\ mm^{-1}$.

3.1.5.3 Tissue pathology

Insufficient reports of the optical properties of pathological tissues exist at present to warrant compilation of these data. Human tumour tissue has been investigated by Svaasand and Ellingsen, 1985 (brain), Profio and Doiron, 1981 (lung) and Navarro and Profio, 1988 (breast). Animal tumours have been studied by Arnfield *et al*, 1988 and 1989, and van Gemert *et al*, 1985. The optical properties of atheromatous plaque have been reported by Kaminow *et al*, 1984, and Oraevsky *et al*, 1988.

3.1.6 *Optical properties of skin*

Melanin is the main constituent of the epidermis which controls the transmission of ultraviolet wavelengths, and is the pigment which determines the discernible colour of the skin. Two to three orders of magnitude difference in transmittance through the epidermis between fair and dark–skinned individuals has been reported (Anderson and Parrish, 1982). A five–fold difference in reflectance was reported by Kuppenheim and Heer (1952) between white and negro skin. For this reason only general statements can be made about the transmission properties of skin at wavelengths up to about 1000 nm. Since melanin absorption is negligible above about 1100 nm, neither transmission nor diffuse remittance are affected by skin colour at infrared wavelengths (Hardy *et al*, 1956; Jacquez *et al*, 1955).

Epidermal transmission is characterised by a broad absorption band at about 275 nm (due to absorption by aromatic chromaphores). At higher wavelengths, in the visible and infrared bands, water absorption lines are seen at about 760, 1000, 1200, 1450 and 2000 nm. (Jacquez *et al*, 1955; Jacquez and Kuppenheim, 1955). Additionally, absorption lines associated with oxyhaemoglobin occur at about 420, 540 and 580 nm. Anderson *et al* (1981) have reported Kubelka–Munk coefficients for human dermis which show a broad transmission band between 700 nm and 1200 nm, with an absorption peak at about 1900 nm. The attenuation coefficients are always lower in the dermis than in the epidermis, over a broad range of frequencies (van Gemert *et al*, 1989; Everett *et al*, 1966). For skin as a whole, Anderson and Parrish (1981) have given estimates of penetration depths over the range 250 nm to 1200 nm, and these values are presented in Figure 3.2.

Findlay (1970) reported an investigation of the apparent blue colour of some skin, and its relationship to the presence of melanin, a deep brown pigment. He concluded that the epidermis was little involved, and the blue colour resulted from the scattering characteristics of the dermis causing subtractive colour mixing. Light scattering in the dermis is such that scattered light at shorter wavelengths has shorter path–lengths, encounters less melanin and therefore suffers less absorption.

Apart from water, melanin and haemoglobin, other components which may influence the reflectance spectrum of the skin are fibrous protein, collagen, fat, bilirubin and carotene. The main absorption band from carotenoids in blood is at approximately 460 nm (Buckley and Grum, 1961).

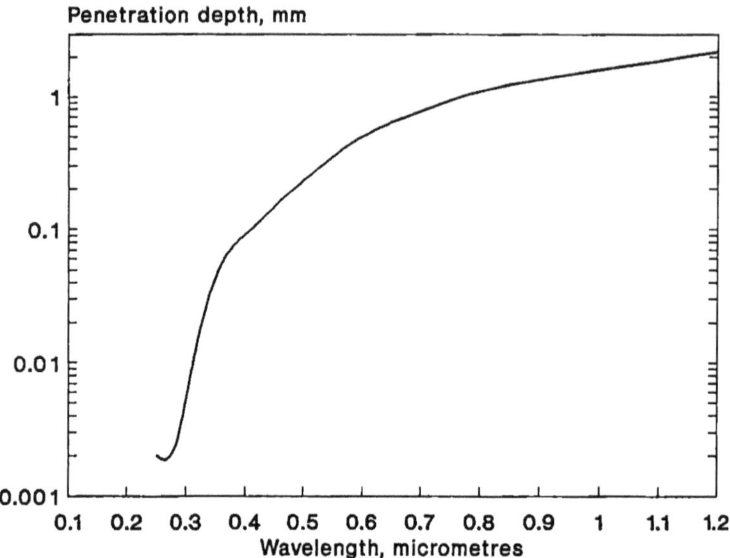

Figure 3.2 Estimated penetration depths in fair caucasian skin; from Anderson and Parrish (1981).

3.1.7 *Optical properties of blood*

Spectrophotometric studies of whole blood and of its components were intensively reported by Drabkin and co–workers from the early 1930s onwards (eg Drabkin and Austin, 1932, 1935a and 1935b for *in–vitro* studies, and Drabkin and Schmidt (1945) and Gordy and Drabkin (1957) for *in–vivo* measurements). The absorption coefficient of blood components is usually given in terms of molar concentration. The variation of absorption of reduced haemoglobin and oxyhaemoglobin over the range of wavelengths from 200 to 1100 nm is shown in Figure 3.3, from Gordy and Drabkin (1957). Buckley and Grum (1961) give absorption bands at 273, 343, 415, 542 and 575 nm in diluted, oxygenated whole blood, lysed or heparinised, and in solutions of oxyhaemoglobin. Reduced haemoglobin shows absorption bands at 425 and 555 nm.

Transmittance through whole blood is also influenced strongly by scattering from cells, an effect which is clearly dependent on haematocrit. As with other tissues, the simple exponential loss with distance predicted by Beer's law does not hold (Kramer *et al*, 1951). The theory of light transmission through blood has subsequently been investigated by several authors (Twersky, 1970; Pedersen *et al*, 1976; Steinke and Shepherd, 1986). Anderson and Sekelj (1967a, 1967b) have investigated the absorbing and

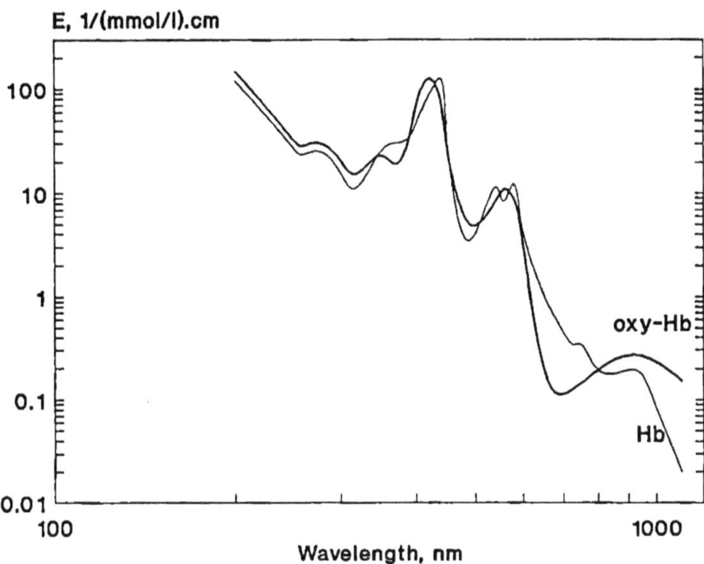

Figure 3.3 Optical attenuation in haemoglobin (Hb) and oxyhaemoglobin (oxy–Hb) solutions over the range of wavelengths from 200 nm to 1100 nm. Gordy and Drabkin (1957).

scattering properties of whole blood. It was shown that Twersky's theory, separating the contributions to optical density into independent absorption and scattering components, was well supported by experiment. Absolute measurements of absorption and scatter coefficients for blood were not reported. However it was demonstrated that the relationship between light scattering and red–cell concentration was parabolic, and that the absorption of light within the erythrocyte was shown to be the same as in a solution of haemoglobin. Reflectance and transmittance were found to be linearly related in thin films of blood.

In a theoretical analysis, Takatani and Grayham (1979) predicted that, for instance, the penetration depth, δ, in whole blood at 500 nm is 0.06 mm at 45% haematocrit and 0.505 mm at 4.5% haematocrit, for an oxygen saturation of 100%. Penetration depth increases slightly to 0.067 mm and 0.560 mm respectively for an oxygen saturation of 40%. At 955 nm and 100% saturation the predicted penetration depths are 0.403 mm and 2.475 mm at 45% and 4.5% haematocrit, whilst at 40% oxygen saturation the values are 0.441 mm and 2.630 mm.

3.1.8 *Optical transmission through the eye*

Transmission and absorption of optical wavelengths in the media composing the eye has been widely studied and reported. Early studies by Ludvigh and McCarthy (1938) have been subsequently criticised, giving somewhat higher absorption than later studies. Reviews by Wyszecki and Stiles (1967) and Norren and Vos (1974) are useful. The most widely quoted values are those from Boettner and Wolter (1962) and Geeraets and Berry (1968) for values of transmission through typical human eye components. The variation with wavelength is presented in Figure 3.4 and Table 3.6 as percentage transmission as far as a particular structure. Absorption coefficients are generally not given in the literature.

For the young adult the cornea transmits radiation between about 300 nm and 2500 nm, reaching over 80% transmission by 400 nm. Water absorption bands appear at 1430 nm and 1950 nm. The aqueous humor exhibits a strong absorption band at 265 nm (Kinsey, 1948; Cooper and Robson, 1969), and behaves in the infrared as water. Lens transmission extends from the ultraviolet to an upper limit of 1900 nm in the infrared. Ageing strongly affects transmission at shorter wavelengths. A transmission band in the lens at about 320 nm under 5 years of age, and in some animals, is largely absent in the adult human. The yellowing of the lens with age has been widely studied (for instance, Said and Weale, 1959; Cooper and Robson, 1969). Boettner and Wolter (1962), report that the transmission through a lens of a young eye reaches 90% at 450 nm, whereas through a

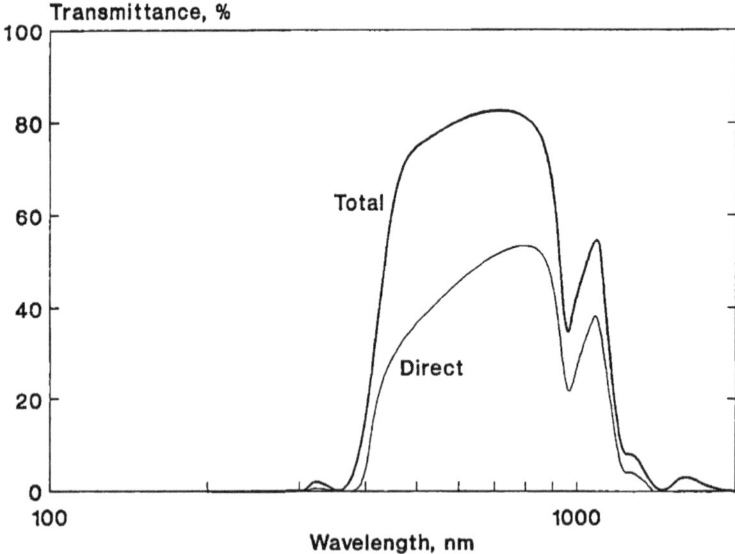

Figure 3.4 Transmittance reaching the retina expressed as a percentage of the incident radiation reaching the eye. For direct (collimated) and total (uncollimated) light. Boettner and Wolter (1962).

Table 3.6 Percentage transmission through the eye reaching the anterior surface of a particular structure

λ	Cornea	Aqueous humor		Lens		Vitreous humor		Retina	
nm	total	direct	total	direct	total	direct	total	direct	total
230	2.7%		–		–		–		–
250	26.0%		–		–		–		–
270	28.7%		–		–		–		–
290	51.5%		2.0%		0.4%		–		–
300	70.0%		27.0%		14.3%		–		–
310	78.0%		64.0%		50.5%		–		–
320	81.0%		78.0%		74.0%		0.3%		0.3%
350	86.5%		86.0%		82.5%		1.8%		1.6%
370	89.5%		90.0%		87.0%		12.1%		11.0%
400	94.5%		94.0%		93.0%		68.5%		63.5%
450			96.0%		96.0%		84.0%		80.5%
500		63%	88%	60%	83%	47%	78%	37%	75%
600		70%	91%	67%	88%	58%	84%	46%	81%
700		75%	92%	73%	90%	64%	87%	52%	83%
800		78%	93%	76%	91%	67%	87%	54%	82%
900		80%	94%	76%	88%	68%	82%	52%	75%
1000		80%	92%	69%	80%	54%	69%	23%	40%
1100		83%	93%	75%	81%	63%	76%	40%	55%
1200		79%	89%	51%	59%	34%	39%	4%	9%
1300		81%	89%	55%	60%	37%	42%	5%	7%
1400		45%	67%	–	10%	–	–	–	–
1700		63%	70%	10%	11%	–	3%	–	–
2000		–	2%	–	–	–	–	–	–
2200		25%	30%	–	–	–	–	–	–

Values for 230–450 nm for albino rabbit eyes, from Kinsey (1948): values for 500–2200 nm for young human eyes, from Boettner & Wolter (1962). Direct means collimated, and total means uncollimated light. Also see Pitts (1959), cow; Boettner & Wolter (1962), rabbit and monkey.

63 year old lens this transmission was reached only at 540 nm. However, a wide variation in ageing characteristics were reported. The vitreous humor transmits from 300 nm to 1400 nm with strong water absorption bands at 980 nm and 1200 nm.

The scattering contribution to the total loss was also evaluated by Boettner and Wolter, showing scatter to be contributory in all components except the aqueous humor. Figure 3.4 shows separately the transmittance due to the direct and total radiation at the retina. Others have investigated scattering in the sclera (Fine *et al*, 1985), the cornea (Maurice, 1957; Feuk and McQueen, 1971) and in the lens (Mellerio, 1971). Bettelheim and Kaplan (1973) noted that the birefringence of the cornea (Maurice, 1957) affects the small angle light scattering. In general scatter is of greater importance at shorter wavelengths and becomes less significant towards the infrared end of the spectrum.

3.1.9 *Optical properties of tooth enamel*

Boehm and Cunnington (1987) have reported a broad transmittance peak for both bovine and human dental enamel between about 500 and 2000 nm. Reflectance peaked at about 700 nm, and was low and constant at about 5% above 2000 nm. Spitzer and Ten Bosch (1975) report an absorption peak for human enamel at 270±5 nm.

3.1.10 *Optical properties of some materials other than tissue*

3.1.10.1 Water

The attenuation of a wide range of wavelengths by pure water, associated with the refractive index of water (Section 3.3) has been reviewed by Hale and Querry (1973), amongst others. Values are given in table 3.9, and the general variation of the absorption of water over the range of wavelength from 200 nm to 100 μm is shown in Figure 3.5.

3.1.10.2 Tissue substitute materials

Several materials and mixtures have been reported whose optical properties are intended to match those of tissue. Marynissen and Star (1984) used a liquid phantom composed of polyvinyl acetate paticles ('pure scatterers') and india ink ('pure absorber'). A similar mixture using a suspension of fat particles ('Intralipid') and ink has been investigated by Flock *et al* (1989). The optical properties of Intralipid suspensions have also been reported by Driver *et al* (1989). Wilson *et al* (1985) reported the use of a gel phantom consisting of agar and blood. They report penetration depths in pure 2% agar gel increasing from 3.5 mm at a wavelength of 400 nm to 18 mm at 700 nm. In order to simulate the optical properties of the breast in the range 500 to 950nm, Linford *et al* (1986) used a mixture of non-dairy

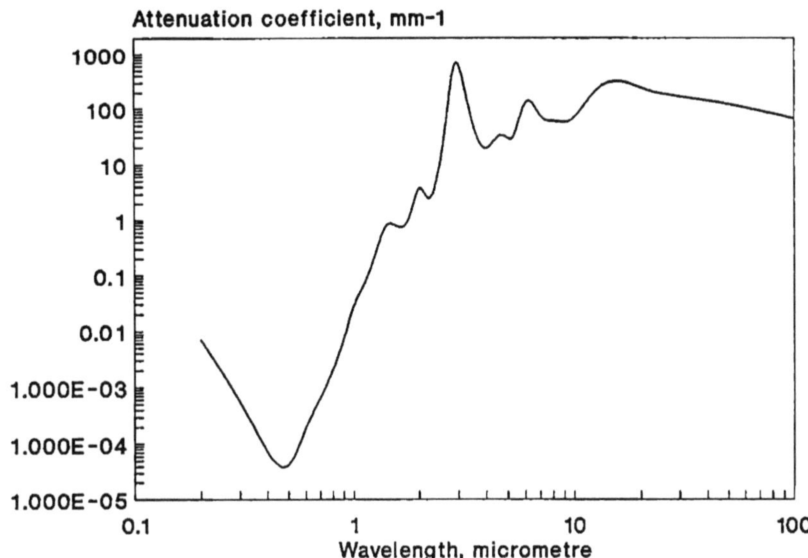

Figure 3.5 Optical absorption coefficient for water from 200 nm to 100 μm.

creamer in Methocel and saline as the scattering component and haemolised blood as the absorbing component. A suitable mixture which completely matches tissue and is stable with time does not appear to have been reported.

3.2 Reflectance and emissivity

According to Anderson and Parrish (1981) the reflectance of normally incident radiation over the entire spectrum from 250 nm to 2000 nm lies between 4% and 7% for both caucasian and negroid skin. Hardy (1973) also reports measurements on the normal reflectance of both wet and dry skin extending from 4 μm to 30 μm, giving a value of about 5% over the whole range with almost no difference between wet and dry skin.

The emissivity of human skin above 1000 nm has been demonstrated by a number of authors to approximate to unity. Together with the observations of reflectivity, this consitutes good evidence that skin may be considered as a black–body radiator over normal temperature ranges. Exposed surfaces of

Table 3.7 **Emissivity of tissue**

Tissue	Wavelength band, μm	Emissivity	Reference
Epicardium,*in-vitro*	1–14	0.83	Steketee 1973
Fat,cow	0.4–1.5	0.86±0.05	Guerassimov & Runyatsev
Muscle,cow	0.4–1.5	0.90±0.05	1972
–,pig	0.4–1.5	0.92±0.05	–
Skin, *in-vivo*	2–5.4	≥0.98	Watmough & Oliver 1968
–,*in-vivo/vitro* normal or burnt	1–14	0.98±0.01	Steketee 1973
–,*in-vitro*,white	up to 20	0.997±0.001	Mitchell *et al* 1967
–,–,Bantu,hydrated	up to 20	0.997±0.006	–
–,–,–,10% water	up to 20	0.919±0.014	–

other tissues are reported as demonstrating slightly lower emissivity (Steketee, 1973; Guerassimov and Rumyantsev, 1972), and lower values are also found from dehydrated skin (Mitchell *et al*, 1967). Some values of tissue emissivity are compiled in Table 3.7.

3.3 Refractive index

3.3.1 *Terminology and definitions*

The **refractive index**, n(λ), and the **extinction coefficient**, k(λ), of a material are the real and imaginary parts of the spectral complex refractive index $\tilde{n}(\lambda)$:

$$\tilde{n}(\lambda) = n(\lambda) - ik(\lambda) \tag{3.9}$$

where λ is the wavelength of the electromagnetic wave in vacuum. The extinction coefficient k at any wavelength is related to the absorption coefficient, α, by the relationship;

$$\alpha = 4\pi k/\lambda \tag{3.10}$$

3.3.2 *Refractive index of tissue*

The measured refractive index for soft tissues lies in the range 1.33–1.55 (Table 3.8). The refractive index of tooth enamel is higher, reaching 1.73 at 220 nm.

Table 3.8 Refractive index of tissues and tissue components

Tissue substance	Refractive index	Reference
CSF,dog	1.3349–52	Krause & Yudkin 1930
Cytoplasm	1.350–1.367	Barer *et al* 1953
Eye:aqueous humor,dog	1.3348–53	Krause & Yudkin 1930
–,man	1.336	Westheimer 1972
Eye:cornea,man	1.376	Westheimer 1972
–,ox	1.382	Aurell & Holmgren 1953
–,pig	1.373	–
–,fibrils	1.47	Maurice 1957
–,ground substance	1.35	–
Eye:lens,man,surface	1.386	Westheimer 1972
–,–,centre	1.406	–
–,cat	1.554	–
Eye:vitreous humor,man	1.336	Westheimer 1972
Mesentary,rat,*in–vitro*	1.52±0.01	Gahm & Witte 1986
–,–,*in–vivo*	1.38±0.1	–
Stratum corneum	1.55	Scheuplein 1964
Tears, human	1.3361–1.3379	Altman & Dittmer 1972
Tooth enamel, 220nm	1.73	Spitzer & Ten Bosch 1975
–,400–700nm	1.62	–
–, apatite	>1.623	Houwink 1974

3.3.3 *Refractive index of water*

The refractive index and attenuation coefficient of pure water has been widely studied. Selected values from a survey by Hale and Querry (1973) are given in Table 3.9.

3.4 Fluorescence of tissue

Several authors have reported on the autofluorescence of skin tissues. Light absorbed at one wavelength is re–emitted at a different, longer wavelength. Anderson and Parrish (1982) have noted that autofluorescence may be an important source of error in the measurement of dermal transmittance at lower wavelengths, and suggest the use of a solar–blind photomultiplier tube to overcome the problem. Anderson reports a broad emission band at 330 nm to 360 nm from an excitation band near 280 nm, consistent with tryptophan or tyrosine fluorescence. A green autofluorescence in human epidermal cells was observed by Fellner (1976), *in–vitro*, showing optimum fluorescence in the range 500 nm to 530 nm from absorption at 300 nm to

Table 3.9 Refractive index and extinction coefficient for water

Wavelength λ, μm	Extinction coefficient, $k(\lambda)$	Refractive index, $n(\lambda)$	Attenuation coeff, α mm^{-1}
0.2	1.1×10^{-7}	1.396	6.9×10^{-3}
0.3	1.6×10^{-8}	1.349	6.7×10^{-4}
0.4	1.86×10^{-9}	1.339	5.84×10^{-5}
0.5	1.00×10^{-9}	1.335	2.51×10^{-5}
0.6	1.09×10^{-8}	1.332	2.28×10^{-4}
0.7	3.35×10^{-8}	1.331	6.01×10^{-4}
0.8	1.25×10^{-7}	1.329	1.96×10^{-3}
0.9	4.86×10^{-7}	1.328	6.79×10^{-3}
1.0	2.89×10^{-6}	1.327	3.63×10^{-2}
1.2	9.89×10^{-6}	1.324	0.104
1.4	1.38×10^{-4}	1.321	1.24
1.6	8.55×10^{-5}	1.317	0.671
1.8	1.15×10^{-4}	1.312	0.803
2.0	1.10×10^{-3}	1.306	6.91
2.2	2.89×10^{-4}	1.296	1.65
2.4	9.56×10^{-4}	1.279	5.00
2.6	6.7×10^{-3}	1.219	31.8
2.8	1.15×10^{-1}	1.142	516
3.0	2.72×10^{-1}	1.371	1139
3.5	9.4×10^{-3}	1.400	33.7
4.0	4.60×10^{-3}	1.351	14.5
4.5	1.34×10^{-2}	1.332	37.4
5.0	1.24×10^{-2}	1.325	31.1
5.3	9.8×10^{-3}	1.312	23.2
6.1	1.31×10^{-1}	1.319	270
7.0	3.20×10^{-2}	1.317	57.4
8.0	3.43×10^{-2}	1.291	62.7
9.0	3.99×10^{-2}	1.262	55.7
10.0	5.08×10^{-2}	1.218	63.8
12.5	2.59×10^{-1}	1.123	260
15.0	4.02×10^{-1}	1.270	337
17.5	4.29×10^{-1}	1.401	308
20.0	3.93×10^{-1}	1.480	247
25.0	3.56×10^{-1}	1.531	179
50.0	5.14×10^{-1}	1.587	129
100.0	5.32×10^{-1}	1.957	66.8
200.0	5.04×10^{-1}	2.130	31.7

Source: Hale and Querry (1973).

330 nm. At longer wavelengths still, Pottier *et al* (1986) report red 700 nm fluorescence from mouse skin *in-vivo*, from excitation at 410 nm, suggesting the presence of a porphyrin.

References

Altman P.L. and Dittmer D.S., 1972, *Biology Data Book,* 2nd edition, Federation of American Societies for Experimental Biology.

Anderson N.M and Sekelj P., 1967a, Light-absorbing and scattering properties of non-haemolysed blood, Phys Med Biol, 12, 173-184.

Anderson N.M. and Sekelj P., 1967b, Reflection and transmission of light by thin films of non-haemolysed blood, Phys Med Biol, 12, 185-192.

Anderson R.R and Parrish J.A., 1981, The optics of the human skin, J Invest Dermatol, 77, 13-19.

Anderson R.R. and Parrish J.A., 1982, Optical properties of human skin. In *The Science of Photomedicine*, J.D. Regan and J.A. Parrish (eds), Plenum, NY, pp 147-194.

Anderson R.R., Hu J. and Parrish J.A., 1981, Optical radiation transfer in the human skin and application in *in vivo* remittance spectroscopy. In *Bioengineering and the Skin,* R. Marks and P.A. Payne (eds), MTP, Lancaster, pp.253-265.

Andreola S., Bertoni A., Marchesini R. and Melloni E., 1988, Evaluation of optical characteristics of different human tissues *in vitro*, Lasers Surg Med, 8, 142 (abstract).

Arnfield M.R., Tulip J. and McPhee M.S., 1988, Optical propagation in tissue with anisotropic scattering, IEEE Trans Biomed Eng, BME-35, 372-381: erratum, IEEE Trans Biomed Eng, BME-36, 572.

Arnfield M.R., Tulip J., Chetner M. and McPhee M.S., 1989, Optical dosimetry for interstitial photodynamic therapy, Med Phys, 16, 602-608.

Aurell G. and Holmgren H., 1953, On the metachromic staining of corneal tissue and some observations on its transparency, Acta Ophthal Kbh, 31, 1-27.

Bachem A. and Reed C.I., 1931, The penetration of light through human skin, Am J Physiol, 97, 86-91.

Ballowitz L. and Avery M.E., 1970, Spectral reflectance of the skin, Biol Neonate, 15, 348-360.

Barer R., Ross K.F.A. and Tkaczyk S., 1953, Refractometry of living cells, Nature, 171, 720-724.

Bettelheim F.A. and Kaplan D., 1973, Small angle light scattering of bovine cornea as affected by birefringence, Biochim Biophys Acta, 313, 268-276.

Boehm R.F. and Cunnington G.R., 1987, Measurement of the reflectance of tooth enamel, Trans ASME; J Heat Transfer, 109, 812-815.

Boettner E.A. and Wolter J.R., 1962, Transmission of the ocular media, Invest Ophthalmol, 1, 776-783.

Bolin F.P., Preuss L.E. and Cain B.W., 1984, A comparison of spectral transmittance for several mammalian tissues: effects at PRT frequencies. In *Porphyrin Localization and Treatment of Tumors*, Alan R. Liss Inc., pp.115-132.

Bolin F.P., Preuss L.E. and Tuharsky D.R., 1987, Determination of penetration depths of visible and near IR light (400-1100 nm) in mammalian tissue by transmission spectroscopy, Lasers Surg Med, 7, 72 (abstract).

Browning C.H. and Russ S., 1917, The germicidal action of ultra-violet radiation, and its correlation with selective absorption, Proc R Soc, London, 90 B, 33-38.

Bruls W.A.G. and van der Leun J.C., 1984, Forward scattering properties of human epidermal layers, Photochem Photobiol, 40, 231-242.

Bruls W.A.G., Slaper H., van der Leun J.C. and Berrens L., 1984, Transmission of human epidermis and stratum corneum as a function of thickness in the ultraviolet and visible wavelengths, Photochem Photobiol, 40, 485-494.

Brunsting L.A. and Sheard C., 1929a, The color of the skin as analyzed by spectrophotometric methods II. The rôle of pigmentation, J Clin Invest, 7, 575-591.

Brunsting L.A. and Sheard C., 1929b, The color of the skin as analyzed by spectrophotometric methods III. The rôle of superficial blood, J Clin Invest, 7, 593-613.

Buckley W.R. and Grum F., 1961, Reflection spectrophotometry, Arch Dermatol, 83, 249-261.

Cheong W.F., Motamedi M. and Welch A.J., 1987, Optical modeling of laser photocoagulation of bladder tissue, Lasers Surg Med, 7, 72 (abstract).

Clark C., Vinegar R. and Hardy J.D., 1953, Goniometric spectrometer for the measurement of diffuse reflectance and transmittance in the infrared spectral region, J Opt Soc Am, 43, 993-998.

Commission International de l'Eclarage - International Commission on Illumination, 1970, International Lighting Vocabulary, CIE.

Cooper G.F. and Robson J.G., 1969, The yellow colour of the lens of man and other primates, J Physiol, 203, 411-417.

Crawford M.L.J. and Marc R.E., 1976, Light transmission of cat and monkey eyelids, Vision Res, 16, 323-324.

Crilly R., 1987, MSc thesis, Univ. of Alberta, Edmonton, Canada; cited in Arnfield *et al*, 1988; also see Flock *et al*, 1987.

Dawson J.B., Barker D.J., Ellis D.J., Grassam E., Cotterill J.A., Fisher G.W. and Feather J.W., 1980, A theoretical and experimental study of light absorption and scattering by *in vivo* skin, Phys Med Biol, 25, 695-709.

Derksen W.L. and Monahan T.I., 1952, A reflectometer for measuring diffuse reflectance in the visible and infrared regions, J Opt Soc Am, 42, 263-265.

Doiron D.R., Svaasand L.O. and Profio A.E., 1983, Light dosimetry in tissue: application to photoradiation therapy, *Porphyrin Photosensitization,* Plenum, NY, pp.63-76.

Drabkin D.L. and Austin J.H., 1932, Spectrophotometric studies I. Spectrophotometric constants for common hemoglobin derivatives in human, dog and rabbit blood, J Biol Chem, 98, 719–733.

Drabkin D.L. and Austin J.H., 1935a, Spectrophotometric studies II. Preparations from washed cells; nitric oxide hemoglobin and sulfhemoglobin, J Biol Chem, 112, 51–65.

Drabkin D.L. and Austin J.H., 1935b, Spectrophotometric studies V. A technique for the analysis of undiluted blood and concentrated hemoglobin solutions, J Biol Chem, 112, 105–115.

Drabkin D.L. and Schmidt C.F., 1945, Spectrophotometric studies XII. Observation of circulating blood *in vivo*, and the direct determination of the saturation of hemoglobin in arterial blood, J Biol Chem, 157, 69–83.

Driver I., Feather J.W., King P.R. and Dawson J.B., 1989, The optical properties of aqueous suspensions of Intralipid, a fat emulsion, Phys Med Biol, 34, 1927–1930.

Edwards E.A. and Duntley S.Q., 1939, The pigments and color of living human skin, Am J Anat, 65, 1–33.

Eichler J., Knof J. and Lenz H., 1977, Measurements on the depth of penetration of light (0.35–1.0 μm) in tissue, Radiat Environm Biophys, 14, 239–242.

Ertefai S. and Profio A.E., 1985, Spectral transmittance and contrast in breast diaphanography, Med Phys, 12, 393–400.

Everett M.A., Yeargers E., Sayre R.M. and Olson R.L., 1966, Penetration of epidermis by ultraviolet rays, Photochem Photobiol, 5, 533–542.

Fellner M.J., 1976, Green autofluorescence in human epidermal cells, Arch Dermatol, 112, 667–670.

Feuk T. and McQueen D., 1971, The angular dependence of light scattered from rabbit corneas, Invest Ophthalmol, 10, 294–299.

Findlay G.H., 1966, The measurement of epidermal melanin by reflectance, Br J Dermatol, 78, 528–531.

Findlay G.H., 1970, Blue skin, Br J Dermatol, 83, 127–134.

Fine I., Loewinger E., Weinreb A. and Weinberger D., 1985, Optical properties of the sclera, Phys Med Biol, 30, 565–571.

Flock S.T., Wilson B.C. and Patterson M.S., 1987, Total attenuation coefficients and scattering phase functions of tissues and phantom materials at 633 nm, Med Phys, 14, 835–841.

Flock S.T., Wilson B.C. and Patterson M.S., 1989, Monte Carlo modeling of light propagation in highly scattering tissues–II: Comparison with measurements in phantoms, IEEE Trans Biomed Eng, 36, 1169–1173.

Gahm T. and Witte S., 1986, Measurement of optical thickness of transparent tissue layers, J Microsc, 141, 101–110.

Geeraets W.J. and Berry E.R., 1968, Ocular spectral characteristics as related to hazards from lasers and other light sources, Am J Ophthalmol, 66, 15–20.

Gordy E. and Drabkin D.L., 1957, Spectrophotometric studies XVI. Determination of the oxygen saturation of blood by a simplified technique, applicable to standard equipment, J Biol Chem, 227, 285–299.

Guerassimov N.A. and Rumyantsev Y.D., 1972, Measurement of meat surface emissivity, *Freezing and Storage of Fish, Poultry and Meat*, Bull Inst Int Froid, Annexe 1972.2, 225–227.

Hale G.M. and Querry M.R., 1973, Optical constants of water in the 200–nm to 200–μm wavelength region, Appl Opt, 12, 555–563.

Hardy J.D., 1939, The radiating power of human skin in the infra–red, Am J Physiol, 127, 454–462.

Hardy J.D., 1973, Reflectance of human skin for low temperature radiators, Arch Sci Physiol, 27, A21–A33.

Hardy J.D. and Muschenheim C., 1935, Radiation of heat from the human body. V. The transmission of infra–red radiation through the body, J Clin Invest, 15, 1–9.

Hardy J.D., Hammel H.T and Murgatroyd D., 1956, Spectral transmittance and reflectance of excised human skin, J Appl Physiol, 9, 257–264.

Hasselbalch K.A., 1911, Quantitative Untersuchungen uber die Absorption der menschlichen Haut von ultravioletten Strahlen, Skand Arch Physiol, 25, 5–68.

Houwink B., 1974, The index of refraction of dental enamel apatite, Br Dent J, 137, 472–475.

Ishimaru A., 1978, *Wave Propagation and Scattering in Random Media*, Vols 1 and 2, Academic Press, New York.

Jacques S.L. and Prahl S.A., 1987, Modeling optical and thermal distributions in tissue during laser irradiation, Lasers Surg Med, 6, 494–503.

Jacques S.L., Alter C.A. and Prahl S.A., 1987, Angular dependence of HeNe laser light scattering by human dermis, Lasers Life Sci, 1, 309–333.

Jacquez J.A. and Kuppenheim H.F., 1955, Spectral reflectance of human skin in the region 235–1000 mμ, J Appl Physiol, 7, 523–528.

Jacquez J.A., Huss J., McKeehan W., Dimitroff J.M. and Kuppenheim H.F., 1955, Spectral reflectance of human skin in the region 0.7–2.6 μ, J Appl Physiol, 8, 297–299.

Kaminow I.P., Wiesenfeld J.M. and Choy D.S.J., 1984, Argon laser disintegration of thrombus and atherosclerotic plaque, Appl Optics, 23, 1301–1302.

Kiefhaber P., Nath G. and Moritz K., 1977, Endoscopic control of massive gastrointestinal hemorrhage by irradiation with a high–power Nyodymium–Yag laser, Proc Surg, 15, 140–155.

Kinsey V.E., 1948, Spectral transmission of the eye to ultraviolet radiations, Arch Ophthalmol, 39, 508–513.

Kollias N. and Baqer A., 1986, On the assessment of melanin in human skin *in vivo*, Photochem Photobiol, 43, 49–54.

Kramer K., Elam J.O., Saxton G.E. and Elam W.N., 1951, Influence of oxygen saturation, erythrocyte concentration and optical depth upon the red and near–infrared light transmittance of whole blood, Am J Physiol, 165, 229–245.

Krause A.C. and Yudkin A.M., 1930, The chemical composition of the normal aqueous humor of the dog, J Biol Chem, 88, 471–477.

Kubelka P., 1948, New contribution to the optics of intensly light–scattering materials. Part I., J Opt Soc Am, 38, 457–488; errata: J Opt Soc Am, 38, 1067.

Kubelka P. and Munk F., 1931, Ein Beitrag zür Optik der Farbanstriche , Z Technichse Physik, 12, 593–601.

Kuppenheim H.F. and Heer R.R., 1952, Spectral reflectance of white and negro skin between 440 and 1000 mμ, J Appl Physiol, 4, 800–806.

Linford J., Shalev S., Bews J., Brown R. and Schipper H., 1986, Development of a tissue–equivalent phantom for diaphanography, Med Phys, 13, 869–875.

Lucas N.S., 1931, IX The permeability of human epidermis to ultra–violet irradiation, Biochem J, 25, 57–70.

Ludvigh E. and McCarthy E.F., 1938, Absorption of visible light by the refractive media of the human eye, Arch Ophthalmol, 20, 37–51.

Marynissen J.P.A. and Star W.M., 1984, Phantom measurements for light dosimetry using isotropic and small aperture detectors. In *Porphyrin Localisation and Treatment of Tumors*, Alan R. Liss Inc., pp.133–148.

Maurice D.M., 1957, The structure and transparency of the cornea, J Physiol, 136, 263–286.

McKenzie A.L. and Byrne P.O., 1988, Can photography be used to measure isodose distributions of space irradiance for laser photodynamic therapy?, Phys Med Biol, 33, 113–131.

McKenzie A.L. and Carruth J.A.S., 1984, Lasers in surgery and medicine, Phys Med Biol, 29, 619–641.

Mellerio J., 1971, Light absorption and scatter in the human lens, Vision Res, 11, 129–141.

Mitchell D., Wyndham C.H. and Hodgson T., 1967, Emissivity and transmittance of excised human skin in its thermal emission wave band, J Appl Physiol, 23, 390–394.

Muller P.J. and Wilson B.C., 1986, An update on the penetration depth of 630 nm light in normal and malignant brain tissue *in vivo*, Phys Med Biol, 31, 1295–1299.

Navarro G.A. and Profio A.E., 1988, Contrast in diaphanography of the breast, Med Phys, 15, 181–187.

Norren D.V. and Vos J.J., 1974, Spectral transmission of the human ocular media, Vision Res, 14, 1237–1244.

Oraevsky A.A., Letokhov V.S., Ragimov S.E., *et al*, 1988, Spectral properties of human atherosclerotic blood vessel walls, Lasers Life Sci, 2, 257–270.

Parker S.G. and Diffey B.L., 1983, The transmission of optical radiation through human nails, Br J Dermatol, 108, 11–16.

Pearson A.R. and Norris R.E., 1933, The transmission of infra–red radiation through the horny layer of human skin, Br J Radiol, 6, 480–486.

Pedersen G.D., McCormick N.J. and Reynolds L.O., 1976, Transport calculations for light scattering in blood, Biophys J, 16, 199–207.

Pitts D.G., 1959, Transmission of the visible spectrum through the ocular media of the bovine eye, Am J Opt and Arch Am Acad Opt, 36, 289–298.

Pottier R.H., Chow Y.F.A., LaPlante J.–P., Truscott T.G., Kennedy J.C. and Beiner L.A., 1986, Non–invasive technique for obtaining fluorescence excitation and emission spectra *in vivo*, Photochem Photobiol, 44, 679–687.

Preuss L.E., Bolin F.P. and Cain B.W., 1982, Tissue as a medium for light transport- implications for photoradiation therapy, Lasers in Medicine and Surgery, SPIE, 357, 77–84.

Profio A.E. and Doiron D.R., 1981, Dosimetry considerations in phototherapy, Med Phys, 8, 190–196.

Russell E.H. and Russell W.K., 1933, *Ultro-violet Radiation and Actinotherapy,* 3rd edition, Livingstone, Edinburgh.

Said F.S. and Weale R.A., 1959, The variation with age of the spectral transmissivity of the living human crystalline lens, Gerontologia, 3, 213–231.

Scheuplein R.J., 1964, A survey of some fundamental aspects of the absorption and reflection of light by tissue, J Soc Cos Chem, 15, 111–122.

Schmitt J.M., 1986, Optical measurement of blood oxygen by implantable telemetry, PhD dissertation, Stanford University, cited in Steinke and Shepherd, 1987.

Sliney D. and Wolbarsht M., 1980, *Safety with Lasers and Other Optical Sources*, Plenum, NY.

Spitzer D. and Ten Bosch J.J., 1975, The absorption and scattering of light in bovine and human enamel, Calcif Tissue Res, 17, 129–137.

Steinke J.M. and Shepherd A.P., 1986, Role of light scattering in whole blood oximetry, IEEE Trans Biomed Eng, BME-33, 294–301.

Steinke J.M. and Shepherd A.P., 1987, Diffuse reflectance of whole blood: model for a diverging light beam, IEEE Trans Biomed Eng, BME-34, 826–834.

Steketee J., 1973, Spectral emissivity of skin and pericardium, Phys Med Biol, 18, 686–694.

Svaasand L.O. and Ellingsen R., 1983, Optical properties of human brain, Photochem Photobiol, 38, 293–299.

Svaasand L.O. and Ellingsen R., 1985, Optical penetration in human intracranial tumors, Photochem Photobiol, 41, 73–76.

Takatani S. and Grayham M.D., 1979, Theoretical analysis of diffuse reflectance from a two-layer tissue model, IEEE Trans Biomed Eng, BME-26, 656–664.

Twersky V., 1970, Absorption and multiple scattering by biological suspensions, J Opt Soc Am, 60, 1084–1093.

Van Gemert J.C., Berenbaum M.C. and Gijsbers G.H.M., 1985, Wavelength and light-dose dependence in tumour phototherapy with haemato-porphyrin derivative, Br J Cancer, 52, 43–49.

Van Gemert M.J.C., Jacques S.L., Sterenborg H.J.C.M. and Star W.M., 1989, Skin optics, IEEE Trans Biomed Eng, 36, 1146–1154.

Walsh J.T. and Deutsch T.F., 1989, Pulsed CO_2 laser ablation of tissue: Effect of mechanical properties, IEEE Trans Biomed Eng, 36, 1195–1201.

Wan S., Anderson R.R. and Parrish J.A., 1981a, Analytical modeling for the optical properties of the skin with *in vitro* and *in vivo* applications, Photochem Photobiol, 34, 493–499.

Wan S., Parrish J.A., Anderson R.R. and Madden M., 1981b, Transmittance of nonionising radiation in human tissues, Photochem Photobiol, 34, 679–681.

Watmough D.J. and Oliver R., 1968, Emissivity of human skin in the waveband between 2μ and 6μ, Nature, 219, 622–624.
Westheimer G., 1972, Optical properties of vertebrate eyes. In *Physiology of Photoreceptor Organs*, M.G.F. Fuortes (ed), Springer-Verlag, NY, p.453.
Wilksch P.A., Jacka F. and Blake A.J., 1984, Studies of light propagation through tissue. In *Porphyrin Localization and Treatment of Tumors*, Alan R. Liss Inc., pp.149–161.
Wilson B.C. and Patterson M.S., 1986, The physics of photodynamic therapy, Phys Med Biol, 31, 327–360.
Wilson B.C., Jeeves W.P., Lowe D.M. and Gabriel A., 1984, Light propagation in animal tissues in the wavelength range 375–825 nanometers. In *Porphyrin Localization and Treatment of Tumours*, Alan R. Liss Inc., pp.115–132.
Wilson B.C., Jeeves W.P. and Lowe D.M., 1985, *In vivo* and *post mortem* measurements of the attenuation spectra of light in mammalian tissues, Photochem Photobiol, 42, 153–162.
Wilson B.C., Muller P.J. and Yanch J.C., 1986a, Instrumentation and light dosimetry for intra-operative photodynamic therapy (PDT) of malignant brain tumours, Phys Med Biol, 31, 125–133.
Wilson B.C., Patterson M.S. and Burns D.M., 1986b, The effect of photosensitizer concentration in tissue on the penetration depth of photoactivating light, Laser Med Sci, 1, 235–244.
Wilson B.C., Patterson M.S. and Flock S.T., 1987, Indirect *versus* direct techniques for the measurement of the optical properties of tissues, Photochem Photobiol, 46, 601–608.
Wyszecki G. and Stiles W.S., 1967, *Color Science*, Wiley, New York.

Chapter 4

Acoustic Properties of Tissue at Ultrasonic Frequencies

The mechanical response of tissues may be studied over the complete spectrum of vibrational frequencies, from its response under static loading to its behaviour at frequencies up to and beyond 100 MHz. In general, two regions in this range are of particular interest. Firstly the mechanical properties of tissues when subjected to static loading may be considered. The associated static mechanical properties are discussed in Chapter 5. The second range of frequencies are ultrasonic, those over 20 kHz, with particular emphasis on the frequency range 1–10 MHz, used widely in clinical medicine. This chapter deals almost exclusively with the properties of tissue at ultrasonic frequencies, although there is also a brief mention of transmission properties at audio frequencies.

4.1 Acoustic velocity

The terms acoustic speed and acoustic velocity are often used interchangeably, although strictly velocity is a vector quantity and speed a scalar quantity. In this chapter the term velocity is generally used even though the vector nature of the quantity under discussion may not be explicit.

4.1.1 *Terminology and definitions*

The **phase velocity**, c_p, of a plane progressive acoustic wave is the ratio of angular frequency ω ($2\pi f$) and wave number k:

$$c_p = \omega/k \qquad (4.1)$$

The **group velocity**, c_g, of a broadband wave depends upon the **dispersion**, dc_p/df, and is given by:

$$c_g = \frac{\partial \omega}{\partial k} = c_p \left[1 - \frac{f}{c_p}\frac{dc_p}{df}\right]^{-1} \tag{4.2}$$

For an unbounded fluid medium the phase velocity depends upon the adiabatic compressibility β_a and the density ρ:

$$c = \frac{1}{\surd(\beta_a \rho)} \tag{4.3}$$

In isotropic solids both shear (transverse) and longitudinal waves may be propagated. The **longitudinal wave velocity**, c_L, is given by:

$$c_L = \left[\frac{K + 4G/3}{\rho}\right]^{\frac{1}{2}} = \left[\frac{E(1 - \nu)}{\rho(1 - 2\nu)(1 + \nu)}\right]^{\frac{1}{2}} \tag{4.4}$$

K being the bulk modulus, G the rigidity modulus, E Young's modulus and ν is Poisson's ratio. The **shear wave velocity**, c_s, when the direction of propagation of the wave is perpendicular to the particle displacements, is:

$$c_s = (G\rho)^{\frac{1}{2}} = \left[\frac{E}{2\rho(1 + \nu)}\right]^{\frac{1}{2}} \tag{4.5}$$

The velocity of sound in a gas at moderate pressure is:

$$c = (\gamma P/\rho)^{\frac{1}{2}} \tag{4.6}$$

where γ is the ratio of specific heat at constant pressure to that at constant volume, and P the ambient pressure. The SI unit for velocity is metre second^{-1} (m s^{-1}).

The velocity is related to the **specific acoustic impedance** Z by the relationship

$$Z = \rho c \tag{4.7}$$

4.1.2 *Measurement of acoustic velocity*

The measurement of velocity is conceptually simple, involving only a knowledge of distance and time. Several simple methods have been used to give measurements of the acoustic velocity through tissue with a high degree of accuracy. The method most widely used is the time-of-flight method, using the pulse transmission time between a transmitter and receiver of known separation. The sing-around method uses two parallel transducers bounding the tissue sample, superimposing echo sequences between them. Time domain spectrometry (Heyser and Le Croisette, 1974) has been used to determine the minimun transit time in inhomogenous tissue samples. Multiple time-of-flight measurement from numerous directions have been used to reconstruct two-dimensional velocity maps of soft tissue (for instance Rajagopalan *et al*, 1979). Velocity estimates have also been derived from measurements of Z using Equation 4.7, but such measurements are generally imprecise. Velocity measurements on tissue samples at 100 MHz using interference patterns generated by a scanning laser acoustic microscope have been reported (for instance, O'Brien *et al*, 1981; Edwards and O'Brien, 1985). Such studies allow improved resolution of local variation of velocity through a tissue sample. Critical angle reflection techniques enable both longitudinal and transverse velocities to be measured on prepared samples of bone and tooth (Lees and Rollins, 1972; Yoon and Katz, 1979a). Methods for *in-vivo* velocity measurement have generally used time-of-flight techniques through accessible segments of tissue although pulse-echo methods have been proposed (Kossoff, 1976). Useful reviews of velocity measurement methods applied to tissue samples are given by Wells (1977) and by Bamber (1986).

4.1.3 *Historical background*

Early reports of acoustic velocities for tissue (Ludwig, 1950, Frucht, 1953, Goldman and Richards, 1954) provided a firm experimental basis for further measurements. Subsequent work has broadened the information available, including data on temperature and frequency dependence of acoustic velocity, tissue composition and anisotropy, as well as adding to the total volume of reported measurements. Several extensive literature surveys have presented compilations of velocity and attenuation values for tissue, of which the most extensive are those of Goss *et al* (1978b and 1980), Chivers and Parry (1978), and O'Brien (1977). Many detailed measurements may be found in Linzer (1979). The values included in the tables are drawn partly from these compilations and includ additional, more recent material.

4.1.4 *Values of acoustic velocity through tissue*

Measured values of the velocity of ultrasound through tissue are given in Tables 4.1 to 4.4. Biological fluids have been grouped separately in Table 4.1, with soft tissues in Table 4.2, hard tissues (bone and teeth) in Tables 4.3a and b, and some pathological tissue velocities in Table 4.4. Tissues are entered in alphabetical order. For any tissue, values for human tissue are first. The temperature of measurement is noted, and when measurements were made over a range of temperatures, the velocity closest to body temperature was selected. No distinction is made between phase velocity and

Table 4.1 Velocity of ultrasound through human biological fluids

Fluid	Temp °C	Velocity m s^{-1}	Reference
Amniotic fluid	37	1534±3	Povall *et al* 1984
Blood:whole,			
–,38.7% Hct,25.6–65.8MHz	37	1584.2±1.4	Collings & Bajenov 1987
–,40% packed cell vol.	37	1590	Hughes *et al* 1979
–,10% packed cell vol. (min value)	37	1540	-
–,0.6g/100ml protein, 1.6g/100ml Hb	37	1559.2	Aubert *et al* 1978
–,6.2g/100ml protein 16g/100ml Hb	37	1604.3	-
–,during coagulation,♂	37	1586±5.5	Grybauskas *et al* 1978
–,after retraction,♂	37	1602±7.5	-
–,during coagulation,♀	37	1580±4.6	-
–,after retraction,♀	37	1597±7.3	-
–,fetal,8wk	37	1580	Wladimiroff *et al* 1975
–,17g/100ml Hb*	36	1587	Bronez *et al* 1985
–,extracorporeal†	37	1580	Nasoni 1981
Blood:cells,25.6–65.8 MHz	37	1627.4±0.6	Collings & Bajenov 1987
Blood:plasma			
1.29g/100ml protien	37	1529.6	Aubert *et al* 1978
5.163g/100ml protein	37	1548.4	-
–,25.6–65.8 MHz	37	1550.3±0.8	Collings & Bajenov 1987
Blood:haemoglobin soln.			
1.6g.100ml, 10MHz	15	1476.2	Carstensen & Schwan 1959b
16.5g/100ml, 300kHz	15	1528.8	-
16.5g/100ml, 10MHz	15	1530.4	-
–*			
8g/100ml	25	1523	-
30g/100ml, 1MHz		1607.5	-
30g/100ml, 10MHz		1610	-
Cerebrospinal fluid	21.8 –25	1499– 1515	Van Venrooij 1971
–,ventricular	22	1499–1505	Schiefer *et al* 1968
Milk	30	1540	Kossoff *et al* 1973
–*	40	1543	Fitzgerald *et al* 1961
Pericardial fluid	37	1554.2	Aubert *et al* 1978
Pus	22	1521–1523	Schiefer *et al* 1968

* cow † dog

Measured at between 1 and 10 MHz, except where noted.

Table 4.2 Velocity of ultrasound through normal soft tissues

Tissue	Temp °C	Velocity m s⁻¹	Reference
Blood vessel,			
cow,aortic arch	20	1570±3.3	Geleskie & Shung 1982
-,-,aorta(upper)	20	1572±4.8	-
-,-,aorta(lower)	20	1575±5.3	-
-,-,iliac artery	20	1565±7.1	-
-,-,carotid artery	20	1575±9.8	-
-,-,pulmonary trunk	20	1568±9.1	-
-,-,vena cava	20	1559±4.3	-
-,dog,aorta	37	1586±7	Hughes & Snyder 1980
-,dog and human	37	1560-1660	Rooney *et al* 1982
Brain,human	37	1562	Kremkau *et al* 1981
-,-,corna radiata	40	1573	-
-,-	22	1532	Schiefer *et al* 1968
-,-,cerebellum	22	1537	-
-,-,newborn,*in-vivo*	37	1525	Willocks *et al* 1964
-,-,fetal,			
17-19wk,91-91.2%water	37	1517-1523	Wladimiroff *et al* 1975
24-29wk,90-91.7%water	37	1523-1529	-
38-40wk,88-89%water	37	1540	-
-,cow	30	1548	Law *et al* 1985
-,dog	37	1563	Bowen *et al* 1979
-,pig	26	1550	Sun Yongchen *et al* 1986
-,cat,*in-vivo*		1554,1564	Robinson & Lele 1969
Breast,human,*in-vivo*,			
pre-menopausal		1450-1570 (av 1510)	Kossoff *et al* 1973
-,-,-,post-menopausal		1430-1520 (av 1468)	-
-,-,-		1553±35	Scherzinger *et al* 1989
-,-	38	1564±6.4	Rajagopalan *et al* 1979
-,-,lymph node	15.5	1513.8	Sun Yongchen *et al* 1986
Cervix,human,*in-vivo*			
non-pregnant		1633±2.9	Bakke & Gytre 1974
pregnant		1625±1.6	-
Collagen (calculated)			
longitudinal		1570	Lees & Rollins 1972
shear		585	-
Eye:aqueous,calf	25.5 -29	1481-1525 (av 1497)	Begui 1954
-,pig	37	1530±3.1	Rivara & Sanna 1962
Eye:choroid,human	20	1527	Thijssen *et al* 1985
Eye:cornea,human	37	1586±3.9	Rivara & Sanna 1962
-,pig	37	1588±3.5	-

cont.

Table 4.2 cont. Velocity of ultrasound, soft tissue

Tissue	T,°C	c,m s^{-1}	Reference
Eye:lens,human	37	1647±3.5	Rivara & Sanna 1962
–,–	37	1640±1.2	Jansson & Kock 1962
–,–,child	37	1653–1674	Coleman *et al* 1975
–,–,cateractous	37	1629±38	-
–,–,–	36.5 to37.4	1543–1665	Jansson & Kock 1962
–,pig	37	1673±3.1	Rivara & Sanna 1962
Eye:retina,human	20	1538±20	Thijssen *et al* 1985
–,pig	20	1532±4	-
Eye:sclera,human	37	1647±3.9	Rivara & Sanna 1962
–,pig	37	1656±3.9	-
Eye:vitreous,human	37	1532±0.5	Jansson & Kock 1962
–,–	37	1523±2.1	Rivara & Sanna 1962
–,pig	37	1532±2.9	-
Fat,human,subcutaneous	35	1476	Bullen *et al* 1965
–,–,high water content	24	1487±1.5	Frucht 1953
–,–,low water content	24	1469±2	-
–,–,dehydrated	24	1459±1	-
–,–,breast fat	37	1436±18.4	Rajagopalan *et al* 1979
–,–,orbital fat	37	1462±23.7	Buschmann *et al* 1970
–,human and animal	37	1427±12.7	Errabolu *et al* 1988
–,–,subcutaneous	37	1430	-
–,–,omentum	37	1412	-
–,–,breast fat	37	1420	-
–,–,mesanteric	37	1427	-
–,cow	37	1430	Bamber & Hill 1979
–,cow,*in–vivo*		1460,1471	Lewin & Busk 1982
–,dog,abdominal fat	37	1412	Bowen *et al* 1979
–,–,omentum fat	36.3	1452,1459	Nasoni *et al* 1979
–,–,subcutaneous,100MHz	22	975–1225	O'Brien *et al* 1981
–,pig,*in–vivo*	*iv*	1426	Lewin & Busk 1982
Gall bladder,human	26	1583.6	Sun Yongchen *et al* 1986
Kidney,human	37.2	1560±1.8	Rajagopalan *et al* 1979
–,–,fetus,7mo.	17.3	1513.5	Sun Yongchen *et al* 1986
–,cow	40	1565	Bronez *et al* 1985
–,pig,medulla,100 MHz	23–26	1564±6.1	Frizzell & Gindorf 1981
–,–,cortex,–	23–26	1578±7.2	-
–,–	26	1569	Gong *et al* 1989
–,dog	37	1567–1571	Bowen *et al* 1979
–,–,*in–vivo*,79.1–82.5% water, 4.1–10% lipid	38.5	1567–1580	Nasoni *et al* 1979
–,–,region 1 (cortex)	22	1549–1556	Sarvazyan & Klemin 1983
–,–,region 2	22	1539–1548	-
–,–,region 3	22	1537–1540	-
–,–,region 4 (core)	22	1527–1533	-
–,–,pelvis	22	1487–1499	-

cont.

Table 4.2 cont. Velocity of ultrasound, soft tissue

Tissue	T,ºC	c,m s^{-1}	Reference
Kidney,mouse,100 MHz	23–26	1586±10.7	Frizzell & Gindorf 1981
Liver,human	37.2	1578±2.9	Rajagopalan *et al* 1979
–,–,71%water,2.9%fat	37	1592±6	Sehgal *et al* 1986
–,–,slightly fatty	37	1578,1584	–
–,–,24% fat	37	1522	–
–,–,*in-vivo*	*iv*	1578±5	Robinson *et al* 1987
–,–,*in-vivo*	*iv*	1593±46	Bamber *et al* 1987
–,–	37	1607	Bamber & Hill 1979
–,–,fetus,7mo.	17.3	1541.5	Sun Yongchen *et al* 1986
–,cow	30	1596–1632	Law *et al* 1985
–,–	37	1597,1639	Bamber & Hill 1979
–,–	36	1587	Bronez *et al* 1985
–,sheep,100 MHz	23–26	1565±7.8	Frizzell & Gindorf 1981
–,pig	23–24	1553±15	Ludwig 1950
–,–	26	1588	Gong *et al* 1989
–,dog,*in-vivo*	*iv*	1597,1602	Nasoni *et al* 1979
–,–	37	1592–1604	Bowen *et al* 1979
–,cat,100 MHz	23–26	1567±13.2	Frizzell & Gindorf 1981
–,rabbit	26	1575±9.4	Goldman & Richards 1954
–,–,70→75% water	25	1595→1575	Sarvazyan *et al* 1987
–,rat,100MHz,3%fat	room	1545–1556 (av 1550)	Tervola *et al* 1985
–,–,–,5.2%fat	room	1539–1552 (av 1546)	–
–,–,–,11%fat	room	1525–1539 (av 1532)	–
Lung,cow,0.01–0.8 MHz			
0.72 g cm^{-3}, 32% air	room	948	Pedersen & Ozcan 1986
0.54 g cm^{-3}, 49% air	room	733	–
0.36 g cm^{-3}, 66% air	room	583	–
–,dog,1.0 MHz			
0.7 g cm^{-3}	35	976	Dunn 1986
0.5 g cm^{-3}	35	787	–
0.4 g cm^{-3}	35	644,658	–, Dunn 1974
–,–,3.0 MHz			
0.7 g cm^{-3}	35	1218	Dunn 1986
0.5 g cm^{-3}	35	1027	–
0.4 g cm^{-3}	35	916	–
–,–,5.0 MHz			
0.7 g cm^{-3}	35	1414	Dunn 1986
0.5 g cm^{-3}	35	1266	–
0.4 g cm^{-3}	35	1196,1180	–, Dunn 1974
–,–,7.0 MHz			
0.4 g cm^{-3}	35	1472	Dunn 1986

cont.

Table 4.2 cont. Velocity of ultrasound, soft tissue

Tissue	T,°C	c,m s⁻¹	Reference
Muscle:cardiac,human,			
fetus,7mo.	18	1529	Sun Yongchen *et al* 1986
–,cow	37	1567	Bronez *et al* 1985
–,–	30	1566,1570	Law *et al* 1985
–,pig	26	1572	Gong *et al* 1989
–,dog,perp to fibres	20	1550±6	Mol & Breddels 1982
–,–,parallel to fibres	20	1558±6	-
–,–	37	1595	Nasoni *et al* 1979
Muscle:skeletal,human,			
forearm,*in–vivo*	*iv*	1580,1583	Sollish 1979
–,–,pectoral	24	1568±5.5	Frucht 1953
–,–,psoas	37.2	1576±1.1	Rajagopalan *et al* 1979
–,–,eye	37	1631±15.3	Buschmann *et al* 1970
–,–,fetal,7mo.	18	1547	Sun Yongchen *et al* 1986
–,cow,10%fat,*in–vivo*	*iv*	~1600	Miles *et al* 1984
–,–,30%fat,*in–vivo*	*iv*	1580	-
–,–,*in–vivo*	*iv*	1588,1604	Lewin & Busk 1982
–,dog	37.5	1609,1629	Nasoni *et al* 1979
–,–	37	1589–1603	Bowen *et al* 1979
–,–,perp. to fibres	26	1576±9.6	Goldman & Richards 1954
–,–,parallel to fibres	26	1592±5.0	-
–,pig,*in–vivo*	*iv*	1579	Lewin & Busk 1982
–,–	30	1593,1607	Law *et al* 1985
–,rabbit,thigh,			
perp. to fibres	20	1548±6	Mol & Breddels 1982
parallel to fibres	20	1557±7	-
–,–,– perp. to fibres	26	1587±5.5	Goldman & Richards 1954
parallel to fibres	26	1603±7.9	-
Nerve:optic,human	20	1644±25.4	Buschmann *et al* 1970
–,–	37	1615±3.1	-
Pancreas,Pig,100 MHz	23–26	1591±9.2	Frizzell & Gindorf 1981
Skin,human,fetal,7mo.	17.5	1537	Sun Yongchen *et al* 1986
–,cow,*in–vivo*	*iv*	1591,1630	Lewin & Busk 1982
–,pig,*in–vivo*	*iv*	1503	-
–,–,15 MHz,			
normal and burned	25	1720±120	Cantrell *et al* 1978
–,dog,100MHz	22	1590–2170	O'Brien *et al* 1981
–,–,–,scar at 7 days	22	1540–1575	-
–,–,–,scar at 35 days	22	1700–2000	-
–,–,–	30	1632±34	Riederer–Henderson *et al* 1988
–,mouse,young	20.5	1536–1543	Bhagat *et al* 1980
–,–,old	20.5	1510–1515	-
Spinal chord,human	37.2	1542±3.3	Rajagopalan *et al* 1979

cont.

Table 4.2 cont. Velocity of ultrasound, soft tissue

Tissue	T,°C	c,m s^{-1}	Reference
Spleen,human,*in-vivo*	*iv*	1567±9	Chen *et al* 1987
–,–	37.2	1567±2.3	Rajagopalan *et al* 1979
–,pig	26	1575	Gong *et al* 1989
–,–,100MHz	23–26	1575±9.7	Frizzell & Gindorf 1981
–,dog	38.2	1567–1635	Nasoni *et al* 1979
Tendon,cow, parallel to fibre		1750	Dussik & Fritch 1956
–,mouse,threads,100 MHz	22	1733±56	Goss & O'Brien 1979
–,–,–,–,in 0.9% NaCl*	20–22	1631±36	Edwards & O'Brien 1985
–,–,425MHz	50	1850–2000	Daft & Briggs 1989
–,rat,tail,100 MHz	23	2100±100	Leeş *et al* 1983b
–,–,–,~10GHz,0% water	25	3530	Harley *et al* 1977
–,–,–,–,85% water	25	2620	–
–,–,–,–,parall. to fibres		2640	Cusack & Miller 1979
–,–,–,–,perp. to fibres		1890	–
Testis,human,*in-vivo*	*iv*	1595	Jellins & Barraclough 1978

Measurements at 1–10 MHz except where noted.

* Variation with concentration of bathing media also given.

Table 4.3a Velocity of ultrasound through bone*

Tissue	Temp °C	Velocity m s^{-1}	Reference
Whole bone,human, radius 23–33 yr ♂	*iv*	3406±126	Craven *et al* 1973
–,–,47–71 yr ♀	*iv*	3198±278	–
–,–,femur,transverse	*iv*	3250±190	André *et al* 1980
–,–,radius,transverse	*iv*	3190±146	–
–,–,–,–,♂,20–60yr	*iv*	3311±205	Greenfield *et al* 1981
–,–,–,–,♀,18–61yr	*iv*	3359±198	–
–,–,tibia,20kHz	*iv*	3260–3572 (av 3481)	Anast *et al* 1958
–,–,phalanx,adult	*iv*	2405±63	Fredfeldt 1986
–,–,–,child	*iv*	2109	–
–,–,tibia,150kHz	*iv*	3519±24	Vilks *et al* 1978
–,pig,1MHz		2814	Lehmann & Johnson 1958
–,dog,tibia,3–5MHz,long.	22	3210–3620	Adler & Cook 1975
–,–,–,shear	22	~1760	–

cont.

Table 4.3a cont. Velocity of ultrasound, bone

Tissue		T,°C	c,m s^{-1}	Reference
Cortical bone,human,				
tibia,0.1MHz,16–30 yr		23	3526±100	Abendschein & Hyatt 1970
–,–,femur,5MHz				
1.94–2.01g cm^{-3}	3L*	22	3850–4040	Meunier *et al* 1982
	3T	22	1850–1890	–
	1L	22	3230–3450	–
	2T	22	1630–1720	–
–,–,tibia,10MHz,axial			3760±100	Lees *et al* 1983a
–,–,–,radial			3130±140	–
–,cow,femur,0.5MHz				
1.97–2.06 g cm^{-3}			3380±99	McKelvie & Palmer 1987
–,–,tibia,10MHz,axial			4180±50	Lees *et al* 1983a
–,–,–,–,radial			3320±80	–
–,–,femur,15MHz	3L	40	4170±60	Yoon & Katz 1979a
	3T	40	2660±130	–
–,–,–,surface wave 3S		40	1950±20	–
–,–,phalanx 5MHz	3L		4030±110	Lang 1970
2.25MHz	3T		1660±90	–
5MHz	1L		3160±170	–
2.25MHz	$1T_h$		1390±40	–
2.25MHz	$1T_3$		1640±60	–
–,pig,1MHz			2941±614	Lehmann & Johnson 1958
–,rabbit,femur,radial				
1.98to2.10 g cm^{-3}			3254to3430	Lees *et al* 1987
Trabecular bone,human,				
0.94–1.17g cm^{-3},0.5MHz			1688–2084	McKelvie & Palmer 1987
–,pig,1MHz			2407±554	Lehmann & Johnson 1958
Skull,human,outer,3–4MHz		25	2650–3050 (av 2920)	Martin & McElhaney 1971
–,1.7MHz,fixed	c_g	37	2960±127	Fry & Barger 1978
–,–,–,3–4MHz	c_p	37	2930±151	–
–,–,inner,3–4MHz		25	2903–3258 (av 3098)	Martin & McElhaney 1971
–,1.7MHz,fixed	c_g	37	2590±252	Fry & Barger 1978
–,–,–	c_p	37	2960±168	–
–,diploë		25	3022–3289 (av 3198)	Martin & McElhaney 1971
–,–,fixed,0.5MHz	c_g	37	2240±180	Fry & Barger 1978
–,–,–,–	c_p	37	2190±165	–
–,–,–,1.5MHz	c_g	37	2770±185	–
–,–,–,–	c_p	37	2580±223	–
–,–,–,3.5MHz	c_g	37	3110	–
–,–,–,–	c_p	37	2870±105	–
–,–,flexural,50kHz			1680±260	Dzenis & Purin'sh 1981

* All values are for wet bone. For dry bone see text.

Table 4.3b Velocity of ultrasound through tooth material

Tissue	c,m s^{-1}	Reference
Tooth:cementum,human,9MHz	3300	Reich *et al* 1967
–,fixed,18MHz	3200	Kossoff & Sharpe 1966
Tooth:dentine,longitudinal wave		
Human,incisor,fixed	3800±370	Barber *et al* 1969
–,9MHz	3700	Reich *et al* 1967
–,fixed,18MHz	4000	Kossoff & Sharpe 1966
Cow	3140–4120	Gilmore *et al* 1969
–,parallel to long axis	4140	Lees & Rollins 1972
–,perp. to long axis	3880	–
–,wet	3550±30	Lees *et al* 1983a
–,incisor,fixed	3630±300	Barber *et al* 1969
Dog,incisor,fixed	3200–3240	–
–,Shear wave		
Cow	1300–2050	Gilmore *et al* 1969
–,parallel to long axis	2120	Lees & Rollins 1972
–,perp. to long axis	2000	–
Tooth:enamel,longitudinal wave		
Human,incisor,fixed	6250±410	Barber *et al* 1969
–,9MHz	5600	Reich *et al* 1967
–,fixed,18MHz	4500	Kossoff & Sharpe 1966
Cow	4740–6190	Gilmore *et al* 1969
–,parallel to long axis	6260	Lees & Rollins 1972
–,perp. to long axis	4300–5800	–
–	6030	Lees *et al* 1983a
–,incisor,fixed	5750,5950	Barber *et al* 1969
Dog,incisor,fixed	4800	–
–,Shear wave		
Cow	2810–3530	Gilmore *et al* 1969
–,parallel to long axis	3460	Lees & Rollins 1972
–,perp. to long axis	3050–3340	–

Table 4.4 Velocity of ultrasound through human pathological tissues

Tissue	Temp °C	Velocity m s^{-1}	Reference
Arterial wall			
Calcified lesions	37	1900–2000	Rooney *et al* 1982
Blood			
Sickle cell	22	1551	Shung & Reid 1977
Brain			
Acoustic neurinoma	22	1522–1547	Schiefer *et al* 1968
Arachnoidal sarcoma	22	1530	-
Astrocytoma	27.5	1545	Van Venrooij 1971
Ependymoma	22	1537–1545	Schiefer *et al* 1968
Glioma	22	1525–1547	-
Meningioma	22	1540–1550	Schiefer *et al* 1968
-	19.7	1548–1569	Van Venrooij 1971
Metastatis	22	1535	Schiefer *et al* 1968
Spongioblastoma	22	1532	-
Tumour cyst contents	22	1504–1524	-
Breast			
with carcinoma	*iv*	1437–1547	Kossoff *et al* 1973
with fibrocystic disease	*iv*	1466–1551	-
with fibroadenosis	*iv*	1482–1566	-
with fibroadenoma	*iv*	1502–1556	-
cancer	*iv*	1550±35	Scherzinger *et al* 1989
fibroadenoma	*iv*	1584±27	-
cyst	*iv*	1568±40	-
benign lump	*iv*	1561±32	-
Breast, *in-vitro*			
Carcinoma	24	1573±7	Frucht 1953
Multiple myoloma	37	1556.2	Sehgal *et al* 1984
Gastropathy	24.6	1555.8	Sun Yongchen *et al* 1986
Gallstone		1400–2200	Ludwig & Struthers 1950
Liver			
Cirrhotic	37	1562	Sehgal *et al* 1986
Congested	37	1581±9	-
Diffusely malignant	20±2	1631	Bamber & Hill 1981
Tumour	37	1579±8	Sehgal *et al* 1986
Cirrhotic	*iv*	1558±23	Chen *et al* 1987
Fatty not fibrotic	*iv*	1547±18	-
Metastatic	*iv*	1581±12	-
Muscle			
Thigh,malignancy	24	1564.4	Sun Yongchen *et al* 1986
Spleen			
Splenomegaly,nonfibrotic	*iv*	1566±12	Chen *et al* 1987
Myelofibrosis, Grade I	*iv*	1559±10	-
-,Grade II	*iv*	1551±11	-
-,Grade III	*iv*	1526±17	-

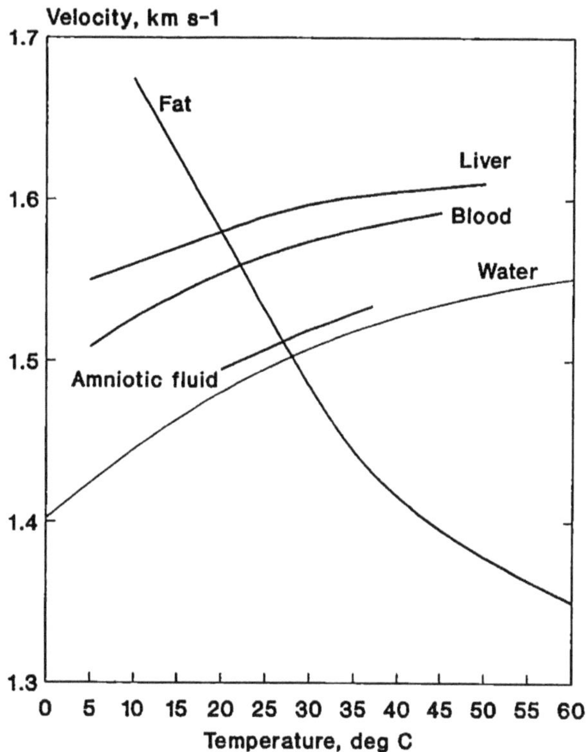

Figure 4.1 Temperature variation of velocity of ultrasound in tissues, body fluids and water. From Bamber *et al* (1979), liver and fat; Povall *et al* (1984), amniotic fluid; Collings and Bajenov (1987), blood.

group velocity in the tabulation of data for soft tissues, since dispersion in most soft tissues is minimal (Section 4.1.5.2). For dispersive tissues (lung and bone) the velocity measured depends upon the measurement technique and method of data analysis (Pedersen and Ozcan, 1986; Fry and Barger, 1978) and care is needed in its interpretation. Generally a measurement of group velocity can be assumed.

4.1.5 *Factors affecting acoustic velocity*

4.1.5.1 Temperature

The variation of velocity with temperature, dc/dT, for most soft tissue follows a similar pattern to that shown by water (Figure 4.1). Unlike most liquids

Table 4.5 Temperature coefficient of acoustic velocity, dc/dT

Tissue	dc/dT m s^{-1} $^oC^{-1}$	Temp ^oC	Reference
Amniotic fluid	2.3	20–37	Povall *et al* 1984
Blood:whole,human	2.0	20–40	Aubert *et al* 1978
–,–	1.3	26–40	Bakke *et al* 1975
–,–,fetal	2.7	0–37	Wladimiroff *et al* 1975
–,cow	1.7	20–45	Bronez *et al* 1985
–,dog	1.6	20–40	Aubert *et al* 1978
Blood:plasma,human	2.1–2.2	20–40	–
Bone,cortical,cow	–6.6	20–40	Yoon & Katz 1979a
Brain,human,fetal	1.5	24–37	Wladimiroff *et al* 1975
–,dog	0.67	37	Bowen *et al* 1979
–,–	0.26	43	–
Breast*	–0.1–2.5	37	Haney & O'Brien 1986
Eye,aqueous,pig	2.0	4–37	Rivara & Sanna 1962
Cornea,human	1.3	4–37	–
Lens,human	0.9	4–37	–
–,pig	1.0	23–37	Jansson & Sundmark 1961
Sclera,human	1.1	4–37	Rivara & Sanna 1962
Vitreous,–	1.9	4–37	–
–,pig	1.8	23–37	Jansson & Sundmark 1961
Fat,human,breast	–3.1	22–37	Rajagopalan *et al* 1979
–,–,orbital	–7.1	20–37	Buschmann *et al* 1970
–,cow	–7.4	37	Bamber & Hill 1979
–,dog,abdomen	–2.9	37	Bowen *et al* 1979
Kidney,cow	1.6	12–40	Bronez *et al* 1985
–,–	1.1–1.2	*in-vivo*	Nasoni *et al* 1979
Liver,human	1.0	37	Bamber & Hill 1979
–,–	1.5	20–37	Sehgal *et al* 1986
–,cow	0.6,0.3	37	Bamber & Hill 1979
–,–	0.9	12–36	Bronez *et al* 1985
–,dog	0.9–1.1	37	Bowen *et al* 1979
–,–	1.1–1.3	38.5	Nasoni *et al* 1979
Muscle,cardiac,cow	1.1	12–37	Bronez *et al* 1985
Muscle,skeletal,dog	1.1–1.2	37	Bowen *et al* 1979
–,–	0.9,0.6	38.5	Nasoni *et al* 1979
Nerve*,optic	–1.7	37	Haney & O'Brien 1986
–,spinal chord	0.8–2.2	37	–
Spleen,dog	1.3	37	Bowen *et al* 1979
–,–	0.8	43	–

* Values estimated from published data.

for which dc/dT is negative, the velocity in tissue increases with temperature from freezing point up to temperatures greater than body temperature. At temperatures above about 50°C there is no further increase and the temperature coefficient may become slightly negative. Values for dc/dT from several reports are given in Table 4.5. Most of the values are derived from linear least-squares fits to experimental data over a finite temperature range. Since the assumption of a linear dependence upon temperature becomes increasingly invalid as the temperature range increases, some caution should be used in the interpretation of these temperature coefficient data. Other measurements comparing velocities of identical tissue samples at different temperatures have been reported (Buschmann *et al*, 1970; Hughes and Snyder, 1980; Kremkau *et al*, 1981; Law *et al*, 1985; Rajagopalan *et al*, 1979; Wladimiroff *et al*, 1975).

In contrast, for lipids, dc/dT is strongly negative in the temperature range 0–50°C (Hustad *et al*, 1971). Tissues with high fat content may also demonstrate a slight negative temperature dependence (Sehgal *et al*, 1984). Furthermore, unusual variations of velocity with temperature for fatty tissue have been reported. Kremkau *et al* (1981) noted a velocity minimum at 15°C in brain tissue. This minimum was absent in measurements on neo-natal brain. The observation was tentatively related to the solidification point of human depot fat, also 15°C, and to possible membrane phospholipid phase changes. In a second study, Rajagopalan *et al* (1979) observed an abrupt drop in the velocity of human breast fat around 37°C with slight evidence of a similar drop around 27°C. It was noted that Hoyer and Nolle (1956) have reported that the sound velocity of nemantic and cholesteric liquid crystals goes through a minimum at the phase transition temperature. On the other hand it has been demonstrated that sound velocity in cholesteryl esters shows no anomalous changes around the phase trasnsition (Dyro and Edmonds, 1974).

Temperature coefficients for longitudinal, transverse and surface waves in bone are all slightly negative (Yoon and Katz, 1979a).

4.1.5.2 Frequency

Soft tissues and body fluids demonstrate very little dispersion over the frequency ranges reported. Carstensen and Schwan (1959b) reported studies on haemoglobin solutions which demonstrated, for instance, an increase in velocity from 1528.8 m s^{-1} at 0.3 MHz to only 1530.4 m s^{-1} at 10 MHz for human haemoglobin of 16 g/100 cc concentration at 15°C. Bovine serum albumin has been studied over a much wider frequency range by Choi *et al* (1990). The velocity difference, for a solution of concentration 51 g l^{-1} at neutral pH and 20°C, was 2.0 ms^{-1} between 3 MHz and 1 GHz, with no measurable change in velocity over the frequency range 120–1600 MHz. For whole tissues slightly greater dispersions have been observed. Kremkau *et al* (1981) report measurements in human brain tissue showing a dispersion of about 1.2 m s^{-1} in the frequency range 1 to 5 MHz. Using an accurate acoustic resonator method of measurement, Klemin *et al* (1981) have investigated dispersion over the range 1.7 to 17.4 MHz. They report, for example, a dispersion of about 0.5 m s^{-1} MHz^{-1} for rat liver at 2.8 MHz,

Table 4.6 Dispersion of acoustic velocity in tissue, 1–10 MHz

Tissue	Dispersion, $m\ s^{-1}$/MHz	dc_p/df %/MHz	Reference
Bone,cortical(dry)	10–30	0.3–1.5	Yoon & Katz 1979b
Bone,diploë(wet)	300	10	Fry & Barger 1978
Bone,skull ivory	slight		-
Brain	1.2	0.08	Kremkau *et al* 1981
Haemoglobin solutions			
0–30 g Hb/100cc	0–0.2	0–0.013	Carstensen & Schwan 1959b
Lung	120	10–20	Dunn 1986

and 0.22 m s^{-1} MHz^{-1} for rat cardiac muscle at 8.9 MHz. Studies on soft tissues using a scanning laser acoustic microscope (Frizzell and Gindorf, 1981) appear to show velocities at this frequency which differ little from those in the low megahertz frequency range. Frequency has been therefore omitted as an independent parameter from the tables, provided that the measurements were made in the range 1–15 MHz.

Using some simplifying assumptions it is possible to derive Kramers–Kronig relations linking attenuation and dispersion which allow estimates to be made of the variation of phase velocity with frequency. O'Donnell *et al* (1981a) used the expression

$$\alpha(\omega) = \frac{\pi\omega^2}{2\rho_0{}^2} \frac{dc(\omega)}{d\omega} \qquad (4.8)$$

where $\alpha(\omega)$ is the attenuation coefficient at an angular frequency ω, with measured values of $\alpha(\omega)$ for myocardium, to give values for dispersion falling from 0.7 m s^{-1} to 0.2 m s^{-1} with frequency increase from 1 to 10 MHz. The near linear dependence of attenuation coefficient on frequency predicts a nearly logarithmic dependence for dispersion.

The two tissues in which velocity clearly varies with frequency are bone and lung. Dunn (1986) reported a velocity increase in a modestly inflated lung (~60% air) from 644 m s^{-1} at 1 MHz to 1472 m s^{-1} at 7 MHz. Dispersion in lung tissue at lower frequencies is not as marked (Pedersen and Ozcan, 1986).

Dispersion in wet diploë from human skull (Fry and Barger, 1978) is significantly greater than that reported in dried bone (Yoon and Katz, 1979b) at similar frequencies. In dry bone Yoon and Katz observed that shear wave dispersion was independent of direction whilst longitudinal wave dispersion was anisotropic. Fry and Barger also reported evidence that there is little dispersion in the outer ivory table of the skull. Results from the inner table were uncertain because of the thinness of the specimens.

There is no good information about acoustic dispersion in fat or fatty tissues.

4.1.5.3 Anisotropy

The anisotropy of velocity in muscle has been investigated by Mol and Breddels (1982) who report a slight (~0.5%) but statistically significant increase in velocity in the direction parallel to the fibres of dog myocardium, and rabbit thigh muscle, over that in a perpendicular direction. A slightly higher difference (~1%) was reported by Goldman and Richards (1954) from observations in both dog and rabbit skeletal muscle. A comparable difference (10 m s^{-1}) was also reported by Ludwig (1950).

Anisotropy of velocity through articular cartilage has been studied at 100 MHz by Agemura *et al* (1990). They found slight (about 5%) variations of velocity with direction, the highest velocity (1720±99 m s^{-1}) being across the fibrils.

The hard tissues (bone and tooth) are clearly anisotropic in acoustic velocity (Table 4.3). It has been suggested that the anisotropy in elastic properties (and hence in sound velocity) may be characterised by a pseudohexagonal elastic model (Lang, 1970; Lees and Rollins, 1972; Yoon and Katz, 1976). Orthogonal axes are commonly used to define direction, with the 3-direction parallel to the bone long axis. On the other hand Lees *et al* (1979) have argued that, however appealing, such a simplification is inappropriate as a general concept to allow absolute comparison between bone sample measurements. They suggest the term 'plesio-velocity' to describe propagation in this anisotropic and inhomogenous material, and limit reported velocities to those along and across the bone, and to intermediate directions.

Longitudinal velocities are greatest in cortical bone in a direction parallel to the long axis of the bone (3L), velocities in the perpendicular directions (1L and 2L) being about 20% lower. Similar anisotropy has been noted in the transverse (shear) wave component where the perpendicular velocities (1T, 2T) may be 10% lower than the velocity along the bone axis (3T). Velocities in other directions take intermediate values.

Hard dental tissue is also anisotropic in velocity, most notably in the enamel (Lees and Rollins, 1972).

4.1.5.4 Tissue composition

An increase in either water or fat content results in a decrease in the velocity of ultrasound in soft tissue. Variations in acoustic velocity through fetal brain have been related to water content (Wladimiroff *et al*, 1975). Also it has been observed that the velocity in brain white matter exceeds that for grey matter by about 6 m s^{-1} (Kremkau *et al*, 1981), and this was explained on the basis of the water content of the tissues. Sehgal *et al* (1986) report a study of velocity and B/A for human liver *in-vitro*, giving several empirical expressions for the dependence of both quantities on water and fat fractions x_w and x_f, and prediction equations for x_w and x_f using measured values of c and B/A. At 37°C the expression for velocity in tissue c_t is:

$$10^6/c_t = 5.138 + 1.403x_w + 2.121x_f \qquad (4.9)$$

The corresponding expression for B/A is given in Equation 4.23.

From measurements on fatty rat liver at 100 MHz, O'Brien *et al* (1988) derived the following expression:

$$c = 1580.2 - 2.4L - 1.6P + 0.4W \qquad (4.10)$$

where L, P and W are percentage wet weight of lipid, protein and water respectively. It was concluded that whilst lipid and protein significantly affected the velocity, water content had little effect.

Comparable mixture rules which include density have been proposed theoretically (Apfel, 1986). Sarvazyan *et al* (1987) have proposed an expression based on an analysis of tissue as a 'solution' whose velocity is independent of structure, depending only upon water fraction, and densities and compressibilities of its components.

The presence of water in fat modifies its acoustic velocity (Frucht, 1953). Increasing the level of hydration in fat at body temperature would be expected to increase the velocity since at this temperature sound travels more slowly through fat than through water (Figure 4.1). At room temperature only small differences are to be expected, whilst at temperatures below about 10°C a decrease in velocity should occur with the level of hydration. Fatty tissue *in-vivo* generally has a lower velocity than comparable tissue without fat, an effect studied particularly for liver tissue (Bamber and Hill, 1981; Tervola *et al*, 1985; Sehgal *et al*, 1986) but also for muscle *in-vivo* (Miles *et al*, 1984).

At a more fundamental level, Hustad *et al* (1971) demonstrated that the velocity in fatty acids and triglycerides depended on the number of carbon atoms per molecule. For instance at 41°C fatty acid velocity increased from 1124 to 1290 m s^{-1} on increasing the carbons from 4 to 10 per molecule.

The collagen content of some collagen-rich tissues, notably skin, tendon cartilage and arterial wall, appears to affect the velocity significantly. O'Brien (1977) derived an expression using velocity measurements on 9 tissues in the 1-10 MHz range in which an exponential dependence of velocity on collagen was modelled. Empirical expressions assuming a linear dependence have been postulated by others. For arterial wall, with collagen content in the range 1% to 5%, Rooney *et al* (1982) have reported a positive correlation between velocity and collagen content expressed as:

$$c = 17.8x_c + 1560 \qquad r = 0.77 \qquad (4.11)$$

Studies at 100 MHz (Goss and O'Brien, 1979; Edwards and O'Brien, 1985) have demonstrated velocities in mouse tendon threads higher than those for most soft tissue. The wide range of velocity values measured in skin may be associated with the relative proportions of water and collagen in the different samples measured. Olerud *et al* (1990) report a positive linear correlation between velocity and x_c in the range 5-26%, and a negative correlation for percentage water, W (55-79%). The tissues studied were from dog skin and wounds. The following empirical expressions were derived from their measurements:

$$c = 1516 + 5.34x_c \qquad r = 0.68 \qquad (4.12)$$

$$c = 1893 - 4.43W \qquad r = -0.75 \qquad (4.13)$$

Most measurements have investigated large tissue samples, so that small-scale variability may be masked. Sarvazyan and Klemin (1983) have noted some variability in the acoustic properties of dog kidney. They found, for example, a typical reduction of 1.4% in parenchymal velocity on moving from cortex to pelvis.

4.1.5.5 Tissue fixation

Apart from some data for hard tissues, literature values from fixed tissues have been omitted from the tables. Several authors have reported a slight drop in velocity due to formalin fixation (Kremkau *et al* 1981; Van Venrooij, 1971; Bamber *et al*, 1979). The percentage reduction lies in the range 0.7% to 2.0% and increases with increase in temperature (Hughes and Snyder, 1980). Fixation in ethyl alcohol results in a reduction in velocity of over 10% (Bamber *et al*, 1979).

4.1.5.6 Changes following death

Kremkau *et al* (1981) report a reduction of up to 10 m s^{-1} in velocity for human brain tissue during the 24 hours following death, and that the velocity remained constant thereafter. However, no significant change was detected in velocity for the vitreous humor or lens of the human eye during a period up to 70 hours after death (Jansson and Kock, 1962). In an early study, Frucht (1953) demonstrated that bleeding of tissue prior to excision resulted in an increase in c. Many of his measurements were from bled animals and these data have therefore been omitted from the tables. A slight increase in velocity (0.2%) in human biceps *in-vivo* during their contraction has also been related to blood being squeezed from the tissue (Mol and Breddels, 1982).

4.1.5.7 Animal species

There is no evidence for consistent variation of sound velocity by species. The tables are organised by animal species for cataloguing convenience only.

4.1.6 *Shear wave velocity*

Transverse, or shear waves propagate in hard tissues at velocities around 30–50% lower than the equivalent longitudinal wave velocity (Table 4.3). Shear wave velocity anisotropy has been reported.

Frizzell *et al* (1976) have investigated the propagation of shear waves in soft tissues. An order-of-magnitude range for c_s of between 9 m s^{-1} and 100 m s^{-1} was given for a range of tissues including red blood cells, liver, liver homogenate, kidney and muscle, measured at 2.0, 6.5 and 14 MHz.

4.1.7 *Acoustic velocity through blood*

Several authors have analysed the dependence of velocity c on the constituents of blood, and have given empirical fits to their data. For blood plasma, Aubert *et al* (1978) gave the following dependence expressions on plasma crystal concentration Φ:

$$\begin{aligned} c &= 3.36\Phi + 1523.9 \quad (37^{\circ}C) \\ c &= 4.01\Phi + 1483.82 \quad (20^{\circ}C) \end{aligned} \tag{4.14}$$

For whole blood at 37°C the expression:

$$c = 3.39\Theta + 1553.42 \tag{4.15}$$

is given as the variation of velocity with haemoglobin concentration, Θ, in g/100 ml. This is in line with a value of 3.33 m s^{-1} per g/100 ml given by Urick (1947). Bradley and Sacerio (1972) report a variation 1.69 m s^{-1} per g/100 ml total protein at 37°C in a study on whole blood. These authors also give a variation with haematocrit, H, over a wide range as 0.59 m s^{-1} per unit hematocrit change. This value contrasts with that given by Bakke *et al* (1975) also at 37°C, of 0.98 m s^{-1} per unit hematocrit change, measured over the narrower hematocrit range of 30% to 60%; the fitted expression is:

$$c = 1541.82 + 0.98H \tag{4.16}$$

An explaination of this difference may lie in the observations of Hughes *et al* (1979) who give a parabolic fit to velocity measurements in the range 0–55% hematocrit, based on Urick's theory of the dependence of sound velocity in a fluid suspension of particles:

$$1/c^2 = 0.418(1 + 5.0\times10^{-4}H - 4.19\times10^{-5}H^2) \tag{4.17}$$

where c is in mm μs^{-1}. The minimum velocity of about 1540 m s^{-1} was observed at 10% haematocrit and the predicted velocity of pure plasma was 1546 m s^{-1}. In view of these observations it is to be expected that estimates of the dependence of velocity on haematocrit based on a linear model will vary with the haematocrit range observed, so explaining the discrepancy between the values given by Bakke *et al* and Bradley and Sacerio.

4.1.8 *Acoustic velocity through bone*

The anisotropy and shear wave characteristics of bone have been noted above (Sections 4.1.5.3 and 4.1.6). Additionally, the variation of bone velocity upon bone density has been widely reported (for instance, Abendschein and Hyatt,

1970). Lees *et al* (1987) give an expression for the density dependence of radial velocity for longitudinal waves (1L,2L) in rabbit femur as:

$$c = (1.40\rho_b + 0.461)\times 10^3 \qquad (4.18)$$

where ρ_b is the bone density in g cm^{-3}. Table 4.3 includes some density values where given by these authors. The reduction in velocity observed in osteoporotic bone (Anast *et al*, 1958; Abendschein and Hyatt, 1970; Meunier *et al*, 1982) has been related to reduced density. For instance Meunier *et al* (1982) reported reductions in longitudinal and shear wave velocities both along and across the bone of about 3% in osteoporotic bone

The acoustic velocity through cancellous (trabecular) bone is significantly lower than through cortical bone.

Lakes *et al* (1983) have reported anomalously slow wave propagation in wet bone additional to the expected longitudinal wave propagation, and apparently distinct from shear wave propagation. Velocities of about 1530 m s^{-1} and 2300 m s^{-1} were reported for circumferential and longitudinal velocities respectively. It was suggested that these waves are associated with the dynamics of fluid motion in the pores of the bone, propagating in a similar manner to waves in other porous media.

The velocity in prepared specimens of dried bone is higher than that for wet bone. Lang (1970) and Lees *et al* (1979) report consistently higher values for all modes of propagation in dry bone, with velocities commonly about 10% above those in comparable wet bone samples. The longitudinal and shear wave velocities in apatite have been calculated by Lees and Rollins (1972) as 7370 m s^{-1} and 3530 m s^{-1}.

4.1.9 *Acoustic velocity through lung*

Inflated lung tissue is strongly dispersive in the low megahertz frequency range (Section 4.1.5.2, and Table 4.2). In addition, velocity in lung changes markedly with inflation. For example an increase from 32% to 66% air (density change from 0.72 to 0.36 g cm^{-3}) caused a decrease in velocity from 948 to 583 m s^{-1} according to Pedersen and Ozcan (1986). These measurements were made at 0.01–0.8 MHz and parallel those in the range 1–7 MHz reported by Dunn (1986). Velocities of audio frequency sound in the airways of the lung are controlled in part by their diameter (Rice and Rice, 1987; Table 4.7).

4.1.10 *Acoustic velocity through skin*

Values of velocity for skin vary widely. Those for mouse skin (Bhagat *et al*, 1980) lie in the range associated with other soft tissues, showing a slight reduction in older animals. Velocities for cattle skin (Lewin and Busk, 1982) lie at the upper end of the soft tissue range. However, two reports suggest that velocity through skin can take values intermediate between soft tissue

Table 4.7 Acoustic velocity at audio frequencies

Tissue	Frequency Hz	Velocity m s^{-1}	Reference
Lung parenchyma			
Horse	Audio pulse	25–70	Rice 1983
Lung airway			
Horse			
trachea	252(mean)	282±23.5	Rice & Rice 1987
airway, >25mm dia	252(mean)	150±83	-
-,1-25mm dia	252(mean)	268±44	-
Dog,upper airway	Audio pulse	349±3.5	Rice 1980

and bone. Cantrell *et al* (1978) report a value of 1720±120 m s^{-1} for the group velocity of a 15 MHz pulse through pig skin, a value which showed no alteration with burn necrosis. At 100 MHz, O'Brien *et al* (1981) give velocities for normal dog skin in the range 1590 to 2170 m s^{-1}. This latter study, using an acoustic microscope, also reports increasing velocity with time in a repairing skin wound, changing from 1540–1575 m s^{-1} at 7 days to 1700–2000 m s^{-1} at 35 days following the damage. The velocity in skin and other collagen-rich tissues has been related to the percent collagen content (see Section 4.1.5.4).

4.1.11 *Acoustic velocity through some materials other than tissue*

4.1.11.1 Water

The acoustic velocity in water is given in Table 4.8, including its temperature dependence. Del Grosso and Mader (1972) give a 5th order polynomial fit to data for pure water, for temperature T in the range 0–100°C;

$$c_{water} = \sum_{i=1}^{5} k_i T^i \qquad (4.19)$$

where the coefficients are:

i	k_i (m s^{-1})
0	1402.39
1	5.03711
2	-5.80852×10^{-2}
3	3.34199×10^{-4}
4	-1.47800×10^{-6}
5	3.14643×10^{-9}

Table 4.8 Acoustic velocity and attenuation, and non-linearity parameter B/A for pure water at atmospheric pressure

Temp °C	Velocity $m\ s^{-1}$	Attenuation* 10^{-3} neper m^{-1} MHz^{-2}	B/A
0	1402.4	57	4.16
10	1447.3	36	4.63
20	1482.3	25	4.96
30	1509.1	19	5.22
40	1528.9	15	5.38
50	1542.6	12	5.55
60	1551.0	10.5	5.67
80	1554.5	8.0	5.96
100	1543.1	–	6.11

Sources: Kaye & Laby (1973), Del Grosso & Mader (1972), Beyer (1960) and Hagelberg *et al* (1966).

* Attenuation in $dB/cm/MHz^2$ may be calculated by multiplying by $8.686x10^{-2}$.

The values have been rounded to six significant figures. Pure water is non-dispersive.

4.1.11.2 Aqueous mixtures

The acoustic velocity through saline mixtures of several concentrations and at several temperatures is given in Table 4.9. Coppens (1981) has given a empirical fit to data for saline water at T°C:

$$\begin{aligned} c_{saline} &= 1449.05 + 45.7t - 5.21t^2 + 0.23t^3 \\ &+ (1.333 - 0.126t + 0.009t^2)(10S - 35) \end{aligned} \quad (4.20)$$

Table 4.9 Acoustic velocity in water as a function of temperature and salinity at 1 atmosphere

Salinity	Temperature				
	0°C	10°C	20°C	30°C	40°C
0%	1402.4	1447.3	1482.3	1509.1	1528.9
0.5%	1409.2	1453.5	1488.0	1514.4	1533.9
1.0%	1416.0	1459.6	1493.6	1519.7	1538.9
2.0%	1429.2	1471.6	1504.7	1530.0	1548.2
3.0%	1442.6	1483.7	1516.0	1540.4	1558.6
4.0%	1455.9	1495.9	1527.0	1550.8	1568.2

Sources: Millero & Kubinski (1975) and Del Grosso & Mader (1972).

Table 4.10 Acoustic velocity in ethanol–water mixtures at 20°C

% ethanol by weight	Velocity m s^{-1}	Temperature coefficient
100	1162	negative
70.5	1365	negative
50.1	1490	negative
29.8	1630	negative
20.7	1625	~0
15.6	1600	~0
10.4	1560	positive
0.0	1482	positive

Zero temperature coefficient at 17% ethanol.
Source: Giacomini (1947).

where t = T/10 and S is the salt concentration in g/100 cm^3 water. Velocities in ethanol–water mixtures are given in Table 4.10.

4.1.11.3 Tissue substitute materials

The acoustic velocity in some tissue substitutes is included in the tables giving their properties later in the chapter (Tables 4.23 and 4.24).

4.2 Acoustic non-linearity parameter, B/A

4.2.1 *Terminology and definitions*

For an acoustic wave of finite amplitude travelling in an isotropic medium the equation of state can be expressed by the Taylor expansion:

$$P = P_o + A\left[\frac{\rho - \rho_o}{\rho}\right] + \frac{B}{2}\left[\frac{\rho - \rho_o}{\rho}\right]^2 + \ldots \qquad (4.21)$$

where

$$A = \rho_o \left[\frac{\partial P}{\partial \rho}\right]_{o,s} = \rho_o c_o{}^2$$

and

$$B = \rho_o{}^2 \left[\frac{\partial^2 P}{\partial \rho^2}\right]_{o,s}$$

P, P_0 are the instantaneous and hydrostatic pressures, ρ, ρ_0 are the instantaneous and equilibrium densities, and c_0 is the infinitesimal sound velocity. The derivatives are at equilibrium and at constant entropy S (see, for instance, Bjørnø, 1976). The **isentropic non-linearity parameter**, B/A, may be used to characterise the non-linearity of the medium. B/A may be expressed as the sum of two terms:

$$\frac{B}{A} = \rho_0 c_0 \left[\frac{\partial c}{\partial P}\right]_T + \frac{\beta T \rho_0}{C_p}\left[\frac{\partial c}{\partial T}\right]_P = \left[\frac{B}{A}\right]' + \left[\frac{B}{A}\right]'' \qquad (4.22)$$

The suffix T indicates a derivative at constant temperature, and suffix P at constant pressure. T is the absolute temperature, β the volume coefficient of thermal expansion and C_p is the specific heat at constant pressure. For soft tissue, B/A" dominates. Measurement methods for B/A are based either on Equation 4.22 (thermodymic method), or by measuring the second harmonic generated in the wave under controlled conditions (finite-amplitude method).

4.2.2 *Values of B/A for tissue*

Experimentally determined values of B/A for a variety of normal and pathological tissues are given in Table 4.11.

4.2.3 *Factors affecting B/A*

4.2.3.1 Temperature

Sehgal *et al* (1986) report a temperature coefficient for B/A for human liver of 0.026 $^{\circ}C^{-1}$ in the temperature range 20–37 °C.

4.2.3.2 Tissue composition

Investigating the variation of velocity and B/A in human liver, Sehgal *et al* (1986) give the following equation fitting measured data at 37°C:

$$10^{15}N = 1.378 + 0.106x_w + 4.037x_f \qquad (4.23)$$

where $N = (B/A)\rho c^3$, ρ is in g cm^{-3} and c in cm s^{-1}, and x_w and x_f are the volume fractions of the water and fat content, respectively. Equations are also given for 20°C and 30°C, and an equivalent equation expressing c in terms of x_w and x_f is given in Equation 4.9.

Apfel (1983) has suggested a two-part mixture rule for immiscible mixtures:

Table 4.11 Acoustic non-linearity parameter, B/A, for tissue

Tissue	Temp °C	B/A	Reference
Bile,pig	24	6.00	Sun Yongchen *et al* 1985
Blood,human	26	6.0,6.1	Gong *et al* 1989
–,pig	30	6.3±0.1	Dunn *et al* 1981
Blood:plasma,cow	30	5.74±.02	Everbach 1989
Brain,human fetal	17.3	6.55	Sun Yongchen *et al* 1986
–,cow	30	7.6	Law *et al* 1985
Fat/fatty tissue,			
human breast fat	37	9.63	Sehgal *et al* 1984
–,human & animal	37	10.28±0.34	Errabolu *et al* 1988
–,pig fat	24	9.6,9.9	Sun Yongchen *et al* 1985
–,–	26	10.8,10.9	Gong *et al* 1989
–,pig fatty tissue	30	11.0,11.3	Law *et al* 1985
Gall bladder,human	26	6.22	Sun Yongchen *et al* 1986
Kidney,human,fetus	17.3	8.98	Sun Yongchen *et al* 1986
–,pig	24	7.10	Sun Yongchen *et al* 1985
–,–	26	6.3,6.9	Gong *et al* 1986
–,dog	30	7.2±0.7	Cobb 1982
–,cow	25	5.83±0.15	Everbach *et al* 1989
Liver,human	37	6.75±.14	Sehgal *et al* 1986
–,–,fetal	17.3	8.72	Sun Yongchen *et al* 1986
–,cow	30	6.2–8.9	Law *et al* 1985
–,dog	30	7.6,7.9±0.8	Cobb 1982
–,pig	26	6.9,7.1	Gong *et al* 1989
–,–	23	7.7±0.4	Dunn *et al* 1981
Lymph node,human	15.5	8.21	Sun Yongchen *et al* 1986
Muscle:cardiac,			
human fetal	18	5.8	Sun Yongchen *et al* 1986
–,cow	30	6.8,7.4	Law *et al* 1985
–,pig	24	6.83	Sun Yongchen *et al* 1985
–,–	26	6.8,7.1	Gong *et al* 1989
Muscle:skeletal,			
human fetal	18	7.43	Sun Yongchen *et al* 1986
–,pig	30	7.5,8.1	Law *et al* 1985
–,–	24	7.55	Sun Yongchen *et al* 1985
Skin,human fetal	17.5	7.87	Sun Yongchen *et al* 1986
Spleen,human	30	7.8±0.8	Cobb 1982
–,pig	26	6.3,6.9	Gong *et al* 1989
–,dog	30	6.8±0.7	Cobb 1982
–,cow	25	6.59±.08	Everbach *et al* 1989
Tissue water,bound		~0.4	Yoshizumi *et al* 1987
–,free		8.0	–
Urine,human	24	6.14	Sun Yongchen *et al* 1985

$$\beta_{eff}^2\left[\frac{B}{2A}\right] = \beta_1^2\left[\frac{B}{2A}\right]_1(1 - x) + \beta_2^2\left[\frac{B}{2A}\right]_2 x \qquad (4.24)$$

where $\beta_{eff} = \beta_1(1 - x) + \beta_2 x$ is the effective compressibility of the mixture, β_1 and β_2 are the compressibilities of the two components and x, 1−x their volume fractions.

4.2.4 *B/A for some materials other than tissue*

The non-linearity parameter B/A for water over the temperature range 0–100°C is included in Table 4.8. B/A for various saline solutions is given in Table 4.12.

Table 4.12 B/A for saline water

Salinity	Temperature			
	0°C	10°C	20°C	30°C
3.3%	4.89	5.06	5.21	5.37
3.5%	4.92	5.09	5.25	5.41
3.7%	4.96	5.12	5.27	5.42

Source: Coppens *et al* (1965).

4.3 Ultrasonic attenuation: absorption and scatter

4.3.1 *Terminology and definitions*

The attenuation of plane sound waves is given by

$$\alpha = \frac{1}{2d}\ln\left[\frac{I_o}{I_d}\right] \qquad (4.25)$$

where the initial intensity, I_o, has decreased to I_d after a distance d. α is the **amplitude attenuation coefficient** with units of neper metre^{-1} (Np m^{-1}), and a practical unit Np cm^{-1}. Expressed in decibels per unit length the value is 8.686α. For pure water α is proportional to f^2 in the range 3–70 MHz. At lower frequencies water has a higher attenuation than the f^2 law predicts.

Simple fluids show single relaxations characterised by the equation

$$\frac{\alpha_R}{f^2} = \frac{A}{1 + (f/f_R)^2} \tag{4.26}$$

where $A = (\alpha_R\lambda)_{max}/cf_R$ is a constant defined by the maximum value of the absorption per cycle $(\alpha\lambda)$, the velocity c and the relaxation frequency f_R.

For fluids absorbing through both viscous and multiple relaxation mechanisms the attenuation is given by the classical 'frequency-free' absorption coefficient α_a/f^2:

$$\frac{\alpha_a}{f^2} = \frac{2\pi^2}{\rho_o c^3}\frac{4}{3}\eta_s + \sum_i \frac{A(i)}{1 + [f/f_R(i)]^2} \tag{4.27}$$

where η_s is the coefficient of shear viscosity, and the summation represents the presence of i relaxation processes. The thermal conductivity term has been omitted since it is negligible for biological media. A more detailed discussion of underlying absorption mechanisms, with particular reference to blood, is given by Horak (1977).

The total attenuation in tissue results from the combined losses due to absorption and scattering. Thus

$$\alpha = \alpha_a + \alpha_s \tag{4.28}$$

where α_a is the **amplitude absorption coefficient** and α_s is the **amplitude scattering coefficient.** The **intensity attenuation coefficient,** μ, is 2α. Equivalently the intensity coefficients μ_a and μ_s are related by

$$\mu = \mu_a + \mu_s = 2\alpha \tag{4.29}$$

The terms μ_a, μ_s are also called **absorption** and **scattering cross-section per unit volume** respectively, with common unit cm^{-1} (Np cm^{-1}).

The scattering of sound from tissue is anisotropic. In order to characterise this variation with angle it is necessary to define a differential cross-section (per unit volume) as the power scattered into unit solid angle in any defined direction. For many applications it is the backscatter returning to a source also acting as a receiver which is important. Under these circumstances the differential cross-section at 180°, or **backscatter cross-section,** μ_{bs}, is used. The common unit for μ_{bs} is centimetre^{-1} steradian^{-1} (cm^{-1} Sr^{-1}).

4.3.2 *Measurement of attenuation, absorption and scatter*

The measurement of attenuation has commonly been achieved using a substitution method, recording the reduction in signal strength at a receiver following the introduction of a tissue sample of known thickness into an acoustic path. The path may traverse the sample once or twice. Considerable difficulties arise from the use of piezoelectric receivers, particularly if large, because of phase cancellation when a disturbed wavefront is detected by a phase-sensitive receiver of finite size (Marcus and Carstensen, 1975). Preferred methods use the measurement of total acoustic power with a radiation force balance, or a phase insensitive acousto-electric receiver such as cadmium sulphide (Busse and Miller, 1981). For measurements using phase-sensitive receivers, greatest accuracy results from using the smallest size compatable with sensitivity. Similar problems arise in the measurement of scattering cross-sections.

Absorption coefficients have been measured from the observation of the temperature rise during exposure, either by the transient thermoelectric method (Dunn, 1962, Goss *et al*, 1979), or the rate-of-heating method (Parker, 1983). Values of α_a estimated from the difference between total attenuation and scatter have been omitted from the tables.

The measurement of attenuation *in-vivo* of large homogeneous tissue masses may be achieved from the analysis of the frequency content in backscattered pulses (Lin *et al*, 1987; Taylor *et al*, 1986).

4.3.3 *Historical background*

In one of the earliest papers, Pohlman (1939) established the anomalously near-to-linear frequency dependence of the attenuation of ultrasound in tissue. Full texts of this and other seminal investigations may be found in a single volume (Dunn and O'Brien, 1976). Also included are important early studies of the ultrasonic properties of protein and other biological solutions, data from which has not been included here. Very full surveys of attenuation data have been given by Goss *et al* (1978b, 1980) and Chivers and Parry (1978). A useful overview of attenuation, absorption and scattering, including a thorough discussion of the problems of measurement, is given by Bamber (1986).

4.3.4 *Values of ultrasonic attenuation coefficients for tissue*

Values of attenuation coefficients for ultrasound in tissue are given in Tables 4.13 to 4.17. Tables 4.13, 4.14 and 4.15 list measurements of attenuation coefficient made at specific frequencies for biological fluids, soft tissues and bone. Tables 4.16 and 4.17 express attenuation in tissues and biological fluids as a function of frequency, using the expression $\alpha = af^b$, with normal tissues in Table 4.16, and some pathological tissues in Table 4.17. Some selection from the literature has been applied, mostly giving preference to measurements in which phase-insensitive methods methods using small recievers have been used. Where piezoelectric receivers have been used, values have been drawn primarily from sources in which attention has been paid to minimising phase interference effects.

Table 4.13 Ultrasound amplitude attenuation coefficients for biological fluids

Fluid	Temp °C	Freq MHz	Attenuation coeff Np cm^{-1} x10^{-2}	Attenuation coeff dB cm^{-1}	Reference
Amniotic fluid	22	2	0.52	0.045	Zana & Lang 1974
–	22	3	0.54	0.047	–
–	22	5	0.93	0.080	–
–	22	9	2.75	0.24	–
–	22	14	5.49	0.48	–
Blood	25	1	2.4	0.21	Carstensen & Schwan 1959a
–,42% Hct	NR	2.4	6.1 ±.2	0.53±.02	Carstensen *et al* 1953
–	25	10	32	2.8	Carstensen & Schwan 1959a
–	22	10	26	2.3	Shung & Reid 1977
–	37	4.5–4.8	22.1	1.92	Grybauskas *et al* 1978
–,40% Hct	37	10	29.4	2.55	Hughes *et al* 1979
–,packed cells	25	1	3.2	0.28	Carstensen & Schwan 1959a
–,–	25	10	66	5.7	–
–,clotted	37	4.5–4.8	133	11.5	Grybauskas *et al* 1978
Milk	23.5–27	0.34	2.4	0.21	Hueter *et al* 1953
–	23.5–27	1	4.7	0.47	–
–	23.5–27	4	17,23	1.5,2.0	–
–	27.6–29	10.25	44	3.8	–
Plasma	25	1	0.8	0.069	Carstensen & Schwan 1959a
–	25	10	17	1.5	–
–	15–18	0.87	1.8	0.16	Colombati & Petralia 1950
–	15–18	1.7	3.9	0.34	–

Table 4.14 Ultrasound amplitude attenuation coefficient for normal tissue

Tissue	Temp °C	Freq MHz	Attenuation coeff Np cm^{-1}	dB cm^{-1}	Reference
Artery,human,aorta	20	10(pulse) (2–10)	0.65–0.77	5.6–6.7 mean 6.1	Greenleaf *et al* 1974
Articular capsule human and cow	NR	1	0.38	3.3±0.9	Dussik *et al* 1958
–,–	NR	5	1.29	11.2±5.4	–
Brain,human	25	1	0.069	0.6	Kremkau *et al* 1981
–	25	5	0.518	4.5	–
–,–,grey matter	37	2.2	0.072	0.625	–
–,–,white matter	37	2.2	0.121	1.05	–
–,–,infant	37	1	0.018	0.16	–
–,–,cerebellum	30	0.97	0.06	0.5	Oka 1977
–,–,–	30	2.9	0.14	1.2	–
–,–,dura mater	NR	0.97	0.13±.01	1.1±0.1	–
–,cat	*in–vivo*	4.2	0.28–0.50	2.4–4.3	Robinson & Lele 1969
–,dog	*in–vivo*	0.97	0.054	0.47	Yosioka *et al* 1969
Breast,human *	*in–vivo*	1.76	0.06–0.13	0.5–1.1	Chevenko & Yukhananov 1971
– *	23,37	7	1.09–1.45	9.5–12.6	McDaniel 1977
Cartilage,human,cow	NR	1	0.57	5.0±1.5	Dussik et al 1958
–,–	NR	5	2.19	19.0±5.4	–
Eye:lens,human	NR	10	0.9	8	Lizzi *et al* 1976
–,calf	28	3.25	0.59–0.69	5.1–6.0	Begui 1954
–,cow	22	10	2.6	23	Filipczynski *et al* 1967
–,–	22	13.5	3.6	31	–
Eye:vitreous	25–28	30	0.33	2.9	Begui 1954
–	22	6	0.07	0.6	Filipczinski *et al* 1967
–	22	18	0.23	2.0	–
–	22	30	0.33	2.9	–

cont.

Table 4.14 cont. Attenuation coefficient for tissue

Tissue	T,°C	f,MHz	α,Np cm^{-1}	dB cm^{-1}	Reference
Fat,pig	37	2	0.35	3	Gammell *et al* 1979
–,–	37	10	1.98	17.2	–
–,–,subcutaneous	NR	1	0.21	1.8±0.1	Lehmann & Johnson 1958
–,–	37	1.6	0.07	0.6	Schwan *et al* 1953
–,–	37	6	0.56	4.9	–
–,mouse	50	100–500	180	1600	Daft & Briggs 1989
Kidney,pig	37	2	0.23	2	Gammell *et al* 1979
–	37	10	0.98	8.5	–
–,dog,reg. 1,cortex	20–22	8.8	0.52–0.56	4.5–4.9	Sarvazyan & Klemin 1983
–,–,region 2	20–22	8.8	0.42–0.47	3.6–4.1	–
–,–,region 3	20–22	8.8	0.34–0.36	3.0–3.1	–
–,–,region 4,core	20–22	8.8	0.24–0.30	2.1–2.6	–
–,–,pelvis	20–22	8.8	1.23–1.30	10.7–11.3	–
Liver,rat	22	100	15.0	130±9	O'Brien *et al* 1988
Lung,human	NR	1	3.5	30	Dussik *et al* 1958
–,dog,collapsed	27	2.4	4.4	38±12	Bauld & Schwan 1974
–,–,–	27	7.4	10.1	88±12	–
–,–,0.4g cm^{-3}	35	1	4.2	36	Dunn 1986
–,–,–	35	5	11.4	99	–
–,–,0.5g cm^{-3}	35	1	3.3	29	–
–,–,–	35	5	8.6	75	–
–,–,0.7g cm^{-1}	35	1	1.6	14	–
–,–,–	35	5	4.9	43	–
Muscle:skeletal					
human,thigh	*in–vivo*	4.3	0.54	4.71±.44	Ophir *et al* 1982
–,cow	21–23	2	0.21–0.24	1.82–2.08	Marcus & Carstensen 1975

cont.

Table 4.14 cont. Attenuation coefficient for tissue

Tissue	T,°C	f,MHz	α,Np cm^{-1}	dB cm^{-1}	Reference
Muscle,along	20	2	0.32	2.8	Shore *et al* 1986
–,–,–	20	7	0.99	8.61	–
–,–,across fibres	20	2	0.11	0.98	–
–,–,–	20	7	0.43	3.71	–
–,–,along fibres	20–35	0.87	0.18	1.6	Colombati & Petralia 1950
–,–,–	20–35	3.4	0.60	5.2	–
–,–,across fibres	20–35	0.87	0.055	0.48	–
–,–,–	20–35	3.4	0.26	2.3	–
–,pig,across fibre	NR	1	0.09	0.8±0.1	Lehmann & Johnson 1958
–,mouse	50	100–500	80	700	Daft & Briggs 1989
Nerve,human,along	15–18	1.7	0.14	1.2	Colombati & Petralia 1950
–,–,–	15–18	3.4	0.34	3.0	–
–,–,across fibres	15–18	1.7	0.22	1.9	–
–,–,–	15–18	3.4	0.46	4.0	–
–,cow,along fibres	15–18	3.4	0.35	3.0	–
–,–,across fibres	15–18	3.4	0.55	4.8	–
Pancreas,pig	37	2	0.25–0.32	2.2–2.8	Le Croisette *et al* 1979
–,–	37	9	1.27–1.7	11–15	–
Rectum wall,cow	NR	1	0.06	0.6±0.1	Dussik *et al* 1958
–,–	NR	5	0.28	2.4±0.4	–
Skin,human	23	1	0.4	3.5±1.2	Dussik *et al* 1958
–,–	23	5	1.06	9.2±2.2	–
–,–,breast,foot	40	0.97	0.28	2.4	Nakaima *et al* 1976
–,–,–	40	2.9	0.52	4.5	–
–,dog,back	40	0.97	0.14–0.17	1.2–1.5	–
–,–,–	40	4.8	0.75	6.5	–
–,–	22–24	25	13±4	110	Riederer-Henderson *et al* 1988
–,–	30	100	66±12	570	–

cont.

Table 4.14 cont. Attenuation coefficient for tissue

Tissue	T,°C	f,MHz	α,Np cm^{-1}	dB cm^{-1}	Reference
Skin,mouse,young	20.5	2.25	0.60	5.19	Bhagat *et al* 1980
–,–,–	20.5	10	3.92	34.0	–
–,–,old	20.5	2.25	0.46	4.03	–
–,–,–	20.5	10	2.05	17.7	–
Spleen,human	37	2	0.12	1.0	Gammell *et al* 1979
–,–	37	10	1.32	11.5	–
Tendon,human,cow	NR	1	0.54	4.7±1.0	Dussik *et al* 1958
–	NR	5	1.95	16.9±2.4	–
–,cow,across	NR	1	0.73	6.3±.2	Dussik *et al* 1958
–,–	NR	5	2.35	20.4±5.7	–
–,with grain	NR	1	0.41	3.6±0.7	–
–,–	NR	5	2.35	20.4±5.7	–
Testis,rabbit	25	2	0.07	0.6	Frizzell 1975
Uterus,cow	NR	1	0.22	1.9±0.2	Dussik *et al* 1958
Shear wave propagation					
Soft tissue	25	2–14	$2–30x10^3$	$\sim2–30x10^4$	Frizzell & Carstensen 1977

* Values for some pathological tissues also given.

Table 4.15 Ultrasound amplitude attenuation coefficient for bone and tooth material

Tissue	Temp °C	Freq MHz	Attenuation coeff Np cm^{-1}	dB cm^{-1}	Reference
Bone,skull,human	NR	0.8	1.51	13.1	Theismann & Pfander 1949
–,–,–*	37	1	2.5	22	Fry & Barger 1978
–,–,–*	37	3	9	78	–
–,–,–*,outer table	37	1	1.67	14.5	–
–,–,–*,inner table	37	1	2.15	18.7	–
–,–,–*,diploë	37	1	3,5	26,43	–
–,–,–*,–	37	3	12,16	100,140	–
–,–,–*,infant	37	1	0.7	6.1	–
–,–,–*,–	37	3	2.0	17	–
Trabecular,human	NR	0.5	0.22–1.8†	1.9–15.7	McKelvie & Palmer 1987
Tibia,dog	22	3	1.50	13	Adler & Cook 1975
–,–	22	5	2.19	19	–
Long bone,horse	NR	1.43	2.5	22	Kishimoto 1958
–,–	NR	2.86	4.6–5.8	40–50	–
–,–	NR	4.5	9.2	80	–
–,cow,human	NR	1	1.44	12.5	Dussik *et al* 1958
Femur,cow,cortical	NR	1	0.79†	6.9±1.9	McKelvie & Palmer 1987
–,pig	32	1	0.97	8.4±1.2	Lehmann & Johnson 1958
Vertebra,human	NR	1	0.17–4.4†	1.5–38.2	McKelvie & Palmer 1958
Tooth:pulp	NR	18	2.3	20	Kossoff & Sharpe 1966
Tooth:dentine	NR	18	9.2	80	–
Tooth:enamel	NR	18	14	120	–
Tooth:cementum	NR	18	23	200	–

* Fixed tissue. † 0.2–1.0 MHz.

Table 4.16 Ultrasound amplitude attenuation coefficient for normal tissue: $\alpha = a.f^{b}$

Tissue	Temp °C	Freq MHz	a: Np cm^{-1} MHz^{-b}	a: dB cm^{-1} MHz^{-b}	b	Reference
Bile	22	4.8/pulse	0.0015–0.0036	0.013–0.031	1.28–1.34	Narayana *et al* 1984
Blood,human	22	4.8/pulse	0.014–0.018	0.12–0.16	1.19–1.23	–
Blood:plasma,human	25	1.7–15	0.0066	0.057	1.41	Lang *et al* 1978
Brain,human	room	1–6	0.067–0.069	0.58–0.60	1.20–1.46	Bamber 1981
			mean 0.067	mean 0.58	mean 1.3	
–,–,white matter	room	1–6	0.083–0.11	0.72–0.96	0.99–1.16	–
			mean 0.09	mean 0.8	mean 1.1	
–,–	37	1–5	0.050	0.435	1.08	Kremkau *et al* 1981
–,cow	20–22	0.5–13	0.040	0.35	1.17	Lyons & Parker 1988
Breast,human	25,37	0.5–6	0.086	0.75±0.3	1.5	Foster & Hunt 1979
Fat,human,stomach	room	1–6	0.07–0.6	0.6–5.2	0.4–1.4	Bamber 1981
–,mouse adipocyte	50	100–500			1.66	Daft & Briggs 1989
Liver,human ⊗	35.5	1.25–8	0.0459	0.399	1.139	Lin *et al* 1987
–,–	37	0.5–6	0.081	0.7±0.2	1 *	Foster & Hunt 1979
–,– ⊗	*in–vivo*	2.5/pulse	0.041–0.070	0.36–0.61	1.05±.25	Parker *et al* 1988a
			mean 0.052	mean 0.45		
–,– ⊗	*in–vivo*	3/pulse	0.060	0.52±0.10	1 *	Taylor *et al* 1986
			0.057	0.49±0.167	1 *	
–,cow	22–25	1–10	0.043	0.37	1.266	Pohlhammer *et al* 1981
–,–	22–25	1–100	0.043	0.37	1.270	–
–,calf	20–22	0.5–13	0.032	0.28	1.30	Lyons & Parker 1988
–,pig	24	1.4–9.8	0.0778	0.676	1.0	Segal & O'Brien 1983
–,–	20–22	0.5–13	0.034	0.30	1.32	Lyons & Parker 1988
–,sheep	24	1.4–9.8	0.0855	0.743	0.916	Segal & O'Brien 1983
–,rat	23	2.3–3.4	0.078	0.68	1.24	Klemin *et al* 1981
–,–	23	8.5–9.6	0.116	1.01	1.42	–

cont.

Table 4.16 cont. Attenuation coefficient for normal tissue

Tissue	T,°C	f,MHz	Np cm^{-1} MHz^{-b}	dB cm^{-1} MHz^{-b}	b	Reference
Lung,sheep,fetal	*in-vivo*	5/pulse	0.055–0.077	0.48–0.67	1 *	Meyer *et al* 1984
Muscle:cardiac,dog	37	2–10	0.060±0.002	0.52	1 *	O'Donnell *et al* 1979
–,–	20	4–9	0.072–0.075 ±0.002	0.63–0.65	1 *	O'Donnell *et al* 1979
–,rat	23	2.3–3.4	0.054	0.47	1.09	Klemin *et al* 1981
–,–	23	8.4–9.5	0.080	0.69	1.19	–
Muscle:skeletal, human,thigh ⊗	*in-vivo*	5/pulse	0.064	0.57±0.07	1 *	Berger *et al* 1987
			0.062	0.54±0.06		
–,cow,across	20	1–8	0.13	1.1±0.15	1 *	Nassiri *et al* 1979
–,along fibres	20	1–8	0.33	2.9±0.23	1 *	–
–,–	–20	1–7	0.684†	5.94	0.79±.11	Shore *et al* 1986
–,–	40	1–7	0.169†	1.47	1.15±.18	–
–,mouse	50	100–500			1.43	Daft & Briggs 1989
Pancreas,cow	24	1.4–9.8	0.119	1.03	0.780	Segal & O'Brien 1983
Spleen,human	room	1–6	0.036–0.062 mean 0.046	0.31–0.54 mean 0.4	1.14–1.47 mean 1.3	Bamber 1981
–,cow	24	1.4–9.8	0.0687	0.597	1.08	Segal & O'Brien 1983
–,pig	24	1.4–9.8	0.0571	0.496	1.01	–
–,sheep	24	1.4–9.8	0.0370	0.321	1.29	–
Testis,human	room	3–7	0.012–0.029 mean 0.020	0.01–0.25 mean 0.17	1.26–2.04 mean 1.7	Bamber 1981
Uterus ⊗	*in-vivo*	3.5/pulse	0.10	0.89±0.2	1 *	Masaoka *et al* 1987

* b=1 assumed. † SE = ln(A)±~0.14. ⊗ Data for pathological tissue also given.

Table 4.17 Amplitude attenuation coefficient, $\alpha = a.f^{b}$; human pathological tissues

Tissue	Temp °C	Freq MHz	a Np cm^{-1} MHz^{-b}	a dB cm^{-1} MHz^{-b}	b	Reference
Cyst fluid	22	4.8/pulse	0.0058–0.0083	0.050–0.072	1.27–1.32	Narayana *et al* 1984
–	24	1.7–15	0.0088	0.076	1.42	Lang *et al* 1978
Liver						
Diffuse disease,mild	35.5	1.25–8	0.045	0.395	1.212	Lin *et al* 1987
–,moderate & severe	35.5	1.25–8	0.045	0.391	1.325	–
Metastatic	*in–vivo*	2.5/pulse	0.030–0.081	0.26–0.70	1 *	Parker *et al* 1988a
Chemotherapy toxicity	*in–vivo*	2.5/pulse	0.082–0.090	0.71–0.78	1 *	–
Alcoholic disease	*in–vivo*	2.5/pulse	0.062–0.092	0.54–0.80	1 *	–
Cirrhosis	*in–vivo*	3/pulse	0.067	0.58±0.14	1 *	Taylor *et al* 1986
Cirrhosis,fatty	*in–vivo*	3/pulse	0.075	0.65±0.16	1 *	–
Fatty	*in–vivo*	3/pulse	0.079	0.69±0.15	1 *	–
Muscle						
Duchenne dystrophy	*in–vivo*	5/pulse	0.077	0.67±0.07	1 *	Berger *et al* 1987
carriers			0.075	0.65±0.06		
Pus	22	4.8/pulse	0.023–0.027	0.20–0.23	1.11–1.16	Narayana *et al* 1984
Uterus						
Myoma,degenerate	*in–vivo*	3.5/pulse	0.150	1.30±0.27	1 *	Masaoka *et al* 1987
–,not degenerate	*in–vivo*	3.5/pulse	0.097	0.84±0.17	1 *	–
–,liquified	*in–vivo*	3.5/pulse	0.070	0.61±0.19	1 *	–

* b=1 assumed.

Table 4.18 Temperature coefficient for ultrasound attenuation

	Freq MHz	Temp °C	$d\alpha/dT$ dB cm^{-1} $°C^{-1}$
Soft tissue	2	20	−0.014±0.004
	2	37	0.0±0.019
	4	20	−0.052±0.011
	4	37	−0.021±0.021
	7	20	−0.097±0.026
	7	37	−0.032±0.006
Fat	1	20	−0.21
	1	37	0(appr)
	2	20	−0.35
	2	37	0(appr)
	3	20	−0.51
	3	37	0(appr)

Source: Bamber & Hill (1979).

4.3.5 *Factors affecting attenuation*

4.3.5.1 Temperature

The attenuation of all tissues is dependent upon temperature, a variation which is also frequency dependent. Recent data have been reviewed by Haney and O'Brien (1986) and by Bamber (1986). For many soft tissues in the mid-megahertz range, attenuation reaches a minimum at around 20–40°C (Bamber and Hill, 1979; Kremkau *et al*, 1981; Gammell *et al*, 1979). At frequencies around 1 MHz α/f^2 varies in a complex way with temperature. Dunn and Brady (1973) report absorption coefficients in mouse spinal chord showing a reversal of temperature coefficient at about 0.7 MHz, being negative below and positive above this frequency. Fat and fatty tissue show a large reduction in attenuation between room and body temperature (Bamber and Hill, 1979; Gammell *et al*, 1979). At temperatures above 50°C there is some evidence that denaturation of large molecules contributes to a considerable increase in attenuation (Robinson and Lele, 1969). The attenuation of frozen muscle is considerably greater than that of unfrozen tissue (Shore *et al*, 1986). Temperature coefficients for attenuation from Bamber and Hill (1979) are given in Table 4.18 for soft tissue and fat.

A positive temperature coefficient for bone was reported by Kishimoto (1959) in the frequency range 1.4–4.5 MHz. An approximate value of 0.15 dB cm^{-1} $°C^{-1}$ may be derived from the data presented.

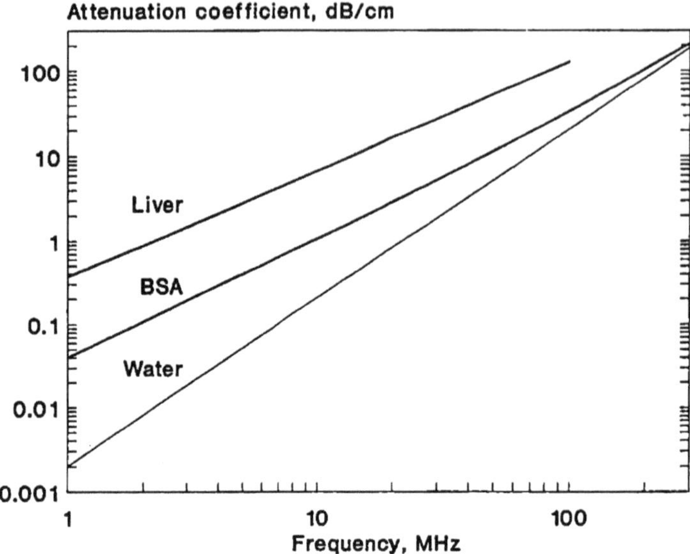

Figure 4.2 Variation of ultrasonic attenuation coefficient with frequency for water, bovine serum albumin (BSA) and liver, at room temperature. BSA data from Choi *et al* (1990), and liver data from Pohlhammer *et al* (1981).

4.3.5.2 Frequency

The frequency dependency of ultrasonic attenuation and absorption can be represented by the expression

$$\alpha = af^b \tag{4.30}$$

where a,b are constants and f is frequency. Several authors have fitted their data to this expression over narrow ranges of frequency. Others have assumed a linear frequency dependence (b=1), particularly when estimating attenuation coefficients from *in-vivo* measurements using the frequency content of backscattered sound. Data in both forms are included in Tables 4.16 and 4.17. Values for b for most soft tissue and biological fluids lie in the range 1.0 to 1.5. There is some evidence that Equation 4.30 may not be appropriate over a wider frequency range. Studying attenuation at low frequencies, Truong *et al* (1978) have demonstrated relaxing elements of muscle structure with characteristic frequencies in the region of 30 kHz. On the other hand at the upper frequency limit there is evidence that, for serum albumin solutions at least, α/f^2 reaches a constant value (27×10^{-15} s^2 m^{-1}; Choi *et al*, 1990). Thus at frequencies above about 500 MHz the relaxation process does not contribute and the absorption becomes almost equal to that of water. At neutral pH the distribution of relaxation times at frequencies below 100 MHz suggest that the dominant relaxation mechanism is from protein hydration, where different relaxations correspond to different degrees of binding of the water molecules. Choi *et al* also reported relaxations at acid pH (200 kHz and 2 MHz) and at alkali pH (200 kHz, 2 MHz and 15 MHz).

A broad comparison between water, serum albumin and liver attenuation over the frequency range 1–100 MHz is shown in Figure 4.2.

4.3.5.3 Anisotropy

Whilst attenuation by most tissues is isotropic, in a few it is anisotropic. Anisotropy has been demonstrated in skeletal muscle, nerve and tendon (Table 4.14). In muscle the attenuation measured along the fibres exceeds that across the fibres by a factor of 2 or 3.

4.3.5.4 Wave amplitude

Goss and Fry (1981) and Carstensen *et al* (1982) have reported on the dependence of tissue attenuation on the amplitude of the interrogating wave. Whilst the attenuation coefficient at any particular frequency is not itself amplitude-dependent, several experimental factors may cause the apparent attenuation to vary with amplitude. Harmonics generated by non-linear transmission in the coupling medium or in the tissue will be more strongly absorbed, raising the measured attenuation coefficient. Conversely, amplitude-dependent loss in the reference water path may be reduced on the introduction of the tissue sample, so lowering the apparent attenuation coefficient (Duck and Starritt, 1987).

4.3.5.5 Tissue fixation

Increases in attenuation following formalin fixation have been reported by several authors (Bamber *et al*, 1979; Kremkau *et al*, 1981; Le Croisette *et al*, 1979). Bamber *et al* measured increases of about 10% at 1 MHz rising to about 50% at 7 MHz. An anisotropic change in attenuation occurs in bovine skeletal muscle after fixation (Nassiri *et al*, 1979). Literature values from fixed tissues have been excluded from the tables, except where noted.

4.3.5.6 Changes following death

The main cause of alteration in attenuation in *post mortem* tissues is the formation of gas bubbles within the tissue, giving erroneously high attenuation measurements. Using carefully prepared samples, Bamber *et al* (1979) report a small but insignificant reduction in attenuation during a 120 hour period following death, in a variety of bovine tissues. However somewhat larger reductions in attenuation have been reported by others (Kremkau *et al*, 1981, Le Croisette *et al*, 1979). Kremkau found a 21% decrease in human brain after 24 hours, and Le Croisette reported similar reductions in several pig soft tissues, but not in fat, after 5 days. The changes were largest at the higher frequencies.

4.3.5.7 Tissue composition

The presence of high fat content in liver tissue is associated with an increase in attenuation both *in-vitro* (O'Brien *et al*, 1988) and *in-vivo* (Taylor *et al*, 1986). Similarly the collagen content in cardiac muscle recovering from infarction has been linked with an increase in attenuation (O'Donnell *et al*, 1979). In view of the predominance of absorption over scatter in attenuation, macromolecular relaxations are important as influencing mechanisms for attenuation (Carstensen, 1979). For instance, Parker *et al* (1988b) have shown that glycogen storage may cause significant variability in liver attenuation *in-vivo* with time.

The local variability in kidney velocity (Section 4.1.5.4) is associated with a similar variation in attenuation (Sarvazyan and Klemin, 1983). At 8.8 MHz the attenuation coefficient of canine kidney was found to reduce to about 0.5 on moving from cortex to pelvis. Similarly local variations in the absorption of ovaries have been reported (Carnes and Dunn, 1988) suggesting a factor of almost 3 between stroma and follicle in some species.

4.3.5.8 Tissue pathology

In view of the wide variability of measured values on pathological tissues, only very few data are presented (Table 4.17). Other data from pathological tissues may be found in the references marked.

4.3.5.9 Animal species

Some reports suggest that a species dependence for absorption and attenuation may exist for some tissues. Segal and O'Brien (1983) report different expressions for the frequency dependence of spleen attenuation coefficient for cow, pig and sheep. The absorption coefficient of ovarian structures is also reported to be species dependent (Carnes and Dunn, 1988). In this case there is also evidence of variability from the estrous cycle. Some caution should therefore be used in applying animal data uncritically to human applications.

4.3.6 *Shear wave attenuation*

Shear waves are very strongly absorbed in soft tissue (Frizzell and Carstensen, 1977). Measurements of shear wave attenuation in bone have not been reported.

4.3.7 *Attenuation by blood*

Hughes *et al* (1979) found that the measured attenuation for blood at 37°C and at 10 MHz fitted the following expression:

$$\alpha = 0.992 + 0.039H \text{ dB cm}^{-1} \tag{4.31}$$

where H is the percentage hematocrit.

4.3.8 *Values of acoustic absorption coefficients in tissue*

Measured values of absorption coefficients for ultrasound in soft tissue are given in Tables 4.19 and 4.20. Values at particular frequencies are included in Table 4.19, and the power–law expression Equation 4.30 used as the basis for the values given in Table 4.20.

Many of the factors discussed for attenuation in Section 4.3.5 are equally relevant to absorption, and reference should be made to this section.

4.3.9 *Values of acoustic scatter from tissue*

Values of scattering cross–sections for normal and pathological tissues measured at particular frequencies are given in Table 4.21. Values using a power–law fit for the data are given in Table 4.22.

4.3.10 *Factors affecting scatter*

4.3.10.1 Anisotropy

The intensity of scattered sound varies with direction. This variability depends upon both the direction of the incident beam and the direction of the scattered sound. For example, according to Nassiri and Hill (1986), back–scattered ultrasound from liver tissue at 6 MHz is about 10 dB below that scattered in a forward direction. However, under similar conditions, back–scatter from blood exceeds forward–scatter by about 6 dB. Muscle has an inherently anisotropic structure. In one of a series of studies of the scattering of ultrasound by canine myocardium, Madaras *et al* (1988) report integrated back–scatter increasing by 5.0±0.4 dB from a parallel to a perpendicular orientation with respect to the fibre axes, measured at end diastole. A smaller variation (3.2±0.4 dB) was observed at end systole.

4.3.10.2 Frequency

Several authors have studied the frequency dependence of scatter in the 1–10 MHz range. Most have presented results in terms of a single power–law fit of the form $\mu = af^b$ and values from several authors are presented in Table 4.22. Scatter from blood shows a 4th power dependence upon frequency, and other tissues somewhat lower. Nicholas (1982) found that a two–term polynomial fit of the form $a + bf^n$ gave a better fit to back–scatter data, especially when data in the frequency range 0.7–2 MHz

Table 4.19 Ultrasound absorption coefficient for soft tissues (i)

Tissue	Temp °C	Freq MHz	Absorption coeff Np cm⁻¹ x10⁻²	Absorption coeff dB cm⁻¹	Reference
Brain,cat	37	0.7	1.4±0.3	0.12±0.02	Goss *et al* 1979
–,–	37	1	2.9±0.4	0.25±0.03	–
–,–	37	7	23±9	2.0±0.8	–
Kidney,cat,cow*	37	0.7	1.7±0.7	0.15±0.06	–
–	37	1	3.3±0.4	0.29±0.03	–
–	37	7	20±2	1.7±0.1	–
Liver,mouse,cow*,	37	0.5	1.0±0.6	0.087±0.05	–
pig*	37	1	2.3±0.4	0.20±0.03	–
–	37	7	24±2	2.1±0.1	–
–,cow	20	1	4.4±0.3	0.38±0.03	Parker 1983
–,–	20	3.4	16.7±1.4	1.45±0.12	–
–,–	20	5.6	27.4±2.2	2.38±0.19	–
–,mouse,*in–vivo*		0.5	1.1	0.10	Frizzell *et al* 1979
–,–,*in–vitro*	37	0.5	0.93	0.08	–
Muscle:cardiac,cat	37	0.7	1.8±0.9	0.16±0.08	Goss *et al* 1979
–	37	1	3.3±0.6	0.29±0.05	–
–	37	7	21±3	1.82±0.2	–
Ovary,cow†,cortex	37	1	4.5±0.1	0.39	Carnes & Dunn 1988
–,–,medulla	37	1	4.4±0.7	0.38	–
–,–,follicle	37	1	1.7±0.2	0.15	–
–,–,corpora lutea, old	37	1	3.4±0.8	0.30	–
–,–,–,active	37	1	4.6±0.4	0.40	–
Spinal chord, mouse,*in–vivo*	20	0.5	8.0	0.7	Dunn & Brady 1973
–,–,–	40	0.5	7.5	0.65	–

cont.

Table 4.19 cont. Absorption coefficient for tissue

Tissue	T,°C	f,MHz	$cm^{-1}x10^{-2}$	dB cm^{-1}	Reference
Spinal chord,rat	*in vivo*	0.98	9.5–11.5	0.83–1.0	Fry & Fry 1953
Tendon,cow*	37	0.5	5±3	0.4±0.2	Goss *et al* 1979
–,–	37	1	11±4	0.96±0.3	–
–,–	37	7	140±50	12±4	–
Testis,mouse,horse	37	0.5	0.78±0.2	0.068±0.01	–
–,–	37	1	1.5±0.3	0.13±0.02	–
–,–	37	7	12±2	1.04±0.1	–

* Previously bled animal. † Data for cat, dog, sheep, mouse and pig also given.

Table 4.20 Ultrasound absorption coefficient (ii); $\alpha = a.f^{b}$

Tissue	Temp °C	a Np cm^{-1} MHz^{-b}	a dB cm^{-1} MHz^{-b}	b	Reference
Brain*	37	0.024	0.21	1.18	Goss *et al* 1979
–,cow,white matter	21	0.064	0.56	1.27	Lyons & Parker 1988
–,–,grey matter	21	0.012	0.10	1.21	–
Kidney*	37	0.028	0.24	1.02	Goss *et al* 1979
Liver*	37	0.026	0.23	1.17	–
–,calf	21	0.030	0.26	1.29	Lyons & Parker 1988
–,pig	21	0.032	0.29	1.30	–
Muscle:cardiac*	37	0.028	0.24	1.04	Goss *et al* 1979
Tendon*	37	0.14	1.2	1.17	Goss *et al* 1979
Testis*	37	0.015	0.13	1.11	–

* See Table 4.20 for details of animals used.

Table 4.21 Acoustic scattering cross-section for tissue (i)

Tissue	Temp °C	Freq MHz	Scattering cross-section: Backscatter cm^{-1} Sr^{-1} $x10^{-3}$	Total cm^{-1}	Reference
Blood,human	NR	4	0.034	$0.28x10^{-3}$	Nassiri & Hill 1986
–,–	NR	7	0.30	$1.8x10^{-3}$	–
Blood cells,40%,cow	23	7.5	0.004		Shung *et al* 1984
Kidney,cow	23	3	0.355±0.032		Fei & Shung 1985
–,–	23	5	1.03±0.01		–
–,–	23	7	3.78±0.36		–
Liver,human	20	4	0.1–1.5		Bamber *et al* 1981
–,–	NR	4	2.2	0.09	Nassiri & Hill 1986
–,–	NR	7	4.6	0.32	–
–,cow	23	3	0.601±0.051		Fei & Shung 1985
–,–	23	5	1.76±0.17		–
–,–	23	7	5.32±0.51		–
Muscle:cardiac,cow,	23	3	0.255±0.024		Fei & Shung 1985
perp. to fibres	23	5	0.932±0.082		–
–,–,–	23	7	3.32±0.38		–
–,–,–	*in-vivo*	2	1.0		Miller *et al* 1983
–,–,–	*in-vivo*	7.5	80		–
Muscle:skeletal,	NR	4	9.2	0.16	Nassiri & Hill 1986
cow	NR	7	11	0.32	–
Pancreas,cow	23	3	0.490±0.062		Fei & Shung 1985
–,–	23	5	1.60±0.26		–
–,–	23	7	5.81±0.75		–
Spleen,cow	23	3	0.165±0.016		–
–,–	23	5	0.632±0.057		–
–,–	23	7	2.50±0.24		–

Table 4.22 Acoustic scattering cross–section for tissue (ii); $\mu = a.f^{b}$

Tissue	Freq MHz	a	b	Reference
Back–scatter		$cm^{-1}\ Sr^{-1}$		
Blood	4–7	$1.27 x 10^{-7}$	3.98	Nassiri & Hill 1986
Brain	0.7–7	$2 x 10^{-5}$	1.24	Nicholas 1982
Liver	0.7–7	$2.7 x 10^{-4}$	1.20	Nicholas 1982
–	4–7	$3.15 x 10^{-4}$	1.38	Nassiri & Hill 1986
Muscle:cardiac	2–10	$3.16 x 10^{-5}$ *	3.1 ±0.6	O'Donnell *et al* 1981b
Muscle:skeletal	4–7	$4.43 x 10^{-3}$	0.53	Nassiri & Hill 1986
Spleen	0.7–7	$1.2 x 10^{-4}$	1.67	Nicholas 1982
Total scatter		cm^{-1}		
Blood	4–7	$(1.2 \pm .3) x 10^{-7}$	3.8±0.3	Nassiri & Hill 1986
Liver	4–7	$(3.7 \pm .6) x 10^{-3}$	2.3±0.2	–
Muscle	4–7	$(3.0 \pm .6) x 10^{-2}$	1.3±0.2	–

* SD $\sim 10^{-(\log a\ \pm 0.5)}$

were included. For liver (μ_l) and brain (μ_b) the following expressions were given:

$$\mu_l = (3.3 + 0.2f^3)\times10^{-4} \text{ cm}^{-1} \text{ Sr}^{-1}$$
$$\mu_b = (0.23 + 0.006f^4)\times10^{-4} \text{ cm}^{-1} \text{ Sr}^{-1} \quad (4.32)$$

4.3.10.3 Tissue composition

The importance of collagen content for back-scatter cross-section from myocardium has been studied by O'Donnell *et al* (1981b) who report a significant increase in back-scatter during the repair of damage due to infarction. A 450% increase in back-scatter by 10 weeks after the infarct was related to the increase in collagen content.

Bamber and Hill (1981) report decreased back-scatter from liver tumour compared with normal tissue. They demonstrated a positive correlation between mean back-scatter cross-section μ_{bs} and fat content, and a negative correlation of μ_{bs} with water content, evaluating data from liver tissue at 4 MHz.

4.3.10.4 Tissue fixation

Following formalin fixation, Bamber *et al* (1979) found a reduction in back-scatter cross-section which became greater with frequency, reaching about 50% at 4 MHz. Other fixatives gave different changes: ethyl alcohol showed an increase of about 450% at 1 MHz, increasing with frequency.

4.3.11 *Scatter from blood*

The back-scatter from blood has been studied by Shung *et al* (1976), demonstrating a dependence on hematocrit which gave a maximum for μ_{bs} in a range of haematocrit from 24% to 30%. The frequency range investigated was 5–15 MHz. A similar dependence on cell concentration was reported by Shung *et al* (1984) for erythrocyte suspensions, although the absolute value of μ_{bs} was about two orders of magnitude lower than that from blood. This was explained by noting that erythrocytes in whole blood form aggregates or rouleaux, which are absent from suspensions of washed erythrocytes. There is also a flow dependence of backscatter, particularly from turbulent flow (Yuan and Shung, 1988a), and a species dependence, apparently related to the tendency of the red corpuscles to aggregate. The scattering from whole bovine blood has been shown to follow very closely to a fourth power law dependence on frequency ($f^{4.05}$ at 44% hematocrit; Yuan and Shung, 1988b). Similarly, bovine erythrocytes in saline showed a frequency dependence of 3.95. Some deviation from the fourth power law was however noted at lower haematocrits and also in flowing pig blood.

4.3.12 *Scatter: proportion of total attenuation*

In the low megahertz range the evidence suggests that the scatter component in soft tissue attenuation accounts for about 10%–15% of the total attenuation. From an early study by Pauly and Schwan (1971), investigating losses in liver homogenate, it can be inferred that μ_s/μ is no more than 30%. More recent studies using both scattering (Nassiri and Hill, 1986) and absorption (Parker, 1983; Lyons and Parker, 1988) measurements suggest lower values. Since the frequency dependence for scattering is slightly higher than that for absorption (Tables 4.20, 4.22), the contribution from scatter loss will increase slightly with frequency.

4.3.13 *Attenuation in some materials other than tissue*

4.3.13.1 Water

The attenuation coefficient of water is given in Table 4.8.

4.3.13.2 Tissue substitute materials

Some properties of oils at room temperature which have been suggested as possible tissue substitutes are given in Table 4.23. Details of some ultrasound tissue substitute materials are given in Table 4.24. Generally these materials consist of a gelatin or agar-based gel, with n-propanol added to control the acoustic velocity, and graphite powder to increase attenuation and give some scatter. The addition of oils can give fat-like materials with lower acoustic velocities.

Table 4.23 Ultrasound properties of some oils at room temperature

Liquid	Temp °C	Freq MHz	Velocity m s^{-1}	Atten Coeff Np cm^{-1}	Reference
Castor oil	23.1	1.4		0.129	Wuensch *et al* 1956
–	22.8	4.25		0.921	–
–	23.0	5.9		1.46	–
–	20		1494		Dunn *et al* 1969
Dow Corning DC–710 Silicone oil	20	1	1378	0.07	–
Rape oil	23	4	1450	0.13	Flesch & Schoknecht 1979
Sandalwood oil	23	4	1450	1.34	–
Shell motor oil					
X–100/10W	23	4	1470	0.16	–
X–100/20W	23	4	1480	0.25	–
BP 10W–40W	23	4	1470	0.22	–
Shell transmission oil	23	4	1690	0.8	–
Shell 969	23	4	1480	0.5	–
Shell 7817	23	4	1460	0.75	–

Table 4.24 Ultrasound tissue substitute materials

Mixture	Temp °C	Velocity* m s^{-1}	Attenuation coeff* $\alpha=af^b$ a, dB $cm^{-1}MHz^{-b}$	b	Reference
20% gelatin, water, 5% n-propanol; graphite powder 0.049 to 0.187 g cm^{-3}	25	1579	0.368 to 1.453	1.24 to 0.99	Madsen *et al* 1978
–, 0% n-propanol	25	1550			
–, 21% n-propanol	25	1650			
77.72% water, 12.65% gelatin 3.27% n-propanol, 5.51% graphite powder	22	1560	0.493	0.961	Madsen *et al* 1986
3% agar, water, 8.6% n-propanol; graphite powder 0.0145 to 0.181 g cm^{-3}	22	1544	0.124 to 1.62	1.18 to 0.94	Burlew *et al* 1980
30% olive oil, 20% castor oil 2.2% n-propanol, gelatin mix	22	1513.2	0.97	1.02	Madsen *et al* 1982
13.6% castor oil, 20.4% olive oil, 8.8% n-propanol, gelatin mix	22	1539.4	0.51	1.08	–
34% olive oil, 3% n-propanol, 0.11 mg cm^{-3} graphite powder, gelatin mix, (breast model)	22	1518.8	0.80	1.01	–
25% olive oil, 25% petrol, 2.2% n-propanol, 0.65% graphite powder, gelatin mix, (fat model)	22	1458.5	0.42	1.02	–
gelatin, sodium alginate, plus 120μm polythene beads	20	1519±1.2	0.12(base gel)	~1	Bush & Hill 1983

* All references give additional information about temperature and mixture dependencies of velocity and attenuation.

References

Abendschein W. and Hyatt G.W., 1970, Ultrasonics and selected physical properties of bone, Clin Orthop, 69, 294–301.

Adler L. and Cook K.V., 1975, Ultrasonic parameters of freshly frozen dog tibia, J Acoust Soc Am, 58, 1107–1108.

Agemura D.H., O'Brien W.D., Olerud J.E., *et al*, 1990, Ultrasonic propagation properties of articular cartilage at 100 MHz, J Acoust Soc Am, 87, 1786–1797.

André M.P., Craven J.D., Greenfield M.A. and Stern R., 1980, Measurement of the velocity of sound in the human femur *in vivo*, Med Phys, 7, 324–330.

Anast G.T., Fields T. and Siegel I.M., 1958, Ultrasonic technique for the evaluation of bone fractures, Am J Phys Med, 37, 157–159.

Apfel R.E., 1983, The effective nonlinearity parameter for immiscible liquid mixtures, J Acoust Soc Am, 74, 1866–1868.

Apfel R.E., 1986, Prediction of tissue composition from ultrasonic measurements and mixture rules, J Acoust Soc Am, 79, 148–152.

Aubert A.E., Kesteloot H. and de Geest H., 1978, Measurement of high frequency sound velocity in blood, Biosigma 78, 1, 421–425.

Bakke T. and Gytre T., 1974, Ultrasonic measurement of sound velocity in the pregnant and the non-pregnant cervix uteri, Scand J Clin Lab Invest, 33, 341–346.

Bakke T., Gytre T., Haagensen A. and Giezendanner L., 1975, Ultrasonic measurement of sound velocity in whole blood, Scand J Clin Lab Invest, 35, 473–478.

Bamber J.C., 1981, Ultrasonic attenuation in fresh human tissues, Ultrasonics, 19, 187–188.

Bamber J.C., 1986, Speed of sound, *Physical Principles of Medical Ultrasonics*, C.R. Hill (ed.), Ellis Horwood, Chichester, pp.200–224.

Bamber J.C. and Hill C.R., 1979, Ultrasonic attenuation and propagation speed in mammalian tissues as a function of temperature, Ultrasound in Med & Biol, 5, 149–157.

Bamber J.C. and Hill C.R., 1981, Acoustic properties of normal and cancerous human liver–I. Dependence on pathological condition, Ultrasound in Med & Biol, 7, 121–133.

Bamber J.C., Fry M.J., Hill C.R. and Dunn F., Ultrasonic attenuation and backscattering by mammalian organs as a function of time after excision, Ultrasound in Med & Biol, 3, 15–20.

Bamber J.C., Hill C.R., King J.A. and Dunn F., 1979, Ultrasonic propagation through fixed and unfixed tissues, Ultrasound in Med & Biol, 5, 159–165.

Bamber J.C., Hill C.R. and King J.A., 1981, Acoustic properties of normal and cancerous human liver–II Dependence on tissue structure, Ultrasound in Med & Biol, 7, 135–144.

Bamber J.C., Cosgrove D.O., Page J. and Bossi C., 1987, *In-vivo* sound speed in normal liver by whole body transit-time measurements using a real-time scanner, Euroson '87, Proc 6th European Congr on Ultrasound in Med & Biol, p 306.

Barber F.E., Lees S. and Lobene R.R., 1969, Ultrasonic pulse-echo measurements in teeth, Archs Oral Biol, 14, 745–760.

Bauld T.J. and Schwan H.P., 1974, Attenuation and reflection of ultrasound in canine lung tissue, J Acoust Soc Am, 56, 1630–1637.

Begui Z.E., 1954, Acoustic properties of the refractive media of the eye, J Acoust Soc Am, 26, 365–368.

Berger G., Laugier P., Fink M. and Perrin J., 1987, Optimal precision in ultrasound attenuation estimation and application to the detection of Duchenne muscular dystrophy carriers, Ultrasonic Imaging, 9, 1–17.

Beyer R.T., 1960, Parameter of nonlinearity in fluids, J Acoust Soc Am, 32, 719–721.

Bhagat P.K., Kerrick W. and Ware R.W., 1980, Ultrasonic characterization of aging in skin tissue, Ultrasound in Med & Biol, 6, 369–375.

Bjørnø L., 1976, Nonlinear acoustics. In *Acoustics and Vibration Progress* 2, R.W.B. Stevens and H.G. Leventhall (eds), Chapman and Hall, London, pp.103–203.

Bowen T., Connor W.G., Nasoni R.L. Pifer A.E. and Sholes R.R., 1979, Measurement of the temperature dependence of the velocity of ultrasound in soft tissues. In Linzer (1979), pp.57–61.

Bradley E.L. and Sacerio J., 1972, The velocity of ultrasound in human blood under varying physiologic parameters, J Surg Res, 12, 290–297.

Bronez M.A., Shung K.K., Heidary H. and Hurwitz D., 1985, Measurement of ultrasound velocity in tissues utilising a microcomputer-based system, IEEE Trans Biomed Eng, BME-32, 723–726.

Burlew M.M., Madsen E.L., Zagzebski J.A., Banjevic R.A. and Sum S.W., 1980, A new ultrasound tissue-equivalent material, Radiology, 134, 517–520.

Bullen B.A., Quaade F., Olesen E. and Lund S.A., 1965, Ultrasonic reflections used for measuring subcutaneous fat in humans, Human Biology, 37, 375–384.

Buschmann W., Voss M. and Kemmerling S., 1970, Acoustic properties of normal human orbit tissues, Ophthal Res, 1, 354–364.

Bush N.L. and Hill C.R., 1983, Gelatine-alginate complex gel: a new acoustically tissue-equivalent material, Ultrasound in Med & Biol, 9, 479–484.

Busse L.J. and Miller J.G., 1981, Response characteristics of a finite aperture, phase insensitive ultrasonic receiver based upon the acoustoelectric effect, J Acoust Soc Am, 70, 1370–1376.

Cantrell J.H., Goans R.E. and Roswell R.L., 1978, Acoustic impedance variations at burn-nonburn interfaces in porcine skin, J Acoust Soc Am, 64, 731–735.

Carnes K.I. and Dunn F., 1988, Absorption of ultrasound by mammalian ovaries, J Acoust Soc Am, 84, 434–437.

Carstensen E.L., 1979, Absorption of sound in tissues. In Linzer (1979), pp.29–36.

Carstensen E.L. and Schwan H.P., 1959a, Absorption of sound arising from the presence of intact cells in blood, J Acoust Soc Am, 31, 185–189.

Carstensen E.L. and Schwan H.P., 1959b, Acoustic properties of hemoglobin solutions, J Acoust Soc Am, 31, 305–310.

Carstensen E.L., Li K. and Schwan H.P., 1953, Determination of the acoustic properties of blood and its components, J Acoust Soc Am, 25, 286–289.

Carstensen E.L., McKay N.D., Delecki D. and Muir T.G., 1982, Absorption of finite amplitude ultrasound in tissues, Acustica, 51, 116–123.

Chen C.F., Robinson D.F., Wilson L.S., *et al*, 1987, Clinical sound speed measurement in liver and spleen *in vivo*, Ultrasonic Imaging, 9, 221–235.

Chevenko A.A and Yukhananov I.Kh., 1971, Two-frequency echo recording method for differential tumour diagnosis, Med. Tekhn., 3, 40–41.

Chivers R.C. and Parry R.J., 1978, Ultrasonic velocity and attenuation in mammalian tissues, J Acoust Soc Am, 63, 940–953.

Choi P.-K., Bae J.-R. and Takagi K., 1990, Ultrasonic spectroscopy in bovine serum albumin solutions, J Acoust Soc Am., 87, 874–881.

Cobb W.N., 1982, Measurement of the acoustic nonlinearity parameter for biological media, PhD dissertation, Yale University.

Collings A.F. and Bajenov N., 1987, Temperature dependence of the velocity of sound in human blood and blood components, Australasian Phys Eng Sci Med, 10, 123–127.

Coleman D.J., Lizzi F.L., Franzen L.A. and Abramson D.H., 1975, A determination of the velocity of ultrasound in cateractous lenses, *Ultrasonography in Ophthalmology*, Bibl. ophthal., No. 83, Karger, Basel, 246–251.

Colombati S. and Petralia S., 1950, Assorbimento di ultrasuoni in tessuti animali, La Ricerca Scientifica, 20, 71–78.

Coppens A.B., 1981, Simple equations for the speed of sound in Neptunian waters, J Acoust Soc Am, 69, 862–863.

Coppens A.B., Beyer R.T., Seiden M.B., *et al*, 1965, Parameter of nonlinearity in fluids. II, J Acoust Soc Am, 38, 797–804.

Craven J.D., Costantini M.A., Greenfield M.A. and Stern R., 1973, Measurement of the velocity of ultrasound in human cortical bone and its potential clinical importance. An *in vivo* preliminary study, Invest Radiol, 8, 72–77.

Cusack S. and Miller A., 1979, Determination of the elastic constants of collagen by Brillouin light scattering, J Mol Biol, 135, 39–51.

Daft C.M.W and Briggs G.A.D, 1989, The elastic microstructure of various tissues, J Acoust Soc Am., 85, 416–422.

Del Grosso V.A. and Mader C.W., 1972, Speed of sound in pure water, J Acoust Soc Am, 52, 1442–1446.

Duck F.A. and Starritt H.C., 1987, Non-linear losses in the measurement of attenuation, *Ultrasonic Tissue Characterization and Echographic Imaging* 6, Commission of the European Communities, Luxembourg, 137–144.

Dunn F., 1962, Temperature and amplitude dependence of acoustic absorption in tissue, J Acoust Soc Am, 34, 1545–1547.

Dunn F., 1974, Attenuation and speed of ultrasound in lung, J Acoust Soc Am, 56, 1638–1639.

Dunn F., 1986, Attenuation and speed of ultrasound in lung: dependence upon frequency and inflation, J Acoust Soc Am, 80, 1248–1250.

Dunn F. and Brady J.K., 1973, Absorption of ultrasound in biological media, Biophysics, 18, 1128–1132.

Dunn F. and O'Brien W.D. Jr., 1976, *Ultrasonic Biophysics*, Dowden, Hutchinson & Ross Inc., Stroudsburg, Pa.

Dunn F., Edmonds P.D. and Fry W.J., 1969, Absorption and dispersion of ultrasound in biological media, *Biological Engineering*, H.P. Schwan (ed.), McGraw-Hill, NY, p.205.

Dunn F., Law W.K. and Frizzell L.A., 1981, Nonlinear ultrasonic wave propagation in biological materials, IEEE 1981 Ultrasonics Symp Proc, 527-532.

Dussik K.T. and Fritch D.J., 1956, Determination of sound attenuation and sound velocity in the structure constituting the joints, and of the ultrasonic field distribution within the joints on living tissues and anatomical preparations, both in normal and pathalogical conditions, Public Health Service, Natl Inst Health project A454, Progr Rep 15 Sept 1956 (cited in Goss *et al*, 1978b).

Dussik K.T., Fritch D.J., Kyriazidou M. and Sear R.S., 1958, Measurements of articular tissue with ultrasound, Am J Phys Med, 37, 160-165.

Dyro J.F. and Edmonds P.D., 1974, Ultrasonic absorption and dispersion in cholesteryl esters, Mol Cryst Liq Cryst, 25, 175-193.

Dzenis V.V. and Purin'sh Yu.I., 1981, Study of human skull bones by flexural ultrasonic waves, from Mekhanika Kompozitnykh Materialov, 3, 1979, translated in Russian Ultrasonics, 11, 30-36.

Edwards C.A. and O'Brien W.D., 1985, Speed of sound in mammalian tendon threads using various reference media, IEEE Trans Son Ultrason, SU-32, 351-354.

Errabolu R.L., Sehgal C.M., Bahn R.C. and Greenleaf J.F., 1988, Measurement of ultrasonic nonlinear parameter in excised fat tissues, Ultrasound in Med & Biol, 14, 137-146.

Everbach E.C., 1989, Tissue composition via measurement of the acoustic nonlinearity parameter, Tech Memo No.6, Dept of Mech Eng, Yale University.

Fei D.Y. and Shung K.K, 1985, Ultrasonic backscatter from mammalian tissues, J Acoust Soc Am, 78, 871-876.

Filipczynski L., Etienne J., Lypacewicz G. and Salkowski J., 1967, Visualising internal structures of the eye by means of ultrasonics, Proc Vib Probl Warsaw, 8, 357-368.

Fitzgerald J.W., Ringo G.R. and Winder W.C., 1961, An ultrasonic method for measurement of solids-not-fat and butter fat in fluid milk 1. Acoustic properties, Proc 56th Ann Meeting Am Dairy Sc Assn, (cited in Goss *et al*, 1980).

Flesch U. and Schoknecht G., 1979, Phantomsubtsanzen zur Simulation von Weichteilgewebe in der Ultraschalldiagnostik, Röntgen, 32, 103-108.

Foster F.S. and Hunt J.W., 1979, Transmission of ultrasound beams through human tissue- focussing and attenuation studies, Ultrasound in Med & Biol, 5, 257-268.

Fredfeldt K.E., 1986, Sound velocity in the middle phalanges of the human hand, Acta Radiologica Diagnosis, 27, 95-96.

Frizzell L.A., 1975, Ultrasonic heating of tissues, PhD thesis, Univ. of Rochester, NY (cited in Goss *et al*, 1978b).

Frizzell L.A. and Gindorf J.D., 1981, Measurement of ultrasonic velocity in several biological tissues, Ultrasound in Med & Biol, 7, 385-387.

Frizzell L.A., Carstensen E.L. and Dyro J.F., 1976, Shear properties of mammalian tissues at low megahertz frequencies, J Acoust Soc Am, 60, 1409–1411.

Frizzell L.A., Carstensen E.L. and Davis J.D., 1979, Ultrasonic absorption in liver tissue, J Acoust Soc Am, 65, 1309–1312.

Frucht A.H., 1953, Die Schallgeschwindigkeit in menschlichen und tierischen Geweben, Zeit ges exp Med, 120, 526–557.

Fry F.J. and Barger J.E., 1978, Acoustical properties of the human skull, J Acoust Soc Am, 63, 1576–1590.

Fry W.J. and Fry R.B., 1953, Temperature changes produced in tissue during ultrasonic irradiation, J Acoust Soc Am, 25, 6–11.

Gammell R.M., Le Croisette D.H. and Heyser R.C., 1979, Temperature and frequency dependence of ultrasonic attenuation in selected tissues, Ultrasound in Med & Biol, 5, 269–277.

Geleskie J.V. and Shung K.K, 1982, Further studies on acoustic impedance of major bovine blood vessel walls, J Acoust Soc Am, 71, 467–470.

Giacomini A., 1947, Ultrasonic velocity in ethanol–water mixtures, J Acoust Soc Am, 19, 701–702.

Gilmore R.S., Pollack R.P. and Katz J.L., 1969, Elastic properties of bovine dentine and enamel, Archs Oral Biol, 15, 787–796.

Goldman D.E. and Richards J.R., 1954, Measurement of high–frequency sound velocity in mammalian soft tissues, J Acoust Soc Am, 26, 981–983.

Gong X., Zhu Z., Shi T. and Huang J., 1989, Determination of the acoustic nonlinearity parameter in biological media using FAIS and ITD methods, J Acoust Soc Am, 86, 1–5.

Goss S.A. and Fry F.J., 1981, Nonlinear acoustic behavior in focused ultrasonic fields: observations of intensity dependent absorption in biological tissue, IEEE Trans Son Ultrason, SU-28, 21–26.

Goss S.A. and O'Brien W.D., 1979, Direct ultrasonic velocity measurements of mammalian collagen threads, J Acoust Soc Am, 65, 507–511.

Goss S.A., Frizzell L.A. and Dunn F., 1978a, Frequency dependence of ultrasonic absorption in mammalian testis, J Acoust Soc Am, 63, 1226–1229.

Goss S.A., Johnston R.L. and Dunn F., 1978b, Comprehensive compilation of empirical ultrasonic properties of mammalian tissues, J Acoust Soc Am, 64, 423–457.

Goss S.A., Frizzell L.A. and Dunn F., 1979, Ultrasonic absorption and attenuation in mammalian tissues, Ultrasound in Med & Biol, 5, 181–186.

Goss S.A., Johnston R.L. and Dunn F., 1980, Compilation of empirical ultrasonic properties of mammalian tissues. II, J Acoust Soc Am, 68, 93–108.

Greenfield M.A., Craven J.D., Huddleston A., *et al*, 1981, Measurement of the velocity of ultrasound in human cortical bone *in vivo*, Radiology, 138, 701–710.

Greenleaf J.F., Duck F.A., Samayoa W.F. and Johnson S.A., 1974, Ultrasonic data acquisition and processing system for atherosclerotic tissue characterization, 1974 IEEE Ultrasonic Symp Proc, 738–743.

Grybauskas P., Kundrotas K., Sukackas V. and Yaronis E., 1978, Ultrasonic digital interferometer for investigation of blood clotting, Ultrasonics, 16, 33–36.

Hagelberg M.P., Holton G. and Kao S., 1966, Calculation of B/A for water from measurements of ultrasonic velocity versus temperature and pressure to 10 000 kg/cm^2, J Acoust Soc Am, 41, 564–567.

Haney M.J. and O'Brien W.D., 1986, Temperature dependency of ultrasonic propagation properties in biological materials, *Tissue Characterization by Ultrasound*, J.F. Greenleaf (ed.), CRC Press, Florida, 15–55.

Harley R., James D., Miller A. and White J.W., 1977, Phonons and the elastic moduli of collagen and muscle, Nature, 267, 285–287.

Heyser R.C. and Le Croisette D.H., 1974, A new ultrasonic system using time delay spectrometry, Ultrasound in Med & Biol, 1, 119–131.

Horak G., 1977, Real-time ultrasonic-spectroscopy in suspensions, Acustica, 37, 11–20.

Hoyer W.A. and Nolle A.W., 1956, Behaviour of liquid crystal compounds near the isotropic–anisotropic transition, J Chem Phys, 24, 803–811.

Hueter T.F., Morgan H. and Cohen M.S., 1953, Ultrasonic attenuation in biological suspensions, J Acoust Soc Am, 25, 1200–1201.

Hughes D.J. and Snyder B., 1980, Speed of 10 MHz sound in canine aortic wall: effects of temperature, storage and formalin soaking, Med & Biol Eng & Comput, 18, 220–222.

Hughes D.J., Geddes L.A., Babbs C.F., Bourland J.D. and Newhouse V.L., 1979, Attenuation and speed of 10 MHz ultrasound in canine blood of various packed cell volumes at 37°C, Med & Biol Eng & Comput, 17, 619–622.

Hustad G.O., Richardson T., Winder W.C. and Dean M.P., 1971, Acoustic properties of some lipids, Chem Phys Lipids, 7, 61–74.

Jansson F. and Kock E., 1962, Determination of the velocity of ultrasound in the human lens and vitreous, Acta Ophthalmol, 40, 420–433.

Jansson F and Sundmark E., 1961, Determination of the velocity of ultrasound in ocular tissues at different temperatures, Acta Ophthalmol, 39, 899–910.

Jellins J. and Barraclough B.H., 1978, Ultrasonic imaging of the scrotum, *Ultrasound in Medicine* 4, D.N. White and E.A. Lyons (eds), Plenum, New York, pp.151–154.

Kaye G.W.C. and Laby T.H., 1973, *Tables of Physical and Chemical Constants*, Longman, London.

Kishimoto T., 1958, Ultrasonic absorption in bones, Acustica, 8, 179–180.

Klemin V.A., Maiorov E.A., Ruchkin V.V. and Sarvazyan A.P., 1981, Investigation of the frequency dependence of the acoustical characteristics of biological tissue by the resonator method, Sov Phys Acoust, 27, 495–498.

Kossoff G., 1976, Reflection techniques for measurement of attenuation and velocity, Proc Seminar on Ultrasonic Tissue Characterization, NBS Spec Publ 453, 135–139.

Kossoff G. and Sharpe C.J., 1966, Examination of the contents of the pulp cavity in teeth, Ultrasonics, 4, 77–83.

Kossoff G., Fry E.K. and Jellins J., 1973, Average velocity of ultrasound in the human female breast, J Acoust Soc Am, 53, 1730–1736.

Kremkau F.W., Barnes R.W. and McGraw C.P., 1981, Ultrasonic attenuation and propagation speed in normal human brain, J Acoust Soc Am, 70, 29–38.

Lakes R., Yoon H.S. and Katz J.L., 1983, Slow compressional wave propagation in wet human and bovine cortical bone, Science, 220, 513–515.

Lang J., Zana R., Gairard B., Dale G. and Gros Ch.M., 1978, Ultrasonic absorption in the human breast cyst liquids, Ultrasound in Med & Biol, 4, 125–130.

Lang S.B., 1970, Ultrasonic method for measuring elastic coefficients of bone and results on fresh and dried bovine bones, IEEE Trans Biomed Eng, BME-17, 101–105.

Law W.K., Frizzell L.A. and Dunn F., 1985, Determination of the nonlinearity parameter B/A of biological media, Ultrasound in Med & Biol, 11, 307–318.

Le Croisette D.H., Heyser R.C., Gammell P.M., Roseboro J.A. and Wilson R.L., 1979, The attenuation of selected soft tissue as a function of frequency; in Linzer, 1979, 101–108.

Lees S. and Rollins F.R., 1972, Anisotropy in hard dental tissues, J Biomechanics, 5, 557–566.

Lees S., Ahern J.M. and Leonard M., 1983a, Parameters influencing the sonic velocity in compact calcified tissues of various species, J Acoust Soc Am, 74, 28–33.

Lees S., Barnard S.M. and Churchill D., 1987, The variation of sonic plesio-velocity in dose dependent lathyritic rabbit femurs, Ultrasound in Med & Biol, 13, 19–24.

Lees S., Cleary P.F., Heeley J.D. and Gariepy E.L., 1979, Distribution of sonic plesio-velocity in a compact bone sample, J Acoust Soc Am, 66, 641–646.

Lees S., Heeley J.D., Ahern J.M. and Oravecz M.G., 1983b, Axial phase velocity in rat tail tendon fibres at 100 MHz by ultrasonic microscopy, IEEE Trans Sonics Ultrason, 30, 85–90.

Lehmann J.F. and Johnson E.W., 1958, Some factors influencing the temperature distribution in thighs exposed to ultrasound, Arch Phys Med Rehab, 39, 347–356.

Lewin P.A. and Busk H., 1982, *In vivo* ultrasonic measurements of tissue properties, 1982 IEEE Ultrasonics Symp Proc, 709–712.

Lin T., Ophir J. and Potter G., 1987, Frequency-dependent ultrasonic differentiation of normal and diffusely diseased liver, J Acoust Soc Am, 82, 1131–1138.

Linzer M., 1979, *Ultrasonic Tissue Characterization II*, M. Linzer (ed.), NBS Spec. Publ. 525, US Govt. Print. Office, Washington, DC.

Lizzi F., Katz L., St Louis L. and Coleman D.J., 1976, Applications of spectral analysis in medical ultrasonography, Ultrasonics, 14, 77–80.

Ludwig G.D., 1950, The velocity of sound through tissues and the acoustic impedance of tissues, J Acoust Soc Am, 22, 862–866.

Ludwig G.D. and Struthers F.W., 1950, Detecting gallstones with ultrasonic echoes, Electronics, 23, 172–178 (from Naval Research Laboratory report NM 004001, No.4, June 16, 1949).

Lyons M.E. and Parker K.J., 1988, Absorption and attenuation in soft tissues II– experimental results, IEEE Trans Ultrasonics, Ferroelectrics & Freq Contr, 35, 511–521.

Madaras E.I., Perez J., Sobel B.E., Mottley J.G. and Miller J.G., 1988, Anisotropy of the ultrasonic backscatter of myocardial tissue: II. Measurements *in vivo*, J Acoust Soc Am, 83, 726–769.

Madsen E.L., Zagzebski J.A, Banjavie R.A. and Jutila R.E., 1978, Tissue mimicking materials for ultrasound phantoms, Med Phys, 5, 391–394.

Madsen E.L., Zagzebski J.A. and Frank G.R., 1982, Oil–in–gelatin dispersions for use as ultrasonically tissue–mimicking materials, Ultrasound in Med & Biol, 8, 277–287.

Madsen E.L., Frank G.R., Carson P.L., *et al*, 1986, Interlaboratory comparison of ultrasonic attenuation and speed measurements, J Ultrasound Med, 5, 569–576.

Marcus P.W. and Carstensen E.L., 1975, Problems with absorption measurements of inhomogenous solids, J Acoust Soc Am, 58, 1334–1335.

Martin B. and McElhaney J.H., 1971, The acoustic properties of human skull bone, J Biomed Mater Res, 5, 325–333.

Masaoka H., Akamatsu N., Numato A., *et al*, 1987, Ultrasonic attenuation coefficient of uterine myomas, Euroson '87, Finnish Soc for Ultrasound in Med & Biol, p.309.

McDaniel G.A., 1977, Ultrasonic attenuation measurements on excised breast carcinoma at frequencies from 6 to 10 MHz, 1977 IEEE Ultrasonics Symp Proc, 234–236.

McKelvie M.L. and Palmer S.B., 1987, The interaction of ultrasound with cancellous bone, *Ultrasonic Studies of Bone*, S.B. Palmer and C.M. Langton (eds), IoP Short Meetings Ser. No.6, IoP Publishing, Bristol, pp.1–13.

Meunier A., Yoon H.S. and Katz J.L., 1982, Ultrasonic characterization of some pathological human femora, 1982 IEEE Ultrasonics Symp Proc, 713–717.

Meyer C.R., Herron D.S., Carson P.L., *et al*, 1984, Estimation of ultrasonic attenuation and mean backscatter size via digital signal processing, Ultrasonic Imaging, 6, 13–23.

Miles C.A., Fursey G.A.J. and York R.W.R., 1984, New equipment for measuring the speed of ultrasound and its application in the estimation of body composition of farm livestock. In *In–vivo Measurement of Body Composition in Meat Animals,* D. Lister (ed.), Elsevier Applied Science, London, pp.93–105.

Miller J.G., Perez J.E., Mottley T.G., *et al*, 1983, Myocardial tissue characterisation. An approach based on quantitative backscatter and attenuation, 1983 IEEE Ultrasonics Symp Proc, 782–793.

Millero F.J. and Kubinski T., 1975, Speed of sound in sea water as a function of temperature and salinity at 1 atm, J Acoust Soc Am, 57, 312–319.

Mol C.R. and Breddels P.A., 1982, Ultrasound velocity in muscle, J Acoust Soc Am, 71, 455–461.

Nakaima N., Aoyama H. and Oka M.J., 1976, Supplementary study on the ultrasonic absorption of human soft tissues, J. Wakayama Med Soc, 27, 107–115; (in Japanese, cited in Goss *et al*, 1978b)

Narayana P.A., Ophir J. and Maklad N.F., 1984, The attenuation of ultrasound in biological fluids, J Acoust Soc Am, 76, 1–4.

Nasoni R.L., 1981, Temperature corrected speed of sound for use in soft tissue imaging, Med Phys, 8, 513–515.

Nasoni R.L., Bowen T., Connor W.G. and Sholes R.R., 1979, *In vivo* temperature dependence of ultrasound speed in tissue and its application to noninvasive temperature monitoring, Ultrasonic Imaging, 1, 34–43.

Nassiri D.K. and Hill C.R., 1986, The differential and total bulk acoustic scattering cross sections of some human and animal tissues, J Acoust Soc Am, 79, 2034–2047.

Nassiri D.K., Nicholas D. and Hill C.R., 1979, Attenuation of ultrasound in skeletal muscle, Ultrasonics, 17, 230–232.

Nicholas D., 1982, Evaluation of backscattering coefficients for excised human tissues: results, interpretation and associated measurements, Ultrasound in Med & Biol, 8, 17–28.

O'Brien W.D., 1977, The relationship between collagen and ultrasonic attenuation and velocity in tissue, Ultrasonics International 1977, pp 194–205.

O'Brien W.D., Olerud, J., Shung K.K. and Reid J.M., 1981, Quantitative acoustical assessment of wound maturation with acoustic microscopy, J Acoust Soc Am, 69, 575–579.

O'Brien W.D., Erdman J.W. and Hebner T.B., 1988, Ultrasonic propagation properties (@100 MHz) in excessively fatty rat liver, J Acoust Soc Am, 83, 1159–1166.

O'Donnell M., Mimbs J.W., Sobel B.E. and Miller J.G, 1977, Ultrasonic attenuation of myocardial tissue: dependence on time after excision and on temperature, J Acoust Soc Am, 62, 1054–1057.

O'Donnell M., Mimbs J.W. and Miller J.G., 1979, The relationship between collagen and ultrasonic attenuation in myocardial tissue, J Acoust Soc Am, 65, 512–517.

O'Donnell M., Jaynes E.T. and Miller J.G., 1981a, Kramers-Kronig relationship between ultrasonic attenuation and phase velocity, J Acoust Soc Am, 69, 696–701.

O'Donnell M., Mimbs J.W. and Miller J.G., 1981b, Relationship between collagen and ultrasonic backscatter in myocardial tissue, J Acoust Soc Am, 69, 580–588.

Oka M., 1977, Progress in studies of the potential use of medical ultrasonics, Wakayama Med Rep, 20, 1–50 (cited in Goss *et al*, 1978b)

Olerud J.E., O'Brien W.D., Reiderer-Henderson M.A., *et al*, 1990, Correlation of tissue constituents with the acoustic properties of skin and wound, Ultrasound in Med & Biol, 16, 55–64.

Ophir J., Maklad N.F. and Bigelow R.H., 1982, Ultrasonic attenuation measurements of *in vivo* human muscle, Ultrasonic Imaging, 4, 290–295.

Parker K.J., 1983, Ultrasonic attenuation and absorption in liver tissue, Ultrasound in Med & Biol, 9, 363–369.

Parker K.J., Asztely M.S., Lerner R.M., Schenk E.A. and Waag R.C., 1988a, *In-vivo* measurements of ultrasound attenuation in normal or diseased liver, Ultrasound in Med & Biol, 14, 127–136.

Parker K.J., Tuthill T.A. and Baggs R.B., 1988b, The role of glycogen and phosphate in ultrasonic attenuation of liver, J Acoust Soc Am, 83, 374–378.

Pauly H. and Schwan H.P., 1971, Mechanism of absorption of ultrasound in liver tissue, J Acoust Soc Am, 50, 692–699.

Pedersen P.C. and Ozcan H.S., 1986, Ultrasound properties of lung tissue and their measurements, Ultrasound in Med & Biol, 12, 483–499.

Pohlhammer J.D., Edwards C.A. and O'Brien W.D., 1981, Phase insensitive ultrasonic attenuation coefficient determination of fresh bovine liver over an extended frequency range, Med Phys, 8, 692–694.

Pohlman R., 1939, On the absorption of ultrasound in human tissues and their dependence upon frequency, Physik Z, 40, 159–161: English translation in Dunn & O'Brien, 1976, pp 14–18.

Povall J.M., Aindow J.D., Chivers R.C. and Driscoll A.M., 1984, Speed of ultrasound in amniotic fluid, Acoust Lett, 7, 181–186.

Rajagopalan B., Greenleaf J.F., Thomas P.J., Johnson S.A. and Bahn R.C., 1979, Variation of acoustic speed with temperature in various exised human tissues studied by ultrasound computerized tomography. In Linzer (1979), pp.227–233.

Rice D.A., 1980, Sound speed in the upper airways, J Appl Physiol, 49, 326–336.

Rice D.A., 1983, Sound speed in the pulmonary parenchyma, J Appl Physiol, 54, 304–308.

Rice D.A. and Rice J.C., 1987, Central to peripheral sound propagation in excised lung, J Acoust Soc Am, 82, 1139–1144.

Riederer-Henderson M.A., Olerud J.E., O'Brien W.D., *et al*, 1988, Biochemical and acoustical parameters of normal canine skin, IEEE Trans Biomed Eng, 35, 967–972.

Rivara A. and Sanna G., 1962, Determinazione della velocità degli ultrasuoni nei tessuti oculari di uomo e di maiale, Annali Ottalmologia e Clinica Oculistica, 88, 678–682.

Robinson T.C. and Lele P.P., 1969, An analysis of lesion development in the brain and in plastics by high-intensity focused ultrasound at low-megahertz frequencies, J Acoust Soc Am, 51, 1333–1351.

Robinson D.E., Chen C.F., Wilson L.S., *et al*, 1987, Experience in sound speed measurement in liver and spleen, Proc Euroson '87, Finnish Society for Ultrasound in Medicine and Biology, p.305.

Rooney J.A., Gammell P.M., Hestenes J.D., Chin H.P. and Blankenhorn D.H., 1982, Velocity and attenuation of sound in arterial tissues, J Acoust Soc Am, 71, 462–466.

Sarvazyan A.P. and Klemin V.A., 1983, Study of ultrasonic topography of the kidney. In *Ultrasonic Interactions in Biology and Medicine*, R.Millner, E.Rosenfeld and U.Cobet (eds), Plenum, London, pp. 99–104.

Sarvazyan A.P., Lyrchikov A.G. and Gorelov S.E., 1987, Dependence of ultrasonic velocity in rabbit liver on water content and structure of the tissue, Ultrasonics, 25, 244–247.

Scherzinger A.L., Belgam R.A., Carson P.L., *et al*, 1989, Assessment of ultrasonic computed tomography in symptomatic breast patients by discriminant analysis, Ultrasound in Med & Biol, 15, 21–28.

Schiefer W., Kazner E. and Kunze S. (eds), 1968, *Clinical Echo-Encephalography,* Springer-Verlag, New York, pp.67–68.

Schwan H.P., Carstensen E.L. and Li K., 1953, Heating of fat-muscle layers by electromagnetic and ultrasonic diathermy, Trans Am Inst Elec Eng., I 72, 483–488.

Segal L.A. and O'Brien W.D., 1983, Frequency dependent ultrasonic attenuation coefficient assessment in fresh tissue, IEEE 1983 Ultrasonics Symp Proc, 797–799.

Sehgal C.M., Bahn R.C. and Greenleaf J.F., 1984, Measurement of the acoustic nonlinearity parameter B/A in human tissues by a thermodynamic method, J Acoust Soc Am, 76, 1023–1029.

Sehgal C.M., Brown G.M., Bahn R.C. and Greenleaf J.F., 1986, Measurement and use of acoustic nonlinearity and sound speed to estimate composition of excised livers, Ultrasound in Med & Biol, 12, 865–874.

Shore D., Woods M.O. and Miles C.A., 1986, Attenuation of ultrasound in *post rigor* bovine skeletal muscle, Ultrasonics, 24, 81–87.

Shung K.K. and Reid J.M., 1977, The acoustical properties of deoxygenated sickle cell blood and hemoglobin S solution, Ann Biomed Eng, 5, 150–156.

Shung K.K., Sigelmann R.A. and Reid J.M., 1976, Scattering of ultrasound by blood, IEEE Trans Biomed Eng, BME-23, 460–467.

Shung K.K., Yuan Y.W., Fei D.Y. and Tarbell J.M., 1984, Effect of flow disturbance on ultrasonic backscatter from blood, J Acoust Soc Am, 75, 1265–1272.

Sollish B.D., 1979, A device for measuring ultrasonic propagation velocity in tissue. In Linzer (1979), pp.53–56.

Sun Yongchen, Dong Yanwu and Zhao Hengyuan, 1985, Study of the acoustic nonlinearity parameter in highly attenuating biological media, IEEE 1985 Ultrasonics Symp Proc, 891–894.

Sun Yongchen, Dong Yanwu, Tong Jie, and Tang Zhensheng, 1986, Ultrasonic propagation parameters in human tissues, IEEE 1986 Ultrasonics Symp Proc, 905–908.

Taylor K.J.W., Riely C.A., Hammers L., *et al*, 1986, Quantitative US attenuation in normal liver and in patients with diffuse liver disease: importance of fat, Radiology, 160, 65–71.

Theismann H. and Pfander F., 1949, Über die Durchlässigkeit des Knochens für Ultraschall, Strahlentherapie, 80, 607–610.

Tervola K.M.U., Gummer M.A., Erdman J.W. and O'Brien W.D., 1985, Ultrasonic attenuation and velocity properties in rat liver as a function of fat concentration: a study at 100 MHz using a scanning laser acoustic microscope, J Acoust Soc Am, 77, 307–313.

Thijssen J.M., Mol H.J.M. and Timmer M.R., 1985, Acoustic parameters of ocular tissues, Ultrasound in Med & Biol, 11, 157–161.

Truong X.T., Jarrett S.R. and Rippel D.V., 1978, Longitudinal pulse propagation characteristics in striated muscle, J Acoust Soc Am, 64, 1298–1302.

Urick R.J., 1947, A sound velocity method for determining the compressibility of finely divided substances, J Appl Phys, 18, 983–987.

Van Venrooij G.E.P.M., 1971, Measurement of ultrasound velocity in human tissue, Ultrasonics, 9, 240–242.

Vilks Yu.K., Pfafrod G.O., Yansan Kh.A. and Saulgozis Yu.Zh., 1978, Experimental study of the effect of fracture and surgical intervention on the acoustic properties of human tibia, from Mekhanika Polimerov, 1; translated in Russian Ultrasonics, 8, 167–178.

Wells P.N.T., 1977, *Biomedical Ultrasonics*, Academic Press, London, pp.111–117.

Willocks J., Donald I., Duggan T.C. and Day N., 1964, Foetal cephalometry by ultrasound, J Obstet Gyn Brit Comm, 71, 11–20.

Wladimiroff J.W., Craft I.L. and Talbert D.G., 1975, *In vitro* measurements of sound velocity in human fetal brain tissue, Ultrasound in Med & Biol, 1, 377–382.

Wuensch B.J., Hueter T.F. and Cohen M.S., 1956, Ultrasonic absorption in castor oil: deviations from classical behaviour, J Acoust Soc Am, 28, 311–312.

Yoon H.S. and Katz J.L., 1976, Ultrasonic wave propagation in human cortical bone –II. Measurements of elastic properties and microhardness, J Biomech, 9, 459–464.

Yoon H.S. and Katz J.L., 1979a, Temperature dependence of the ultrasonic velocities in bone, 1979 IEEE Ultrasonics Symp Proc, 395–398.

Yoon H.S. and Katz J.L., 1979b, Ultrasonic properties and microtexture of human cortical bone. In Linzer (1979), pp.189–196.

Yoshizumi K., Sato T. and Ichida N., 1987, A physicochemical evaluation of the nonlinear parameter B/A for media predominantly composed of water, J Acoust Soc Am, 82, 302–305.

Yosioka K., Oka M., Omura A. and Hasegawa T., 1969, Absorption coefficient of ultrasound in soft tissues and their biological conditions, Mem Inst Scient Ind Res Osaka U, 26, 55–59 (cited in Goss *et al*, 1978b).

Yuan Y.W. and Shung K.K., 1988a, Ultrasonic backscatter from flowing whole blood. I: dependence on shear rate and hematocrit, J Acoust Soc Am, 84, 52–58.

Yuan Y.W. and Shung K.K., 1988b, Ultrasonic backscatter from flowing whole blood. II: dependence on frequency and fibrinogen concentration, J Acoust Soc Am, 84, 1195–1200.

Zana R. and Lang J., 1974, Interaction of ultrasound and amniotic liquid, Ultrasound in Med & Biol, 1, 253–258.

Chapter 5

Mechanical Properties of Tissue

The mechanical properties of tissue which are reviewed in this chapter include their densities, moduli of elasticity and ultimate strengths. In addition to the mass density, values of electron density and proton density are also included. Whilst these latter two properties relate primarily to the interaction of ionising radiation with tissue and the magnetic resonance of tissue, they are nevertheless grouped here for convenience. Lastly, values of the viscosity of body fluids are given.

5.1 Density and mass

5.1.1 *Mass density*

The **mass density**, ρ, is defined as the mass of unit volume of a material. The SI unit is kilogram metre^{-3} (kg m^{-3}), with a common alternative suitable for most tissues being g cm^{-3}.

The density of tissue may be measured by comparing the mass of a sample in air with the apparent mass measured in water. The density is calculated from the ratio of the apparent masses and a knowledge of the density of water at the measurement temperature. For biological fluids early methods using a falling drop method (eg Phillips *et al*, 1950) have been replaced by the use of a mechanical oscillator (Hinghofer–Szalkay, 1985).

Values of mass density for a range of biological fluids, soft tissues and hard tissues are given in Table 5.1. Where sufficient measurements have been reported, both an average value and an absolute range of values have been included. This gives some indication of the variability in density for a particular organ or tissue. The density of most soft tissue is slightly higher than water, lying typically in the range 1.00 to 1.07 g cm^{-3}. The presence of fat lowers the average density of tissue. Most pure animal fat has a density of about 0.9 g cm^{-3}, although the density of brain fat from a dog has been reported as being greater that 1.0 g cm^{-3} (Méndez *et al*, 1960).

Table 5.1 Mass and densities of adult human organs and tissues

Tissue	Average mass ICRP,1975 g	Mass density kg m^{-3}	Average electron density $m^{-3}x10^{26}$	Average ^{1}H density $kg^{-1}x10^{25}$
Whole body	70000	1070,♂; 1040,♀		
Fluids				
Bile,liver		995–1015 (1010)		
–,gallbladder	62	1010–1032 (1026)	3430	3.60
Blood,whole	5500	1052–1064 (1060)	3510	3.38
erythrocytes	2400	1089–1097 (1093)	3590	3.13
plasma	3100	1025–1030 (1027)	3420	3.60
Breast milk		1026–1037 (1031)		
Cerebrospinal fluid	120	1006–1008	3370	3.71
Gastric juice		1004–1010		
Lymph		1030		
Saliva		1002–1012		
Semen		1020–1040		
Sweat		1001–1008		
Synovial fluid	1	1008–1015		
Urine	(100)	1001–1050	3400	3.67
Soft tissues				
Adrenal glands	14	1016–1033	3425	3.52
Blood vessels				
arteries	200	1050–1075*	3470	3.27
vein		1056±3*		
Brain	1400	1030–1041	3460	
–,grey matter		1039	3460	3.56
–,white matter		1043	3460	3.52
Breasts	♀360	990–1060 (1020)	3300–3510	3.52
Eyes	15	1022–1030		
aqueous humor		1002–1004		
cornea		1076		
lens	0.4	1034–1121	3525	3.16
–,20yr		1034		
–,50yr		1071		
–,90yr		1113		
vitreous humor		1009		
Fatty,adipose	15000	916	3120–3240	3.82
Fat (breast)†		917–939 (928)		
–,brain,dog⊗		1026		
Intestine	1000	1041–1047	3425	3.52
Kidneys	310	1050	3480	3.41
–,cortex		1049		
–,medulla		1044		
Larynx	28	1060–1103		
Liver	1800	1050–1070 (1050)	3480–3540	3.38

Table 5.1 cont. Mass and density

Tissue	Mass,g	Density,kg m^{-3}	e^- dens $m^{-3}x10^{26}$	^1H dens $kg^{-1}x10^{25}$
Lungs,defl	1000	1040–1092 (1050)	3481	3.41
–,inflated		230–290 (260)	862	
Muscle:cardiac	330	1060	3480	3.45
Muscle:skeletal	28000	1038–1056 (1041)	3475–3480	3.38
Oesophagus	40	1040		
Ovary	11	1048±1.4	3490	3.49
Pancreas	100	1040–1050	3455	3.52
Pineal gland	0.18	1047–1050		
Pituitary gland	0.6	1066		
Placenta		995		
Prostate gland	16	1045	3455	3.49
Salivary glands		1041–1055		
Skin	2600	1093–1190 (1100)	3605	3.31
Horny layer		1500		
Epidermis		1110–1190		
Dermis		1116		
Hypodermis		971		
Spinal chord, nerve	30	1038	3460	3.56
Spleen	180	1054	3515	3.41
Stomach	150	1048–1052	3485	3.45
Tendon,ox		1110,1220		
Testes	35	1044	3455	3.52
Thymus	20	1026		
Thyroid gland	20	1036–1066 (1050)	3485	3.45
Trachea	10	1060–1100 (rings)	3510	3.34
Urinary bladder	45	1040	3455	3.49
Uterus	80	1052		
Skeleton and other hard tissues				
Bone,cortical	4000	1990±27	5950	1.05
–,trabecular	1000	1080 (inc. marrow) 1920±20 (bone alone)	3844	2.77
Cartilage	1100	1092–1104	3620	3.16
Red marrow	1500	992–1047 (1027)	3420	3.49
Yellow marrow	1500	923–1027 (980)	3280	3.85
Hair	20	1310		
Nails	3	1300		
Teeth (32)	46	2090–2240 2380(crown) 1950(root)		
Enamel	10	2890–3020 (2970)⊕		
Dentin	35	2030–2350 (2140)⊕		

Sources: ICRP (1975); ICRU (1989); Woodard & White (1986); Diem & Lentner (1970). Also * Geleskie & Shung (1982), † Johns & Yaffe (1987), ⊗ Mendez *et al* (1960) and ⊕ Manly *et al* (1939). Proton density calculated from data given by Woodard & White (1986).

Table 5.2 Average human bone densities

Bone	Mass density kg m^{-3}	Electron density	
		electrons $kg^{-1}x10^{26}$	electrons $m^{-3}x10^{26}$
Clavicle,scapula	1460	3.181	4644
Cranium	1610	3.148	5068
Femur,total	1420	3.191	4544
–,head	1330	3.214	4291
–,trochanter	1360	3.207	4363
–,shaft	1750	3.125	5471
Humerus,total	1460	3.181	4644
–,head	1330	3.214	4291
–,shaft	1490	3.175	4737
Mandible	1680	3.137	5270
Pelvis,male	1410	3.191	4499
–,female	1460	3.181	4644
Rib,male,2nd 6th	1410	3.192	4501
–,–,10th	1520	3.168	4815
Sacrum,male	1290	3.225	4160
–,female	1390	3.198	4445
Sternum,male	1250	3.237	4046
Vertebral column,male,			
excluding cartilage,C4	1420	3.190	4530
–,–,–,D6 L3	1330	3.212	4272
–,–,including cartilage,C4	1380	3.201	4417
–,–,–,D6 L3	1300	3.221	4187
–,–,–,whole column	1330	3.214	4275

Source: White *et al* (1987).

The density of cortical bone may reach 2.0 g cm^{-3}. However, the average density of whole bone is lower than this value. This results from the presence of low-density bone marrow in the spongiosa, which has a density close to that of soft tissue. White *et al* (1987) report a range of whole bone densities ranging from 1.29 g cm^{-3} (sacrum) to 1.75 g cm^{-3} for the proximal shaft of the femur (Table 5.2).

5.1.1.1 Thermal expansion

Density varies as a function of temperature. For solids and liquids the volumetric expansion may be described by the **coefficient of cubical expansion,** α:

$$\alpha = \frac{1}{V_T}\frac{dV_T}{dT} \tag{5.1}$$

where V_T is the volume of a particular mass of the material at a temperature T.

The cubical expansion coefficient has been investigated for some tissues in the temperature range 0°C to 40°C in several studies (Table 5.3). The expansion coefficient is itself a function of temperature and values of α are are listed together with the temperature range associated with the measurement. For most soft tissues, α increases with temperature. Generally the cubical expansion of fat exceeds that of muscle or blood. The difference is greater at lower temperatures, since anomalously, α for fat decreases with increase in temperature. The expansivity of tissue is greater than that of water at the same temperature (see Table 5.4).

Table 5.3 Thermal expansion coefficients of tissues

	Temp °C	Expansion coeff $°C^{-1}x10^{-4}$	Reference
Cubical expansion			
Blood:whole	15–20	2.5	Phillips *et al* 1950
–	25–30	3.2	–
–	35–40	4.0	–
Blood:red cells	6	1.9	Hinghofer-Szalkay 1985
–	43	4.2	–
Blood:plasma	15–20	2.3	Phillips *et al* 1950
–	25–30	3.1	–
–	35–40	3.9	–
Blood:ultrafiltrate	6	0.7	Hinghofer-Szalkay 1985
–	43	4.2	–
Bone,cow	18–30	1.97±.03	Lang 1969
Fat,human	10–15	19.9	Fidanza *et al* 1953
–,–	15–37	9.2	–
–,dog rat rabbit	25–37	8.8–11.4	–
–,cow pig sheep	25–37	8.7–17.6	–
–,dog,body	24–36	8.8,10.0	Méndez *et al* 1960
–,–,marrow	25–37	6.9	–
–,–,brain	25–37	1.6	–
–,cow	5–30	15.3±0.8	Jarvis 1971
Muscle,cow	5–30	3.75±0.04	–
–,rabbit	22–36	3.44±0.85	Méndez & Keys 1960
–,dog	21–36	3.90±0.57	–
Linear expansion			
Bone,cow,axial	5–30	0.89±.02	Lang 1969
–,human,cortical	25–75	0.27±.05	Ranu 1987

The cellular and fluid components of blood differ in their temperature dependence of α. The have comparable values at about body temperature, but the cubical expansion of blood cells considerably exceeds that of blood fluid at 6°C (Hinghofer-Szalkay, 1985).

For solid tissues such as bone, it is also possible to measure the **coefficient of linear expansion**, α_l. This is expressed approximately by

$$l_1 = l_o[1 + \alpha_l(T_1 - T_o)] \tag{5.2}$$

where l_1 and l_o are linear dimensions at temperatures T_1 and T_o. Some values for l are also given in Table 5.3. Since bone is not an isotropic solid, the coefficient of linear expansion of bone varies with direction (Lang, 1969).

5.1.1.2 Density of water

The variation of the density of water over the temperature range 0–100°C is given in Table 5.4.

5.1.3 *Organ mass*

Included in Table 5.1 are the average values of the mass of each organ or tissue. These average masses are for 'reference man' and are given in ICRP (1975). They are estimates for an average adult human organ, and were intended primarily for use in ionising radiation dosimetry.

Table 5.4 Density and viscosity of water; 1 atm

Temp °C	Density, ρ kg m^{-3}	Viscosity, η mPa s
0	999.84	1.7865
10	999.70	1.3037
20	998.20	1.0019
30	995.65	0.7982
40	992.22	0.6540
50	988.04	0.5477
60	983.20	0.4674
80	971.79	0.3554
100	958.36	0.2829

Source: Kaye & Laby (1973).

5.1.4 *Electron density*

The **electron density** of a medium, the number of electrons per unit mass, or per unit volume, may be derived from a knowledge of the elemental composition of that material (see Chapter 9). Estimates of the average electron density per unit mass $^m\rho_e$, together with ranges where the variation in the composition was known, are included in Table 5.1. The data is from ICRP (1975), updated by Woodard and White (1986). The average electron density per unit volume, $^v\rho_e$, can be estimated using the mass density, ρ; $^v\rho_e = \rho.^m\rho_e$.

Average electron densities, $^m\rho_e$ and $^v\rho_e$, are given for several human bones in Table 5.2.

5.1.5 *Hydrogen proton density*

The density of hydrogen protons in tissue, whether associated with water or any other constituent molecule, may be calculated directly from a knowledge of the percentage concentration of hydrogen by mass (Chapter 8) and the electron density. Proton density is important when carrying out nuclear magnetic resonance studies on tissue. Some estimates of ^{1}H proton density in tissues are included in Table 5.1.

5.2 Elastic moduli of bone and teeth

5.2.1 *Terminology and definitions*

The mechanical properties of bone have been the subject of considerable study and what follows may be considered as only a very brief overview of the important features, concentrating whenever possible on numeric values rather than qualitative description. Several excellent reviews exist in the literature and the reader is referred for example to those by Yamada (1973), Ashman *et al* (1984) and Cowin *et al* (1987). A general discussion of the mechanics of biological materials is given by Vincent (1982).

On the application of an external stress, which may be tensile, compressional or torsional, a solid material deforms. The deformation is expressed as a **strain**, the change in linear dimension per unit length. The general relationship between the longitudinal strain, and the applied longitudinal stress for bone is shown in Figure 5.1. Initially the strain, ϵ, increases linearly with the stress, σ. In this linear region where Hooke's law applies it is possible to define **Young's modulus**, E, as the stress per unit strain: thus $\sigma = E\epsilon$.

Bone is a non-homogeneous anisotropic and non-linearly viscoelastic material, and to make analysis of its mechanical behaviour tractable some simplifying assumptions about its mechanical characteristics are commonly

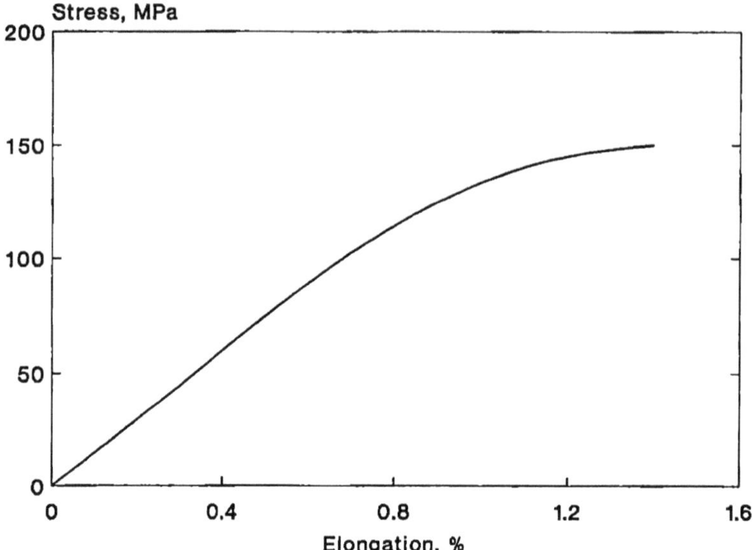

Figure 5.1 Generalised variation of longitudinal strain with stress for bone.

made. Over the range of stresses where Hooke's law may be deemed to hold, bone has been commonly modelled as an anisotropic but linear elastic material with a constitutive matrix equation

$$\sigma_i = C_{ij} \epsilon_j \tag{5.3}$$

Six-fold stresses σ_i are related to six-fold strains ϵ_j through C_{ij}, the **stiffness matrix.** Conversely, the strain-stress relationships can be given in terms of the **compliance matrix,** S_{ij}:

$$\epsilon_i = S_{ij} \sigma_j \tag{5.4}$$

The compliance coefficients can be further expressed in terms of Young's moduli, E_{ij}, Poisson's ratio, ν_{ij}, and shear moduli, G_{ij}. **Poisson's ratio,** is the lateral contraction per unit breadth (i direction) divided by the longitudinal extension per unit length (j direction). For most materials Poisson's ratio lies in the range 0 to 0.5. The **shear modulus,** G_{ij}, relates the strain $2\epsilon_{ij}$ to the stress T_{ij} (strain and stress are written as tensors).

For orthotropic material symmetry, the compliance matrix takes the form

$$[S_{ij}] = \begin{bmatrix} 1/E_1 & -\nu_{21}/E_2 & -\nu_{31}/E_3 & 0 & 0 & 0 \\ -\nu_{12}/E_1 & 1/E_2 & -\nu_{32}/E_3 & 0 & 0 & 0 \\ -\nu_{13}/E_1 & -\nu_{23}/E_2 & 1/E_3 & 0 & 0 & 0 \\ 0 & 0 & 0 & 1/G_{23} & 0 & 0 \\ 0 & 0 & 0 & 0 & 1/G_{31} & 0 \\ 0 & 0 & 0 & 0 & 0 & 1/G_{12} \end{bmatrix} \quad (5.5)$$

Only nine of the twelve components are independent since $\nu_{12}/E_1 = \nu_{21}/E_2$, $\nu_{13}/E_1 = \nu_{31}/E_3$ and $\nu_{23}/E_2 = \nu_{32}/E_3$.

For a transversely isotropic material (an assumption sometimes made for bone) in which by convention the 3-direction is parallel to the bone long axis,

$$E_1 = E_2, \ \nu_{12} = \nu_{21}, \ \nu_{13} = \nu_{31}, \ G_{23} = G_{31}, \qquad (5.6)$$

$$G_{12} = \frac{E_1}{2(1 + \nu_{12})}$$

Values of Young's modulus and shear modulus are expressed in the SI unit newton per metre^2 (N m^{-2}) or pascal (Pa), practical units being MPa or GPa. The unit pounds per square inch (psi or lbf in^{-2}) was commonly used in earlier work; since 1 lbf = 4.448 N, 1 psi = 6.89476×10^3 Pa. Another useful conversion is that 1 kgf = 9.807 N.

The last fundamental elastic constant, which relates the change of volume to external stress, is the **bulk modulus of elasticity**, K. For a homogenous, isotropic material the bulk modulus is

$$K = \frac{E}{3(1 - 2\nu)} \qquad (5.7)$$

The **compressibility**, κ, is the reciprocal of the bulk modulus: $\kappa = 1/K$.

5.2.2 *Measurement of elastic moduli*

There are two main methods for measuring the elastic constants of bone. Mechanical testing involves direct application of known tensile or compressive stresses, and measurement of the associated dimensional changes. Such mechanical testing can be easily extended to provide information about ultimate strength (Section 5.3). Strain rate can be easily controlled and the dependence of elastic modulus on strain rate evaluated. The second method derives elastic constants from measurements of ultrasonic velocity (McSkimin, 1961, Lang, 1970). Ultrasonic methods allow the elastic anisotropy of a single specimen to be more easily studied. Also, since only small specimens are required, the heterogeneity of the elastic properties of a single whole bone may be studied. The elastic constants obtained from each of the two methods of measurement appear to be in broad agreement, although values for Poisson's ratio differ more than those for Young's modulus (Abendschein and Hyatt, 1970; Burris, 1983; Cowin *et al*, 1987).

5.2.3 *Values of elastic moduli for bone*

Table 5.5 presents experimentally determined values of Youngs's modulus, E_i, shear modulus, G_{ij}, and Poisson's ratio, ν_{ij}, for several human and animal cortical bone specimens. All values are for fresh, hydrated specimens measured *in-vitro*. The 3-direction is that of the long axis of the bone. There is broad agreement between values of E_i and G_{ij} for human bone and those for bovine and canine bones. Values of Poisson's ratio have greater variability. Transverse isotropy has been commonly assumed.

The elastic moduli in compression have been found not to differ significantly from those measured under tension (Reilly *et al*, 1974; Bargren *et al*, 1974).

The elastic moduli of trabecular bone have not been so widely investigated as those for cortical bone. Carter and Hayes (1976, 1977) studied bovine and human trabecular bone under compression as a function of both density and strain rate. At a strain rate of 1.0 s^{-1} and density 1.8 g cm^{-3}

Table 5.5 Elastic properties of hydrated cortical bone; Young's modulus, E_i, 10^9 Pa; shear modulus, G_{ij}, 10^9 Pa, and Poisson's ratio, ν_{ij}

	Human femur	Human tibia	Bovine femur,phalanx	Canine femur
E_1	11.5,12.0	6.9	10.2–13.3	12.8
E_2	11.5,13.4	8.5	10.2–14.6	15.6
E_3	17.0,20.0	18.4	18.1–22.6	20.1
G_{12}	3.6,4.3	2.4	3.4–5.3	4.7
G_{13}	3.3,5.6	3.7	3.6–6.3	5.7
G_{23}	3.3,6.2	4.9	3.6–7.0	6.7
ν_{12}	0.38,0.58	0.49	0.30–0.51	0.28
ν_{13}	0.22,0.31	0.12	0.11–0.24	0.29
ν_{23}	0.24,0.31	0.14	0.20–0.22	0.27
ν_{21}	0.42,0.58	0.62	0.38–0.51	0.37
ν_{31}	0.37,0.46	0.32	0.21–0.42	0.45
ν_{32}	0.35,0.46	0.31	0.22–0.40	0.34

Long axis in the 3-direction.

Source: Cowin *et al* (1987). Original data: human femur-Reilly & Burstein (1975) and Ashman *et al* (1984); human tibia-Knets and Malmeisters (1977); bovine phalanx-Lang (1970); bovine femur-Reilly & Burstein (1975), van Buskirk *et al* (1981) and Burris (1983); canine femur-Ashman *et al* (1984).

the value of E was 22.1 GPa. Whilst Carter and Hayes reported that E for trabecular bone appeared to be dependent on the third power of density, others have reported a square-law dependence (Bensusan *et al*, 1983).

5.2.4 *Factors affecting elasticity*

5.2.4.1 Temperature

There is an inverse linear relationship between elastic constants and temperature. The temperature coefficient for bone has been reported as lying in the range −0.17% $^{\circ}C^{-1}$ (Cowin *et al*, 1987) and −0.24% $^{\circ}C^{-1}$ (Bonfield and Tully, 1982).

5.2.4.2 Regional variations

There is evidence that the mechanical properties of individual bones vary with position in the bone. For human tibia, Evans and Vincentelli (1974) found a reduction in the longitudinal modulus of elasticity in compression from 20.7±5.4 GPa in the distal third to 17.3±3.9 GPa in the proximal third of the bone. Comparable variations were also reported in compressive strength, and deformation. The strength correlated positively with the percentage of osteons in the bone segment. Pope and Outwater (1974) found evidence of spatially variable anisotropy in bovine tibia. The bone was found to be highly anisotropic at the mid-diphysis and essentially isotropic at the epiphyses.

5.2.4.3 Viscoelasticity

Whilst many of the mechanical characteristics of bone may be evaluated using a linear elastic model for bone, detailed studies have demonstrated that bone has clear viscoelastic properties. There are two important ways in which bone is found to deviate from the simple elastic model in practice.

Mechanical methods of testing allow the strain-rate dependence of the stress-strain relationship to be measured. McElhaney (1966), examining embalmed human femur in compression, showed that the specimens were both stronger and stiffer at higher strain rates. Carter and Hayes (1976) reported that both strength and elastic modulus were approximately proportional to the 0.06 power of strain rate. Since, during exercise, the strain rate to which bone is subjected may vary by an order of magnitude, the *in-vivo* elastic modulus may vary by as much as 15% during normal activity (Cowin *et al*, 1987).

The second important aspect of the viscoelastic response of bone to applied stress is creep. Although bone always returns to its non-deformed length following the application of stress, this recovery is a time-dependent process. Curry (1965) studied the long-term creep behaviour of wet bovine bone noting a recovery asymptotic to zero strain only after many days. The creep of human cancellous bone in the femoral head has been similarly evaluated by Schoenfeld *et al* (1974), who showed recovery after about 100 s.

Other aspects of the viscoelastic behaviour of bone lie beyond the scope of this review, and readers are referred, for instance, to a discussion by Lakes *et al* (1979) for further detail.

5.2.5 *Elastic moduli for tooth material*

The mechanical behaviour of tooth enamel and dentine have been thoroughly reviewed and reported by Braden (1976). Values of elastic moduli for tooth material included in Table 5.6 are drawn from this review to which reference should be made for access to the original authors. Isotropic properties have been commonly assumed.

Table 5.6 Some mechanical properties of tooth material

	Dentine	Enamel
Young's modulus (E), 10^9 Pa	12.9±2.7	20.0 to 84.2
Shear modulus (G), 10^9 Pa	6.4 to 9.7	29
Bulk modulus (K), 10^9 Pa	3.11 to 4.38	45,65
Poissons ratio, ν	−0.11 to 0.07	0.23,0.30
Compressive strength, 10^9 Pa	0.282±.033	0.095 to 0.386
Tensile strength, 10^9 Pa	0.040 to 0.276	0.030 to 0.035
Shear strength, 10^9 Pa	0.012 to 0.138	0.06
Knoop hardness	64±7	250 to 500

Source: Braden (1976).

5.3 Ultimate strength of bone and teeth

5.3.1 *Terminology and definitions*

Making reference to Figure 5.1, it may be seen that as tensile stresses are increased the elongation no longer remains a linear function of applied stress. Initially the material yields and it is possible to quantify the stress required to cause a deviation from the linear stress–strain relationship as the yield stress. Yield stress is commonly defined by the intersection of an offset yield line and the stress–strain curve (Figure 5.1). Values of yield stresses have not been included here. As the applied stress is further increased the material ultimately breaks. The tensile stress sufficient to cause fracture is termed the **ultimate tensile stress** or **ultimate strength** of the material. The maximum strain which the material can sustain before failure is called its **ultimate strain.** Values of ultimate strength and strain under compression and torsion can also be defined in a similar way. Under compressional stresses cortical bone generally does not yield before failing (Yamada, 1973). In general it may be assumed that ultimate strengths vary with the direction of the applied strain, and that the strengths are different under tensile, compressional and torsional loads.

5.3.2 *Values of ultimate strength and strain for bone and teeth*

The ultimate tensile strength and associated ultimate strain for femoral bone is included in Table 5.10. Ultimate compressive and torsional strengths for bone and cartilage are given in Table 5.7. Both tensile and compressive strengths are anisotropic, bone being stronger in the longitudinal than in the transverse direction. According to Carter and Caler (1983) the ultimate longitudinal tensile strength of cortical bone is given as a function of strain rate $\dot{\epsilon}$ (s^{-1}) by

$$\sigma^{+} = 147\,\dot{\epsilon}^{0.055} \text{ MPa} \qquad (5.8)$$

Trabecular bone is weaker under compression than cortical bone by at least a factor of 10, and has a very wide variation of strength (Schoenfeld *et al*, 1974). The compressive and tensile strengths of trabecular bone are dependent on density, probably with a square–law dependence (Carter and Hayes, 1977; Carter *et al*, 1980; Bensusan *et al*, 1983). Using the values given by Carter, Cowin *et al* (1987) give an expression for the compressive strength σ^{-} for cancellous bone in the form

$$\sigma^{-} = 68\ \dot{\epsilon}^{0.06}\rho^{2} \qquad (5.9)$$

Table 5.7 Ultimate compressive and torsional strength of bone and cartilage

Tissue	Average ultimate compressive strength,MPa	Fractional compressive strain	Average ultimate torsional strength,MPa
Bone,femur,compact*	159	0.018	53
–,–,longitudinal†	195±20	0.022±.006	69±4.41
–,–transverse†	133±17	0.046±.026	
–,vertebra*	6.9	0.052	3.1
–,–,spongy*	1.6	0.024	
–,fem head,cancellous⊗	0.015–13.5		
Costal cartilage*	7.8	0.136	
Vertebral disk*	10.8	0.227	4.4

Sources: *Yamada (1973), †Cowin *et al* (1987) from Reilly & Burstein (1975) and Cezayirlioglu *et al* (1985), ⊗Schoenfeld *et al* (1974).

where σ^- is in MPa, $\dot{\epsilon}$ is in s^{-1} and ρ is in g cm^{-3}.

The shear strength τ of trabecular bone was found by Stone *et al* (1983) to be given by the following density dependence:

$$\tau = 21.6\rho^{1.65} \tag{5.10}$$

where τ is in MPa and ρ in g cm^{-3}.

The density used in Equations 5.9 and 5.10 is that of the total cancellous bone volume, including marrow, and not that of the bony matrix alone.

Bending, impact and cleavage strengths for cortical bone as reported by Yamada (1973) are included in Table 5.8.

Compressive, tensile and shear strengths for human tooth material, from a survey reports made by Braden (1976) are included in Table 5.6.

Table 5.8 Various mechanical properties of human compact bone

Ultimate bending strength	157 MPa *
Impact bending strength	1.4 N cm/mm^2 †
Impact snapping strength,radial	2.5 N cm/mm^2
Cleavage strength	84 N mm^{-1}
Brinell hardness	24

Source: Yamada (1973).

* 189 MPa for femur whole bone. † 15 N cm/mm^2 for femur whole bone.

5.4 Elastic properties of other tissues

The stress–strain relationship of most soft tissues can be simply characterised by three regions (Figure 5.2). At low stress there is a region of relatively low elastic modulus in which large extensions may occur for small increases in tension. At high stresses, below the ultimate strength of the tissue, there is a region of high elastic modulus in which extensions are much smaller for a given stress increment. The elastic properties in both these regions are approximately linear and it is possible in principle to derive a value of an elastic modulus from the slope of the stress–strain response in either of these quasi-linear regions. Viidik (1987) has termed the slope of the high stress region the elastic stiffness to distinguish it from the elastic property of the tissue in the region at low stress. In the middle region there is a constant change in gradient of the stress–strain relationship. For some tissues either or both of the quasi-linear regions are absent; for others the change from low to high modulus occurs over a relatively small range of applied stress (Yamada, 1973).

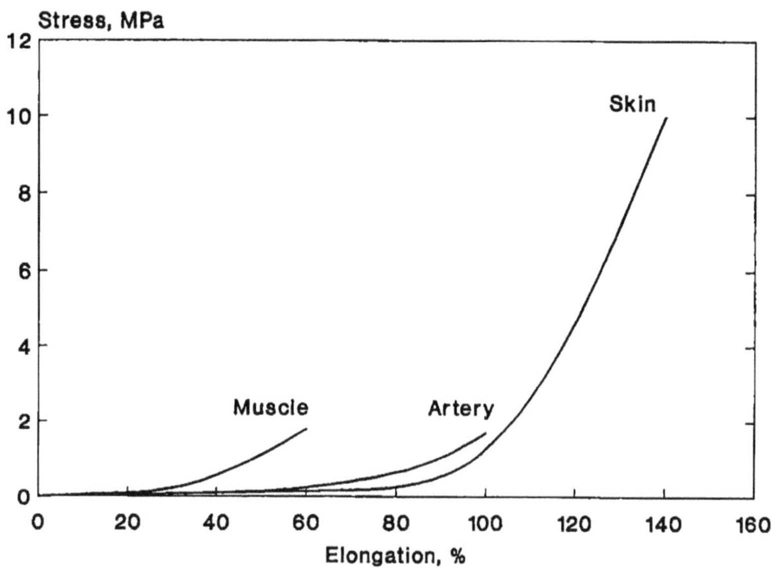

Figure 5.2 Generalised variation of longitudinal strain with stress for three soft tissues: skin, artery and muscle.

5.4.1 *Values of elastic properties of soft tissue*

Some values of Young's modulus for tissues other than bone are given in Table 5.9. Considerable care must be exercised when comparing these values since the measurements have not necessarily been carried out on the same portion of the stress–strain curve. The values summarised by McDonald (1974) for arteries are from his own and earlier work (for instance, Bergel, 1961; Goedhard *et al*, 1973). These values are associated with normal physiological conditions of longitudinal tension and distending blood pressure. They are therefore primarily on the lower portion of the curve. The arterial wall becomes rapidly stiffer at blood pressures in excess of 200 mm Hg and McDonald gives an upper value of 20 MPa for the modulus under these conditions. Similarly the values for hair, nail and *in–vivo* muscle given in Table 5.9 are for the initial linear region. On the other hand, values derived from Yamada's book by Viidik (1987) are for elastic stiffness in the high stiffness upper linear region. Values given for elastic stiffness by Viidik are, for instance, 35 MPa for thorax skin in a transverse direction, and 700 MPa for tendon. Viidik has suggested that tendon *in–vivo* probably operates somewhere in the lower and curved part of the stress–strain relationship. Skin

Table 5.9 Young's modulus, E, of some tissues other than bone

Tissue	E MPa	Reference
Artery		
Thoracic aorta	0.3–0.94	McDonald 1974
Abdominal aorta	0.98–1.42	–
Iliac artery	1.10–3.50	–
Femoral artery	1.23–5.50	–
Cartilage, at equilibrium		
Human,articular,patellar	0.79±0.36	Woo *et al* 1987*
Bovine,articular	0.85±0.21	–
–,–	0.7	Sokoloff 1966
–,condylar,femoral,*in situ*	0.90±0.43	Mow *et al* 1989
–,patellar grove,*in situ*	0.47±0.15	–
–,costal	5.0	Sokoloff 1966
–,nasal	5.64±3.19	Woo *et al* 1987*
Hair,100% humidity	2620±180	Swanbeck *et al* 1970
–,60% humidity	3890	Robbins 1988
–,53% humidity	8800	Goldsmith & Baden 1970
–,6% humidity	11700	–
Menisci,bovine	0.41	Woo *et al* 1987*
Muscle,*in–vivo*,dog		
along fibres	0.50±14%	Gates *et al* 1980
across fibres	0.79±24%	–
–,–,human,relaxed,10% strain	0.0062±8%	Krouskop *et al* 1987
–,–,–,tense,10% strain	0.11±2%	–
Nail	4600±1400	Baden 1970

* Values assembled from other authors.

is normally under slight tension *in–vivo*. It initially exhibits large strains when stressed, becoming increasingly stiff with further strain, in common with most soft tissue, ultimately reaching a linearly elastic region. The elastic properties of skin have been widely studied (see, for instance, Lanir *et al*, 1987).

5.4.2 *Factors affecting elasticity*

5.4.2.1 Humidity

Humidity strongly affects the stress–strain relationships, and is particularly important when analysing the elastic properties of surface tissues such as skin and hair. Young's modulus of skin at 25%–30% relative humidity may be 1000–fold greater than at 100% humidity (Papir *et al*, 1975). A similar, though smaller change with humidity is observed in the elasticity of hair (Robbins, 1988).

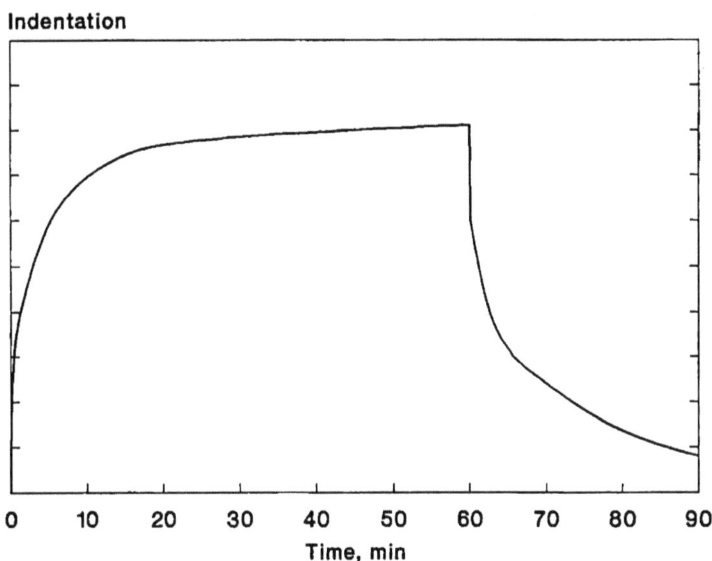

Figure 5.3 Generalised indentation response for cartilage under constant load, showing creep behaviour.

5.4.2.2 Tissue composition: collagen and elastin

The elastic properties of soft tissues, and more particularly those of skin, artery and tendon, have been commonly related to the elastic properties of collagen and elastin. The elastic stiffness of tendon collagen fibre bundles, measured using Brillouin light scattering, has a value of about 5100 MPa (Cusack and Miller, 1979). Harkness (1981) calculated that the strength of a collagen fascicle is about 500 MPa. The equivalent values for elastin, noted by McDonald (1974) are considerably lower; 0.3 MPa for Young's modulus at 100% extension, and 1 MPa for ultimate tensile strength.

5.4.2.3 Viscoelasticity

Many experimental studies on ligament, tendon, skin and other soft tissues have demonstrated aspects of their viscoelastic and plastic behaviour. Cycles of tension and relaxation result in a progressive reduction in the stress in the tissue sample, a process known as preconditioning. In addition, the stress–strain relationships are rate-dependent and exhibit considerable hysteresis. Stress relaxation occurs under constant strain and creep occurs under constant stress (Figure 5.3). The viscoelastic behaviour is also probably anisotropic. No attempt is made here to address these highly complex and, at present, incompletely analysed problems. Reference should be made to the publications cited elsewhere in this chapter.

5.4.3 *Elastic properties of cartilage*

The mechanical properties of cartilage lie broadly between those of bone and of soft tissue. Both articular cartilage and costal cartilage demonstrate creep under stress (Figure 5.3) and it is necessary to distinguish between an equilibrium Young's modulus and an initial Young's modulus when reporting elastic properties. Typically equilibrium is not reached in cartilage under compression until at least 30 minutes after the load is applied (Sokoloff, 1966). Under a static load the elastic deformation is about 25% of the final equilibrium deformation. Under compression the stress–strain relationship is nearly linear until the ultimate strength of the sample is reached (Yamada, 1973; Kenedi *et al*, 1966); there is little yield. Under tension the relationship is generally similar in pattern to that of bone, with a linear region before yielding and then failing. Costal cartilage is more rigid than articular cartilage (Sokoloff, 1966). Both Young's modulus and ultimate strength decrease for the deeper layers of cartilage (Kempson, 1979).

Equilibrium shear modulus for cartilage was reported by Sokoloff as being 0.23 MPa. More recently Arokoski *et al* (1986) have reported an average shear modulus of 0.32±0.05 MPa for canine cartilage. Poisson's ratio has been measured by Mow *et al* (1989) for bovine condyle and patella grove cartilage, who report values of 0.38±0.05 MPa and 0.25±0.07 MPa respectively. Comparable values are given by Kempson (1979) for human femoral condylar cartilage when oriented perpendicular to the cleavage lines (0.42 and 0.47). Values of Poisson's ratio reaching 4.0 may occur when the stress is applied parallel to the cleavage lines, due to the overall reduction in volume under compression.

The mechanical response of cartilage has been described using a biphasic model (Mow *et al*, 1980). This model depicts cartilage as consisting of a soft, permeable elastic solid mixed with water. Recently, a finite deformation theory based on this model has been reported (Kwan *et al*, 1990). A thorough overview of articular cartilage is presented by Freeman (1979).

5.4.4 *Elastic properties of hair*

Hair strains in a linear way on the initial application of a tensile stress. At higher stresses the hair yields, elongating greatly but not fracturing (Figure 5.4). A further stress applied above the yield point is necessary to reach the breaking strain (Robbins, 1988).

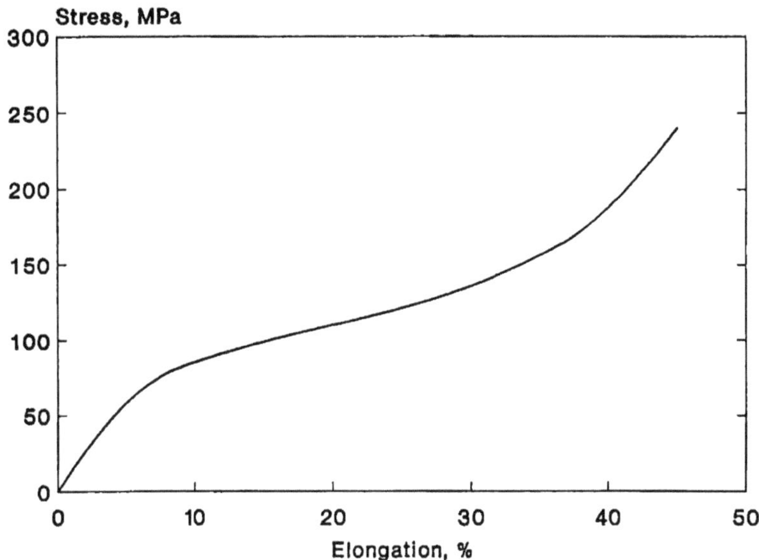

Figure 5.4 Generalised stress–strain relationship for hydrated hair.

5.5 Ultimate strength of soft tissue

5.5.1 *Values of tensile strengths of soft tissue*

The ultimate tensile strength of soft tissue is more amenable to simple numeric evaluation than is the general non–linear stress–strain relationship. In an extensive and broad–ranging study, Yamada (1973) has reviewed the stress–strain relationships and the ultimate strengths of a very wide range of human and animal tissues. All tissue samples were stored in physiological saline between excision and testing to provide for mechanical stabilisation. The storage period was between 1 and 5 days for soft tissues and several months for bone. Values of ultimate tensile strength and breaking strain for adult human tissues are presented in Table 5.10, together with variation at 95% spread. For many tissues their mechanical anisotropy results in slightly different strengths in the longitudinal and transverse directions. These are noted. The greatest variation was for spinal dura mater where a nearly ten–fold difference in strength was observed depending on the direction of the applied stress.

Table 5.10 Ultimate tensile strength and maximum elongation strain of hard and soft tissues (human except where noted)

Tissue (L=longitudinal, T=transverse)	Average ultimate tensile strength MPa	Coeff of variation (95% range)	Maximum fractional strain
Amnion	2.4		0.42
Artery,mixed,L	1.7	±29%	0.87
-,-,T	1.4		0.69
-,coronary,L	1.1		0.64
-,elastic tissue,L	0.78		0.80
-,-,T	1.0		0.82
-,muscle,L	1.4		1.02
-,-,T	1.0		0.75
Bladder,urinary	0.24	±28%	2.26
Bone,femur,L*	133±14		0.029±30%
-,-,T*	51±10		0.0072 ±0.002
Cartilage,costal	2.8	±20%	0.18
Eye,cornea	3.4	±14%	0.15
-,sclera,E	6.8	±15%	0.17
-,-,M	4.7		0.17
Fascia	14		0.16
Gall bladder,rabbit	2.1±0.07		0.53±0.4
Hair	193		0.40
-,100% humidity†	164±10		0.62±0.02
Heart,muscle	0.11	±11%	0.64
-,auricle	3.3		0.26
-,-,cartilage	3.0		0.26
-,chordae tendinae	63		0.33
-,papilliary muscle	0.23		0.30
-,valve,radial	0.92		0.17
-,-,circumference	2.5		0.13
Intestine,large,L	0.68	±20%	1.17
-,-,T	0.44		1.37
-,small,L	0.55	±23%	0.43
-,-,T	0.52		0.89
Kidney,parenchyma	0.05	±45%	0.52
-,calyx,L	1.1		0.35
-,-,T	0.47		0.48
-,fibrous capsule	2.3		0.29
Liver,rabbit	0.024±0.005		0.46
Muscle,skeletal	0.11	±38%	0.61
Nail	18		0.14
Nerve,secondary fibre bundle	13	±10%	0.18
Oesophagus,L	0.59	±10%	0.73
-,T	0.18		1.24

Table 5.10 cont. Strength of tissue

Tissue	Strength,MPa	95% var.	Strain
Peridontal membrane	1.4		
Skin(thorax,neck)	11	±62%	0.47–0.90
–(abdomen,back,arm,			depending
foot)	9.5		on site
–(leg,hand)	7.3		
–(face,head,			
external genitals)	3.7		
Spine,dura mater,L	11	±13%	0.21
–,–,T	1.3		0.34
Stomach,L	0.55	±24%	0.93
–,T	0.43		1.27
Tendon,calcaneal	53	±9%	0.09
Trachea,cartilage	2.4	±19%	0.19
–,membranous wall,L	2.2		0.61
–,–,T	0.35		0.81
–,–,intercartilage			
membrane	1.9		3.38
Umbilicus,mature	1.5		0.59
Ureter,L	1.8	±42%	0.36
–,T	0.44		0.98
Uterus,rabbit	0.18±0.007		1.50±0.07
Vagina,rabbit,T	0.45±0.04		1.94±0.14
–,–,L	0.57±0.04		1.13±0.15
Vein,T	2.9		0.66
–,L	1.7		0.89
Vertebral disk	2.7		0.57
–,annulus fibrosis,L	13		0.14
–,–,T	5.2		0.12

E Direction of equator, M Direction of meridian; * Cowin *et al* (1987) from Reilly & Burstein (1975) and Cezayirlioglu *et al* (1985), † Swanbeck *et al* (1970). Source: mainly from Yamada (1973).

5.5.2 *Factors affecting tensile strength*

5.5.2.1 Age

Ultimate strengths of some fetal tissues are included in Table 5.11. The values given in Table 5.10 are largely average values for adult human tissue. In addition, Yamada gives an extensive evaluation of the variation of human tissue strength with age, using the 20–29 year age group as a reference. Average values taken from this survey are given in Table 5.12. It is concluded that there is an average ageing in the strength of human tissues above 20 years of age at a rate of 0.5% per year. This average trend for

Table 5.11 Ultimate strength of some human fetal tissues

Lunar month	Ultimate strength, MPa 4th	6th	8th	10th	Fractional strain
Tensile strength					
Aorta,thoracic,L	...	0.78	0.94	1.14	1.09
–,–,T	...	1.27	1.35	1.46	0.99
Intestine,small,L	...	0.39	1.04	1.22	0.22→0.42
–,–,T	...	...	0.46	1.04	0.58→0.88
Stomach,L	...	0.15	0.27	0.39	0.48→0.84
–,T	...	0.14	0.21	0.33	0.58→1.11
Umbilicus	0.25 ±0.01	0.61 ±0.04	0.97 ±0.04	1.36 ±0.06	0.36→0.57
Compressive strength					
Femur	24	31	40	53	0.041

L = longitudinal direction. T = transverse direction.
Source: Yamada (1973).

Table 5.12 Ageing rate for the strength of tissue: average percentage change from the strength at 20–29 years

Tissue	30–39 yr	40–49 yr	50–59 yr	60–69 yr	70–79 yr
Artery	–5	–15	–22	–24	...
Bladder,urinary	–17	–17	–30	–30	–30
Bone	–1	–10	–14	–20	–22
Cartilage	–1	–21	–26	–29	–30
Dura mater	–2	–10	–12	–18	–20
Eye;cornea,sclera	0	0	0	0	0
Hair(head)	–2	–10	–12	–19	...
Intestine	–14	–24	–32	–39	–42
Kidney	–6	–8	–12	–17	–17
Muscle,cardiac	–8	–14	–14	–24	–24
–,skeletal	–8	–16	–19	–24	–24
Nerve	–4	–12	–16	–20	–22
Skin	+26	+26	+8	–4	–23
Stomach,oesophagus	–8	–17	–22	–28	–28
Teeth	+11	+22	+20	–3	...
Tendon	–1	–1	–1	–6	–18
Trachea	–8	–19	–24	–35	–40

Source: Yamada (1973).

tissue to weaken with age must be seen in the context of the average ranges within each age band, which, at 95% confidence limits, increase from 21% (coefficient of variation) at 20–29 years to 26% in the 70–79 year band.

5.5.2.2 Humidity

Tensile strength also depends upon humidity. Hair is stronger when dryer, and the ultimate strength of skin drops from about 100 MPa at low humidity (26%) to about 10 MPa in the wet state (Papir *et al*, 1975). There is a similar drop with temperature in the 25–60°C range. The maximum elongation of skin increases with humidity by two orders of magnitude from approximately 2% at 26% relative humidity to 200% if wet.

5.5.3 *Shear strength and bursting strength*

Yamada also reported values for the ultimate expansive (bursting) strength and ultimate shear strength of some organs and tissues. These values are given in Tables 5.13 and 5.14.

Table 5.13 Ultimate expansive strength per unit thickness

Tissue	Ultimate expansive strength, MPa mm^{-1}
Amnion	15
Artery,elastic	44
Fascia	663
Intestine,large	42
–,small	37
Oesophagus	39
Stomach	53

Source: Yamada (1973).

Table 5.14 Ultimate shear strength of some soft tissues

Tissue	Shear strength,MPa
Artery (thoracic aorta),dog,L	6.2±0.3
–,–,T	5.4±0.2
Dura mater,human	19.1
Intestine,small,dog,L	2.2±0.2
–,–,–,T	2.9±0.2
Oesophagus,dog,L	3.4±0.2
–,–,T	3.8±0.2
Skin,dog	6.5±0.2

Source: Yamada (1973). L=longitudinal. T=transverse.

5.6 Compressibility and bulk modulus of elasticity

The adiabatic bulk modulus K_a can be calculated from measurements of the ultrasonic velocity, c, and density, ρ, of the tissue: $K_a = c^2\rho$ (from Equation 4.3). Both these quantities vary with temperature, and therefore so does K_a. Experimentally determined values for K_a for a few soft tissues are listed in Table 5.15. Estimates for other tissues and temperatures may be made using values given in Tables 5.1 for density and 4.1 and 4.2 for acoustic velocity. For liquids and soft tissues also, the isothermal bulk modulus K_i obtained by static measurements is very close to K_a.

Table 5.15 Adiabatic bulk modulus, K_a, of some soft tissues

Tissue	Temp °C	K_a GPa	Reference
Aorta	22	2.61	Shung & Ried 1978
Cartilage		0.25–0.71	Hori & Mockros 1976
Fat,breast	22	2.01	Apfel 1986
–,–	30	1.99	–
–,–	37	1.86	–
–	24	1.96	Frucht 1953
–,50% water	24	2.15	–
Kidney,cortex	22	2.54	Sarvazyan & Klemin 1983
–,medulla	22	2.46	–
Liver	30	2.60	Apfel 1986
–,70% water	24	2.70	Frucht 1953
–,80% water	24	2.53	–
Water	15	2.05	Kaye & Laby 1973
–	24	2.24	–
–	40	2.32	–

5.7 Viscosity

The **dynamic viscosity,** η, of a (Newtonian) fluid is the ratio of the shearing stress between two parallel flowing planes to the velocity gradient perpendicular to the flow. The SI unit is newton second metre^{-2} (N s m^{-2}) or equivalently pascal second (Pa s). A common alternative unit is poise (P), and 1 P = 0.1 Pa s.

The **relative viscosity** is the ratio of the viscosity of a fluid to that of water at the same temperature. The dynamic viscosity of water is strongly temperature dependent (Table 5.4). The relative viscosity depends much less on temperature, varying only as the ratio of the temperature coefficients of the fluid and water. Viscosity also depends upon pressure.

The **kinematic viscosity** is the ratio of the dynamic viscosity to the density. The SI unit is metre^2 second^{-1} (m^2 s^{-1}). A common alternative unit is stokes (St); 1 St = 1 cm^2 s^{-1}.

The viscosity of a fluid may be measured either by observing its flow along a capilliary tube, or by the use of a rotational viscometer. The viscometer applies an almost uniform shear rate across the fluid, and can be used to investigate the variation of viscosity with velocity gradient. Newtonian fluids show a constant relationship between shear stress and velocity gradient. Whilst many body fluids are Newtonian, blood is not. At high rates of shear, above about 100 s^{-1}, the viscosity of blood is approximately constant. Below shear rates of about 50 s^{-1} the viscosity of blood rises sharply, an effect

Table 5.16 Relative viscosity of human body fluids

Fluid	Temp °C	Viscosity mean	Viscosity range	Notes
Bile,hepatic		1.27	1.07–1.75	
–,gall bladder		2.85	1.31–5.42	
Blood*	18	4.75	3.80–5.70	
–		4.5	3.5–5.4	ICRP 1975
–			2.30–2.75	Dynamic viscosity *in-vivo*
Blood:plasma	18	2.01	1.67–2.35†	
Blood:serum	18	1.88	1.58–2.18†	
–	37.5	1.15	1.08–1.22	Kinematic viscosity
Cerebrospinal fluid	38		1.020–1.027	
Semen	20	6.45		
Synovial fluid	37	46.3	26.7–65.7†	Intrinsic
–	37		>300	Relative
Urine			1.0–1.14	

Source: Diem & Lentner (1970).
* At high shear rate. † 95% range.

which is dependent on the concentration of high molecular weight proteins in the blood.

Blood viscosity increases with haematocrit, and in common with other fluids, decreases with temperature. There is an approximately 30% decrease in dynamic viscosity on increasing the temperature from 25°C to 37°C. One anomalous property of flowing blood is that its apparent viscosity decreases when flowing in narrow tubes, relative to that measured using a viscometer (Fahraeus and Lindqvist, 1931). This effect appears to depend on the deformation of the red blood cells within the tube.

Harkness (1981) reports that plasma shows a 2–3% drop in viscosity for each 1°C increase in temperature over the range 15–40°C. The dynamic viscosity is given as 1.70 to 1.915 mPa s at 20°C and 1.16 to 1.35 mPa s at 37°C. The relative viscosity does not vary over this temperature range. Irreversible changes occur in some plasma constituents above about 42°C resulting in increased viscosity upon return to lower temperatures.

Values of the relative viscosity of blood and of other body fluids are given in Table 5.16.

References

Abendschein W. and Hyatt G.W., 1970, Ultrasonics and selected properties of bone, Clin Orthop, 69, 294–301.

Apfel R.E., 1986, Prediction of tissue composition from ultrasonic measurements and mixture rules, J Acoust Soc Am, 79, 148–152.

Arokoski J., Jurvelin J., Kiviranta I. and Helminen H.J., 1986, Combined biochemical and histochemical analysis of articular cartilage in the canine knee, Proc XVth Symp Europ Soc Osteoarthrology, Kuopio, Finland.

Ashman R.B., Cowin S.C., van Buskirk W.C. and Rice J.C., 1984, A continuous wave technique for the measurement of the elastic properties of cortical bone, J Biomech, 17, 349–361.

Baden H.P., 1970, The physical properties of nail, J Invest Dermatol, 55, 115–122.

Bargren J.H., Bassett C.A.L. and Gjelsvik A., 1974, Mechanical properties of hydrated bone, J Biomech, 7, 239–245.

Bensusan J.S., Davy D.T., Heiple K.G. and Verdin P.J., 1983, Tensile, compressive and torsional testing of cancellous bone, Trans Orthop Res Soc, 8, 132.

Bergel D.H., 1961, The static elastic properties of the arterial wall, J Physiol, 156, 445–457.

Bonfield W. and Tully A.E., 1982, Ultrasonic analysis of the Young's modulus of cortical bone, J Biomed Eng, 4, 23–27.

Braden M., 1976, Biophysics of the tooth, Front Oral Physiol, 2, 1–37.

Burris C.L., 1983, A correlation of quasistatic and ultrasonic measurements of the elastic properties of cortical bone, PhD dissertation, Tulane University, New Orleans; cited in Cowin *et al*, 1987.

Carter D.R. and Caler W.E., 1983, Cycle dependent and time dependent bone fracture with repeated loading, J Biomech Eng, 105, 166–170.

Carter D.R. and Hayes W.C., 1976, Bone compressive strength: the influence of density and strain rate, Science, 194, 1174–1176.

Carter D.R. and Hayes W.C., 1977, The compressive behavior of bone as a two–phase porous structure, J Bone Joint Surg, 59A, 954–962.

Carter D.R., Schwab G.H. and Spengler D.M., 1980, Tensile fracture of cancellous bone, Acta Orthop Scand, 51, 733–741.

Cezayirlioglu H., Bahniuk E., Davy D.T. and Heiple K.G., 1985, Anisotropic yield behavior of bone under combined axial force and torque, J Biomech, 18, 61–70.

Cowin S.C., van Buskirk W.C. and Ashman R.B., 1987, Properties of bone, Chapter 2 in *Handbook of Bioengineering*, R.Skalak and S.Chien (eds), McGraw Hill.

Curry J.D., 1965, Anelasticity of inelastic flow in bone, J Exp Biol, 43, 279–292.

Cusack S. and Miller A., 1979, Determination of the elastic constants of collagen by Brillouin light scattering, J Mol Biol, 135, 39–51.

Diem K. and Lentner C, (eds), 1970, *Documenta Geigy Scientific Tables*, 7th edition, Geigy, Macclesfield.

Evans F.G. and Vincentelli R., 1974, Relations of the compressive properties of human cortical bone to its histological structure and calcification, J Biomech, 7, 1–10.

Fahraeus R. and Lindqvist T., 1931, Viscosity of blood in narrow capillary tubes, Am J Physiol, 96, 562–568.

Fidanza F., Keys A. and Anderson J.T., 1953, Density of body fat in man and other mammals, J Appl Physiol, 6, 252–256.

Freeman M.A.R. (ed.), 1979, *Adult Articular Cartilage*, 2nd edition, Pitman Medical, London.

Frucht A.-H., 1953, Die Schallgeschwindigkeit in menchlichen und tierischen Geweben, Zeit ges exp Med, 120, 526–557.

Gates F., McCammond D., Zingg W. and Kunov H., 1980, *In vivo* stiffness properties of the canine diaphragm muscle, Med Biol Eng Comput, 18, 625–632.

Geleskie J.V. and Shung K.K., 1982, Further studies on acoustic impedance of major bovine blood vessel walls, J Acoust Soc Am, 71, 467–470.

Goedhard W.J.A., Knoop A.A. and Westerhof N., 1973, The influence of vascular smooth muscle contraction on elastic properties of pig's thoracic aortae, Acta Cardiol (Brux), 28, 415–430.

Goldsmith L.A. and Baden H.P., 1970, The mechanical properties of hair I. The dynamic sonic modulus, J Invest Dermatol, 55, 256–259.

Harkness J., 1981, Measurement of plasma viscosity, in *Clinical Aspects of Blood Viscosity and Cell Deformability*, G.D.O.Lowe, J.C.Barbanel and C.D.Forbes (eds), Springer–Verlag, New York, pp 79–87.

Hinghofer–Szalkay H., 1985, Volume and density changes of biological fluids with temperature, J Appl Physiol, 59, 1686–1689.

Hori R. and Mockros L.F., 1976, Indentation tests on human articular cartilage, J Biomech, 9, 259–268.

ICRP, 1975, *Report of the Task Group on Reference Man*, ICRP Publication 23, International Commission on Radiological Protection, Pergamon Press, Oxford.

ICRU, 1989, *Tissue Substitutes in Radiation Dosimetry and Measurement*, ICRU Report 44, International Commission on Radiation Units and Measurements, Bethesda, MD, USA.

Jarvis H.F.T., 1971, The thermal variation of the density of beef and the determination of its coefficient of cubical expansion, J Fd Technol, 6, 383–391.

Johns P.C. and Yaffe M.J., 1987, X-ray characterisation of normal and neoplastic breast tissues, Phys Med Biol, 32, 675–695.

Kaye G.W.C. and Laby T.H., 1973, *Tables of Physical and Chemical Constants*, 14th edition, Longman, London.

Kempson G.E., 1979, Mechanical properties of articular cartilage. Chapter 5 in Freeman, 1979.

Kenedi R.M., Gibson T., Daly C.H. and Abrahams M., 1966, Biomechanical characteristics of human skin and costal cartilage, Fed Proc, 25, 1084–1087.

Knets I. and Malmeisters A., 1977, Deformability and strength of human compact bone tissue. In *Mechanics of Biological Solids*, Bulgarian Acad Sci, Sophia, pp.133–141; cited in Cowin *et al*, 1987.

Kwan M.K., Lai W.M. and Mow V.C., 1990, A finite deformation theory for cartilage and other soft hydrated connective tissues–I. Equilibrium results, J Biomech, 23, 145–155.

Krouskop T.A., Dougherty D.R. and Vinson F.S., 1987, A pulsed Doppler ultrasonic system for making noninvasive measurements of the mechanical properties of soft tissue, J Rehab Res Develop, 24, 1–8.

Lakes R.S., Katz J.L. and Sernstein S.S., 1979, Viscoelastic properties of wet cortical bone: I. Tortional and biaxial studies, J Biomech, 12, 657–678.

Lang S.B., 1969, Thermal expansion coefficients and the primary and secondary pyroelectric coefficients of animal bone, Nature, 224, 798–799.

Lang S.B., 1970, Ultrasonic method for measuring elastic coefficients of bone and results on fresh and dried bovine bones, IEEE Trans Bio-med Eng, BME-17, 101–105.

Lanir Y., 1987, Skin mechanics. Chapter 11 in *Handbook of Bioengineering*, R.Skalak and S.Chien (eds), McGraw-Hill, New York.

Manly R.S., Hodge H.C. and Ange L.E., 1939, Density and refractive index studies of dental hard tissues II. Density distribution curves, J Dental Res, 18, 203–211.

McDonald D.A., 1974, *Blood Flow in Arteries*, 2nd edition, Arnold, London.

McElhaney J.H., 1966, Dynamic response of bone and muscle tissue, J Appl Physiol, 21, 1231–1236.

McSkimin H.S., 1961, Notes and references for the measurement of elastic moduli by means of ultrasonic waves, J Acoust Soc Am, 33, 606–615.

Méndez J. and Keys A., 1960, Density and composition of mammalian muscle, Metabolism, 9, 184–188.

Méndez J., Keys A., Anderson J.T. and Grande F., 1960, Density of fat and bone mineral of the mammalian body, Metabolism, 9, 472–477.

Mow V.C., Kuei S.C., Lai W.M. and Armstrong C.G., 1980, Biphasic creep and stress relaxation of articular cartilage, J Biomech Eng, 102, 73–84.

Mow V.C., Gibbs M.C., Lai W.M., Zhu W.B. and Athanasiou K.A., 1989, Biphasic indentation of articular cartilage–II. A numerical algorithm and an experimental study, J Biomech, 22, 853–861.

Papir Y.S., Hsu K.H. and Wildnauer R.H., 1975, The mechanical properties of the stratum corneum: I. Influence of relative humidity and ambient temperature on the tensile properties of newborn stratum corneum, Biochim Biophys Acta, 399, 170–180.
Phillips R.A., van Slyke D.D., Hamilton P.B., *et al*, 1950, Measurement of specific gravities of whole blood and plasma by standard copper sulphate solutions, J Biol Chem, 183, 305–330.
Pope M.H. and Outwater J.O., 1974, Mechanical properties of bone as a function of position and orientation, J Biomech, 7, 61–66.
Ranu H.S., 1987, The thermal properties of human cortical bone: an *in vitro* study, Eng in Med, 16, 175–176.
Reilly D.T. and Burstein A.H., 1975, The elastic and ultimate properties of compact bone tissue, J Biomech, 8, 393–405.
Reilly D.T., Burstein A.H. and Frankel V.H., 1974, The elastic modulus for bone, J Biomech, 7, 271–275.
Robbins C.R., 1988, *Chemical and Physical Behavior of Human Hair*, 2nd edition, Springer–Verlag, New York.
Sarvazyan A.P. and Klemin V.A., 1983, Study of ultrasonic topography of the kidney. In *Ultrasonic Interactions an Biology and Medicine*, R.Millner, E.Rosenfeld and U.Cobet (eds), Plenum Press, London, pp.99–104.
Schoenfeld C.M., Lautenschlager E.P. and Meyer P.R., 1974, Mechanical properties of human cancellous bone in the femoral head, Med Biol Eng, 12, 313–317.
Shung K.K and Reid J.M., 1978, Ultrasound velocity in major bovine vessel walls, J Acoust Soc Am, 64, 692–694.
Sokoloff L., 1966, Elasticity of ageing cartilage, Fed Proc, 25, 1089–1095.
Stone J.L., Beaupre G.S. and Hayes W.C., 1983, Multiaxial strength characteristics of trabecular bone, J Biomech, 9, 743–752.
Swanbeck G., Nyren J. and Juhlin L., 1970, Mechanical properties of hairs from patients with different types of hair diseases, J Invest Dermatol, 54, 248–251.
Van Buskirk W.C., Cowin S.C. and Ward R.N., 1981, Ultrasonic measurements of orthotropic elastic constants of bovine femoral bone, Trans ASME J Biomech Eng, 103, 67–72.
Viidik A., 1987, Properties of tendons and ligaments. Chapter 6 in *Handbook of Bioengineering*, R.Skalak and S.Chien (eds), McGraw–Hill, New York.
Vincent J.F.V., 1982, *Structural Biomaterials*, Macmillan, London.
White D.R., Woodard H.Q. and Hammond S.M., 1987, Average soft–tissue and bone models for use in radiation dosimetry, Br J Radiol, 60, 907–913.
Woo S. L–Y., Mow V.C. and Lai W.M., 1987, Biomechanical properties of articular cartilage. Chapter 4 in *Handbook of Bioengineering*, R.Skalak and S.Chien (eds), McGraw–Hill, New York.
Woodard H.Q. and White D.R., 1986, The composition of body tissues, Br J Radiol, 59, 1209–1219.
Yamada H., 1973, *Strength of Biological Materials*, F.G.Evans (ed.), Robert E. Krieger, New York.

Chapter 6

Electrical Properties of Tissue

The greatest part of this chapter describes the electrical and dielectric properties of tissue, covering the frequency range from d.c. to over 10 GHz. In addition, the propagation through tissue of electromagnetic radiation at frequencies between 1 MHz and 10 GHz is also discussed. Lastly, electro-mechanical properties, and associated pyroelectric and optoelectric properties displayed by some collagenous tissues are briefly reviewed.

6.1 Electrical conductivity and relative permittivity

6.1.1 *Terminology and definitions*

The electrical character of tissue over a wide range of frequencies may best be described by using the two properties **relative permittivity,** ϵ', and **conductivity,** σ. These are respectively the charge and current densities set up in response to an applied electric field of unit amplitude. Together, these properties allow a complete statement to be made of the **complex relative permittivity,** ϵ^*:

$$\epsilon^* = \epsilon' - j\epsilon'' \tag{6.1}$$

where the **dielectric loss,** ϵ'', is given by:

$$\epsilon'' = \frac{\sigma}{2\pi f \epsilon_o} \tag{6.2}$$

where f is the frequency and ϵ_o is the **permittivity of free space** (ϵ_o = 8.854 x 10^{-12} F m^{-1}). The relative permittivity or dielectric constant (previously known as specific inductive capacity) of a material is defined as the ratio of the capacitance of a capacitor with the material as dielectric to that of one with the same linear dimensions, but with vacuum as dielectric. Relative permittivity is a preferred term to dielectric constant for a quantity which may vary greatly, for instance, with temperature and frequency. 'Relative permittivity' is commonly abbreviated to 'permittivity' although the latter is strictly the quantity $\epsilon_0\epsilon'$. The conductivity is defined as the conductance of a unit volume of matter and has units of siemens metre^{-1} (S m^{-1}) equivalent to the older unit, mho m^{-1}. The **resistivity**, ρ, is the inverse of σ and has units of ohm metre (Ω m).

For simple materials it is common to describe the dielectric loss ϵ'' as $\epsilon'\tan\delta$ where δ is the **loss angle**, and $\tan\delta$ is the **loss tangent**, so describing the energy loss in the dielectric. When this is done the symbols ϵ' and ϵ'' are commonly written ϵ and ϵ' respectively and care is needed in identifying which convention is used. Herein the former convention (ϵ' and ϵ'') will be used in line with current usage in reports giving tissue dielectric properties.

Whereas a simple theoretical model may be adequate to describe the dielectric behaviour of, say, non-polar organic materials, materials containing polar molecules, and furthermore tissues with material inhomogeneities due to cellular components, cannot be so described. These substances are dispersive, demonstrating a spread of relaxation characteristics over the full frequency spectrum.

For the situation where a single relaxation is relevant the following Debye equation may suffice:

$$\epsilon^* = \epsilon_\infty - \frac{j\sigma_s}{2\pi f\epsilon_0} + \frac{(\epsilon_s - \epsilon_\infty)}{1+(jf/f_R)} \tag{6.3}$$

However, for the circumstance typical in tissue of multiple relaxations a Cole–Cole equation is required (Schwan, 1957; Cole and Cole, 1941).

$$\epsilon^* = \epsilon_\infty - \frac{j\sigma_s}{2\pi f\epsilon_0} + \frac{(\epsilon_s - \epsilon_\infty)}{1+(jf/f_R)^{1-\alpha}} \tag{6.4}$$

In these equations σ_s is the conductivity at low frequencies (interpreted in tissues as that due to ionic conduction through the extracellular space) and ϵ_s and ϵ_∞ are the limiting permittivity values at frequencies respectively far below and far above the centre frequency f_R. The empirical parameter, α, in Equation 6.4 characterises the spread of relaxation times. For a single relaxation, $\alpha = 0$ in which case Equation 6.4 becomes the Debye Equation 6.3. Such a single relaxation equation may be satisfactory to describe, for instance, the behaviour of water at frequencies up to around 30 GHz, but, in general the full Cole–Cole equation is necessary to describe the dielectric dispersion of tissues. However, it should be recognised that even Equation 6.4 is inadequate in that it leads to predictions that are physically impossible. For instance at high frequencies σ becomes proportional to f^α.

The dielectric properties may, equivalently, be expressed in terms of the **complex conductivity,** σ^*:

$$\begin{aligned}\sigma^* &= \sigma' + j\sigma'' \\ &= \sigma' + j2\pi f\epsilon_0\epsilon''\end{aligned} \tag{6.5}$$

in which case it is possible to write an equivalent Cole–Cole expression:

$$\sigma^* = \sigma_\infty + j2\pi f\epsilon_0\epsilon_\infty + \frac{\sigma_s - \sigma_\infty}{1 + (-jf/f_R)^{1-\alpha}} \tag{6.6}$$

where σ_∞ is the tissue conductivity at very high frequencies.

These equations have been used as models in curve fitting routines to test their general applicability against the measured values for the dielectric properties of tissue. Alternatively, for some tissues simple functions have been suggested which summarise the dielectric properties. Foster *et al* (1979) have modelled their results for brain tissue using functions of the form:

$$\begin{aligned}
\epsilon' &= Af^{-\theta_1} + B && : 0.01\ \text{GHz} \leqslant f \leqslant 10\ \text{GHz} \\
\text{and } \sigma &= Cf^{\theta_2} && : 0.01\ \text{GHz} \leqslant f \leqslant 1\ \text{GHz} \\
\text{or } \sigma &= D + Ef^2/[1 + (f/f_R)^2] && : 1\ \text{GHz} < f < 10\ \text{GHz}
\end{aligned} \tag{6.7}$$

The parameters in these empirical functions cannot necessarily be interpreted physically. They may, however, provide the basis for calculations for microwave energy distribution.

The temperature variation can be represented, in the simplest case, by substituting the relaxation time $\tau = 1/f_R = \text{constant}.e^{v/kT}$ in the Debye Equation 6.3 (v is the activation energy for the material). Within some temperature range, which varies with frequency, ϵ' will increase from ϵ_∞ to ϵ_0 with increasing temperature, and the loss tangent pass through a maximum value given by $(\epsilon_0 - \epsilon_\infty)/2(\epsilon_0\epsilon_\infty)^{\frac{1}{2}}$.

6.1.2 *Measurement of conductivity and permittivity*

Early methods used for the measurement of tissue conductivity were direct, using two electrodes, carefully designed to minimise the effects of electrode polarisation, the electrode system connnected to an impedance bridge. Rush *et al* (1963) criticised this design because of the difficulties of correctly deducing the true polarisation and for boundary effects. They used a four-electrode system in which a controlled current flowed between two electrodes, and the resulting potential difference was measured between two other electrodes. Concern has been expressed by many authors regarding the unknown magnitude of the effects of coupling fluid (typically saline) around the electrodes.

At frequencies up to the radio-frequency range, relatively simple bridge techniques are sufficient to provide accuracy within a few per cent for both ϵ' and σ. A major problem in biological impedance measurement is that of electrode polarisation and much earlier work can be criticised for the lack of precaution taken to minimise errors due to electrode polarisation. Data where such errors are evident have been excluded from the tabulated values.

For measurements up to microwave frequencies, coaxial line reflection methods have been used and a review of methods is given by Stuchly and Stuchly (1980b). Frequency domain studies have been carried out, making measurements at separate frequencies a time-consuming method, or else using frequency sweep methods using a spectrum analyser. Alternatively, time domain methods may be used in which the frequency components in a fast rise time pulse are analysed. The latter method gains in speed and thus time-varying tissue properties (such as the effect of ageing in *post mortem* samples) may perhaps be more accurately investigated. Generally, resonant methods are not suitable because of the low Q of tissue samples; they are also are limited if a range of frequency values is required. The choice of measurement method depends upon a variety of factors including the need for temperature control, whether *in-vivo* or *in-vitro* measurements are intended, the sample size and the range of frequencies to be used.

6.1.3 *Historical background*

Earliest reports of measurements of tissue electrical properties date from the late nineteenth century. A brief historical note added by Schwan (Epstein and Foster, 1983) refers to a study by Hermann (1871) who demonstrated the anisotropy of the d.c. resistance of frog skeletal muscle. In a useful survey of values of specific resistance of tissue, Geddes and Baker (1967) include extensive reference to results of Galeotti (1902), and later work by Philipson (1920) and Crile *et al* (1922). These measurements were extended to higher frequencies by Rajewsky (1938), who studied mascerated tissue samples, and Osswald (1937). Some material from these sources has been used in compiling the tables of data, together with substantial further contributions in other extensive surveys (Schwan, 1957; Schwan and Foster, 1980; Foster and Schwan, 1986; Pethig and Kell, 1987; Stuchly and Stuchly, 1980a). Subsequent work has extended knowledge of anisotropic properities, the alterations of properties following death, and of regional and species variations, in addition to adding to the body of available data and to the understanding of underlying mechanisms.

6.1.4 *Values of conductivity of tissue at low frequencies*

The electrical conductivity of tissues at low frequencies is given in Table 6.1, including values from both *in-vivo* and *in-vitro* samples. Where values have been reported at several temperatures for *in-vitro* studies, those closest to

Table 6.1 Electrical conductivity, σ, of tissue: d.c. to 10 kHz

Tissue	Temp °C	Conductivity mS cm^{-1}	Reference
Amniotic fluid,			
human	37	16.6	Bolte 1961
–,sheep	37.5	20.4	Geddes & Baker 1967
Bile,cow,pig	37	16.7	Osswald 1937
–,rabbit	39	13.9–16.4	Crile *et al* 1922
Blood:whole,human	37	6.2–7.4	Burger & van Dongan 1961
–,–	22–40	Eqn 6.10	Mohapatra & Hill 1975
–,dog	37	6.5	Kinnen *et al* 1964
–,cow	37	7.6	Burger & van Dongan 1961
Blood:plasma,human	37	14.3	Burger & van Milaan 1943
–,cow	40	15.4	Burger & van Dongan 1961
Blood:serum,sheep		12.3	McDougall 1964
–,horse	37	15.7	Mazzoleni *et al* 1986
Bone,cow,pig	37	0.22 av	Osswald 1937
–,–,long.	24	0.21–0.22	Chakkalakal *et al* 1980
–,–,radial	24	0.06–0.07	–
–,dry		$\rho \sim 10^{12} \Omega$ cm	Reinish & Nowick 1979
Brain,cow,pig	37	1.7	Osswald 1937
–,rabbit,cerebrum	39	1.38–1.92	Crile *et al* 1922
–,–,cerebellum	39	1.17–1.64	–
–,–,cortex	body	3.1	Ranck 1963
–,–,white	body	1.0	van Harreveld *et al* 1963
–,cat,cortex	37	4.5	Freygang & Landau 1955
–,–,white	37	2.9	–
–,–,internal capsule			
longitudinal	body	11.8	Nicholson 1965
transverse	body	1.25	–
CSF,human	24.5	15.3–15.6	Geddes & Baker 1967
–,cat	24.5	15.1–15.3	–
–,rabbit	39	16.1–19.6	Crile *et al* 1922
Eye,lens cortex,cow	32	4.8	Pauly & Schwan 1964
Fat	body	0.20–0.67	Schwan & Kay 1956
–,cow,pig	37	0.4 av	Osswald 1937
Fetal tissue,human			
membrane	37	4.0±0.5	Oostendorp *et al* 1989
Whartons jelly	37	11,12	–
vernix caseosa	37	0.36	–
Kidney,cow,pig	37	9.0	Osswald 1937
Liver,cow,pig	37	1.20	–
–,dog	body	1.43	Rush *et al* 1963
–,–	body	0.83–1.19	Schwan 1955

cont.

Table 6.1 cont. Electrical conductivity of tissue

Tissue	T,°C	σ,mS cm^{-1}	Reference
Lung,dog	body	0.42–0.51	Rush *et al* 1963
–,–	body	0.62–1.11	Schwan 1955
–,–,deflated	body	0.91–1.33	Witsoe & Kinnen 1967
–,–,inflated	body	0.42–0.57	–
–,rabbit	37	0.51–0.71	Crile *et al* 1922
Lymph,sheep		12.0–12.7	McDougall 1964
Muscle:cardiac,			
dog,perp	body	1.8	Rush *et al* 1963
–,–,parallel	body	4.0	–
–,–	body	1.04	Schwan & Kay 1957
–,–	body	2.44	van Oosterom *et al* 1979
–,calf,perp	25	0.259±.003	Clerc 1976
–,–,parallel	25	2.49±.17	–
–,sheep,parallel			
intra–cell	37	2.13±.02	Weidmann 1970
extra–cell	37	21.3±5	–
Muscle:skeletal,			
human	37	4.1	Burger & van Dongan 1961
–,dog,perp	body	0.43–0.53	Rush *et al* 1963
–,–,parallel	body	4.9–6.7	–
–,rabbit,perp	37	0.55	Burger & van Dongan 1961
–,–,parallel	37	8.0	–
Pancreas,cow	37	1.3 av.	Osswald 1937
Semen,human		9.3–11.4	Spector 1956
Spinal chord,cat			
longitudinal	body	4.7–7.2	Ranck & Be Merit 1965
transverse	body	0.83	–
Spleen,cow,pig	37	1.4 av	Osswald 1937
Tooth,human			
enamel		ρ=2.6–6.9x10^6 Ω cm	Mumford 1967
dentine		ρ=11–52x$10^3$$\Omega$ cm	–
Urine,cow,pig	37	33.3	Osswald 1937

body temperature have been selected. Electrical resistivity, ρ, may be calculated from the conductivity, σ, in mS cm^{-1} using the expression $\rho = 10^3.(1/\sigma)$ ohm cm.

Some important sources of variation of tissue conductivity are discussed in the following sections. In addition, some comments included in the discussion of the dielectric properties of tissue at higher frequencies are relevant (Section 6.1.9).

6.1.5 *Factors affecting conductivity*

6.1.5.1 Frequency

The conductivity, σ, can be considered as being the sum of two components, that due to ionic conductivity through the intracellular space, σ_0, and that due to dispersive processes. It is normally considered that σ_0 is not frequency dependent and that the total conductivity is asymptotic to σ_0 at low frequencies. However, small effects have been documented in aqueous solutions of ionic salts suggesting σ_0 may vary with frequency (Little and Smith, 1955).

At low frequencies the charging time–constant for a cell membrane is small enough to charge and discharge the membrane completely during a single cycle, resulting in a high tissue capacitance, and hence a high dielectric constant (see tables of dielectric properties, Tables 6.4 to 6.16).

6.1.5.2 Temperature

The temperature coefficient for electrical conductivity in tissue at low frequencies has been reported as lying in the range 1% to 3% $°C^{-1}$. It is lowest for spleen and highest for brain tissue. Experimental values of temperature coefficients are given in Table 6.2.

6.1.5.3 Anisotropy

Marked anisotropy of both conductivity and permittivity occurs in muscle up to 10 kHz, the conductivity along the fibre direction being several times greater than that in the transverse direction (Epstein and Foster, 1983;

Table 6.2 Electrical conductivity of soft tissue: temperature coefficient

Tissue	$(\Delta\sigma/\sigma)\Delta T^{-1}$ % $°C^{-1}$	Reference
Amniotic fluid,sheep	2.6	Geddes & Baker 1967
Bile,cow,pig	1.8	Osswald 1937
Blood:whole,human	1.7	Burger & van Dongen 1961
–,cow	2.6	–
Blood:plasma,human	2.3	Burger & van Milaan 1943
–,cow	1.9	Burger & van Dongan 1961
Blood:serum,horse	2.63	Mazzoleni *et al* 1986
Brain,cow,pig	3.2	Osswald 1937
Kidney,cow,pig	1.7	–
Liver,cow,pig	1.5	–
Pancreas,cow,pig	1.4	–
Spleen,cow,pig	1.0	–
Urine,cow,pig	1.8	–

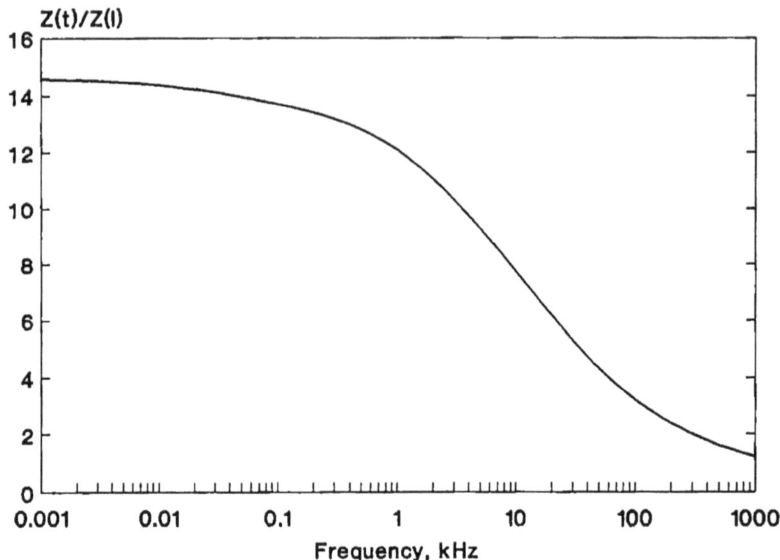

Figure 6.1 The ratio of transverse to longitudinal electrical impedance for skeletal muscle from five animal species: derived from Zheng *et al* (1984).

Zheng *et al*, 1984). Anisotropy of conductivity in skeletal muscle, cardiac muscle, nerve and bone have all been recorded (Table 6.1). At higher frequencies the electrical properties of muscle become essentially isotropic by 1 MHz (Figure 6.1). The measurements of anisotropic properties pose particular problems, and early estimates of the ratio of conductivities in transverse and parallel fibre directions in muscle probably underestimated this ratio.

6.1.5.4 Applied voltage

At high currents, skin impedance reduces. Schwan and Kay (1956) estimated that, for an electrode area of 0.3 cm^2 at 10 Hz the measured skin impedance dropped for applied voltages above about 1 V, or at a current density of about 2 mA cm^{-2}. Yamamoto and Yamamoto (1981) have observed that this threshold for non-linear response may be even lower, lying between 10 and 100 μA cm^{-2} depending upon frequency and true skin impedance. The mechanism of the ionic conduction in the keratin layer may be one cause of the observed non-linearity.

6.1.6 Conductivity of blood

The conductivity of blood was observed to alter both with motion and with haematocrit by Sigman *et al* (1937), and this observation has been confirmed by many subsequent authors. For cow blood at 40% haematocrit, Sigman reported a 1.3% decrease in resistivity when the blood flowed at 15 cm s^{-1} in comparison with the value when stationary.

The low frequency electrical properties of mixtures of materials such as blood, a suspension of cells in plasma, have been evaluated using developments of Maxwell's (1873) expression for the conductivity of a dilute suspension of non-conducting spheres in a conducting medium. Fricke (1924, 1925) extended the expression to include spheriods and these expressions have been used with some success in the analysis of the variation of conductivity of blood with haematocrit, H (Goodwin and Sapirstein, 1957). Extending the Fricke mixture equation they developed the following expression:

$$\sigma_b = \frac{1 - H}{1 + H/F} \sigma_p \tag{6.8}$$

where σ_p is the plasma conductivity and F a form factor whose value depends upon the orientation and shape of the ellipsoidal erythrocytes (Velick and Gavin, 1940). Goodwin and Sapirstein give F = 1.07. Suitable forms for an expression for σ_p are reviewed by Troutman and Newbower (1983) who give:

$$\sigma_p = \frac{100 - 3.46P}{100} \sigma_s \tag{6.9}$$

where σ_s is the saline conductivity and P the concentration of plasma proteins (g/100 ml).

Measurements of resistivity for fresh human blood at 100 kHz over a range of temperatures, and values of percentage haematocrit have been reported by Mohapatra and Hill (1975). Their data fitted the relationship:

$$\rho_b = (6.272H + 75.176) - (0.104H + 1.467)\theta \tag{6.10}$$

where ρ_b is the blood resistivity in Ω cm, and θ is the temperature in °C.

Alternatively, blood resistivity may be represented by the empirical expression

$$\rho = ae^{bH} \tag{6.11}$$

where H = percentage haematocrit and a and b are constants depending on the type of blood. Values of resistivity at body temperature and their

Table 6.3 Resistivity of blood at body temperature
$\rho_b = a.e^{bH}$; H = percentage haematocrit

Species	a	b	Freq, MHz	Reference
Human	62.9	0.0195	1	Rosenthal & Tobias 1948
Human	53.2	0.022	25	Geddes & Sadler 1973
Human	65.6	0.0175	50	Yamakoshi *et al* 1980
Dog	56.6	0.022	100	Kinnen *et al* 1964
Dog	56.8	0.025	24	Geddes & da Costa 1973

variation with haematocrit are given in Table 6.3, based on the empirical expression given in Equation 6.11.

6.1.7 *Conductivity of skin*

Resistance values for skin are not given in Table 6.1. It is well known that skin resistance *in-vivo* is widely variable depending primarily upon the state of hydration of the stratum corneum (Edelburgh, 1977; Campbell *et al*, 1977). Lawler *et al* (1960) have demonstrated that skin resistance decreases, and the skin capacitance increases when the stratum corneum is progressively removed. Dispersion of the absolute value of skin impedance at 10 Hz has been reported by Almasi and Schmitt (1970). Yamamoto and Yamamoto (1978) have demonstrated approximately log-normal distributions of skin impedance, showing a wide range even when the environmental conditions are controlled. In the low frequency limit, resistance values from 10 kΩ cm^2 to 2 MΩ cm^2 and capacitance values lying approximately between 0.02 μF cm^{-2} and 0.2 μF cm^{-2} were found.

6.1.8 *Values of dielectric properties of tissue up to 90 GHz*

The dielectric properties of normal body tissues are presented in Tables 6.4 to 6.16 for frequencies in the range from 100 Hz to 90 GHz. Values for each major organ are included in an individual table as follows: blood and serum in Table 6.4: bone and collagen in Table 6.5: brain in Table 6.6: eye in Table 6.7: fat and bone marrow in Table 6.8: kidney in Table 6.9: liver in Table 6.10: lung in Table 6.11: cardiac muscle in Table 6.12: skeletal muscle in Table 6.13: skin in Table 6.14: spleen in Table 6.15: and other tissues not included elsewhere in Table 6.16. Source references are indicated in each table by a numeric code, and the short-hand list of the codes and references is given immediately following Table 6.16. Although values of σ and ϵ' are separated for convenience it must be recognised that pairs of values constitute a single measurement of the relative permittivity ϵ^* (Equation 6.1). The selection of frequencies allows interpolation to be used

Table 6.4 Blood and serum, relative permittivity and conductivity

Animal	rabbit	pig	rat	cow	human	human	rabbit	human serum	human serum
Temp, °C	37	37	23	27	35	37	37	23	37
Ref	54	19	18	7	5	2	59	50	2
Freq, Hz	Relative permittivity								
100k	2740								
1M	2040	1800							
10M	200	180							
25M		80							
100M		56	86	67*				67.4–81.3	
300M			67	63				60.1–70.7	
1G			62	63*					
3G			62		56	53			70
9.4G			50		47.8	45			57.5
24G					30.2	32			45.5
35G							23.6		
	Conductivity, mS cm^{-1}								
100k	6.8								
1M	7.1	7.7							
10M	11.0	9.7							
25M		10.3							
100M		10.8	9	10.4*				11.4–12.7	
300M			10	11.1				9.3–11.9	
1G			11.5	12.5*					
3G			30		26.5	27.8			41.7
9.4G			155		103	121			126
24G					344	263			380
35G							498		

* Measurements at 200 and 900 MHz.

Table 6.5 Bone* and collagen, relative permittivity and conductivity

Animal Tissue	cow femur radial	cow femur circum- ference	cow femur axial	rat femur cortical	goat femur 72% hu- midity	goat collagen
Temp, °C	21	21	21	37	17	17
Ref	34	34	34	33	55	55
Freq. Hz	**Relative permittivity**					
100				3800		
1k	600	800	1800	1000		
10k	300	300	800	640		
100k	85	100	200	280		
1M	25	30	90	87		
10M				37		
100M				23		
400M					23.8	24.5
1G					22.5	22.5
	Conductivity, mS m^{-1}					
100				12.6		
1k	18.5	27	59	12.9		
10k	19	28	60	13.3		
100k	19	32	67	14.4		
1M	33	43	100	17.3		
10M				23.7		
100M				57.4		
400M					280	80
1G					170	120

* All values are for 'wet' bone.

Table 6.6 Brain: cortex, cerebrum, cerebellum and brain stem, permittivity and conductivity

Animal	cow,pig	cat,rat	rat	mouse,rabbit		mouse	mouse	mouse	human
Tissue		cortex	cortex	cortex		cere-bellum	cerebrum	brain stem	cortex
				adult	birth+8hr				
	in-vitro	*in-vivo*	*in-vivo*	*in-vitro*	*in-vitro*	*in-vitro*	*in-vitro*	*in-vitro*	*in-vitro*
Temp, °C	37	35	32	37	37	37	37	37	37
Ref	1	27,53	18	35,39	35,39	31	31	31	14
Freq, Hz	Relative permittivity								
10M				240	455				
25M				160	310				
50M	110–114								
100M	81–83	73	84	80	110	84.8	74.4	71.8	
300M		60	64	50	65	55.8*	52.3*	50.6*	
1G		55	55	40	60	44.4	41.6	39.9	
3G		51.5	53	37	55	41.5*	39.6*	36.9*	
5G		48	53	36	52	36.0*	35.7*	32.5*	
10G		44	42	35	50				
18G				25	37				
	Conductivity, mS cm^{-1}								
10M						3.2	3.9		
25M	4.55			4.1	5.8				
50M	4.76–5.26			4.7	7.2				
100M	5.13–5.56	9.0	8	6.0	9.2	5.1	4.6	4.5	
300M		10.5	9	7.2	10.1	6.9*	6.3*	6.3*	
1G		12.5	12	10.2	13.0	9.4	9.0	8.9	
3G		22.5	23	18	25	16.8*	17.3*	14.7*	30
5G		40	45	32	45	28.9*	29.6*	26.8*	
10G			120	80	90				
18G				170	250				

* Measured at 268 MHz, 2680 MHz and 5180 MHz.

Table 6.6 cont. Brain: grey matter, pia and dura mater, permittivity and conductivity

Animal	dog	cow	cat	dog	dog	dog	dog	rabbit
Tissue	grey	grey	grey	grey	grey	pia*	grey	grey
	in-vitro	*in-vitro*	*in-vivo*	*in-vitro*	*in-vivo*	*in-vivo*	*in-vitro*	*in-vitro*
Temp, °C	37	24.5	33	20	36	36	37	37
Ref	28	47	52	60	42	42	17	37,59
Freq, Hz	**Relative permittivity**							
100k	3800	2800						
1M	1250	920						
10M	352	270	237–289					
100M	90	83	65–80	77.5				
400M			50–54	63.6†		61		
1G			47–51	58.8		58	45	
2.45G					43	52.7	43	47
10G				39.0†			37	40
18G								32
35G								21.3
	Conductivity, mS cm^{-1}							
100k	1.7	1.7						
1M	2.1	1.85						
10M	3.5	3.0	4.5–6.3					
100M	6.9	5.7	5.2–8.5	6.2				
400M			8.9–11.7	8.5†		10.0		
1G				10.9		13.0	11.0	
2.45G					17.4	19.3	21.0	25
10G				162†			100	100
18G								205
35G								389

* Values for dura mater at 2.45 GHz are also given; ϵ=45.3–53.9, σ=18.4–23.5 mS cm^{-1}.

† Measurements at 500 MHz and 11.8 GHz.

Table 6.6 cont. **Brain: white matter, relative permittivity and conductivity**

Animal	dog *in–vitro*	cow *in–vitro*	cat *in–vivo*	dog *in–vitro*	dog *in–vivo*	dog *in–vitro*	rabbit *in–vitro*
Temp, °C	37	24.5	33	20	37	37	37
Ref	28	47	52	60	42	17	37,59
Freq, Hz	**Relative permittivity**						
100k	1960,3400	1400					
1M	543,827	400					
10M	163,209	115	190–191				
22M	109,135		132–134				
100M	57,66	58	58–64	59.9			
1G			37.5–38.5	41.2		37	
2.45G					32.3	34	36
10G				30.1*		28	31
18G							25
35G							17.8
	Conductivity, mS cm^{-1}						
100k	1.2,1.5	0.8					
1M	1.4,1.9	1.2					
10M	2.1,2.8	1.6	2.9–3.1				
22M	2.5,3.3		3.3–3.5				
100M	3.6,4.8	2.8	4.8–5.1	3.5			
1G			8.1–8.2	6.5		7.5	
2.45G					12.1	12.5	20
10G				67*		67	70
18G							135
35G							280

* Measured at 9.0 GHz.

Table 6.7 Eye, relative permittivity and conductivity

Animal	cow	cow	rabbit	rabbit	rabbit	rabbit	rabbit		
Tissue	lens	lens	retina	iris	cornea	lens	lens	lens	vitreous
in-vitro	cortex	nucleus				cortex	nucleus		humor
Temp, °C	32	32	37	37	37	37	37		
Ref	11	11	24,30	24,30	24,30	24,30	24,30	22	22
Freq, Hz	**Relative permittivity**								
500k	720	110							
1M	580	80							
10M	179	60	600	280	150	205	58		
100M	66		100	83	80	67	46	46	70
1G			65	60	60	50	32	32	70
3G			65	60	55	45	30	30	70
8.5G			60	50	40	40	25		
10G			50	40	35	35	20	28	62
	Conductivity, mS cm^{-1}								
500k	5.06	1.9							
1M	5.15	1.95							
10M	6.09	2.0	9.0	8.5	15	5.5	0.6		
100M	7.25	2.8	13	11	16	6.5	1.8	4.0	16.7
1G			15	12	19	10	6	5.3	20
3G			26	25	25	16	8	15	30
8.5G			80	90	60	35	30		
10G			150					100	178

cont.

Table 6.7 cont. Eye, relative permittivity and conductivity

Animal	human	human	rabbit	rabbit	rabbit	rabbit	rabbit	rabbit
Tissue	lens	lens	retina	cortex†	retina	cortex†	lens	lens
in–vitro	nucleus	cortex	....supercooled....		.......frozen......		cortex	nucleus
Temp, °C	37	37	−9	−9	−9	−9	37	23
Ref	30	30	41	41	41	41	59	59
Freq, Hz	Relative permittivity							
10M			170	100				
100M	48	62	72	61	12	11		
1G	36	47	63	52	7	8		
8.5G	29	40						
10G			16	15	3	4		
35G							17.9	8.9
	Conductivity, mS cm^{-1}							
10M			3.8	1.8	0.16*	0.05*		
100M	3.5	7	4.6	2.3				
1G	8	11	9.7	8.1				
8.5G	35	43						
10G			110	58				
35G							386	173

* Calculated ionic conductivity.

† Values for lens nucleus are also given.

Table 6.8a Low-water-content Tissues: fat, bone-marrow, relative permittivity

Tissue	fat	fatty tissue	fat tissue	fat	fat	bone marrow	bone marrow	fat	fat
Animal	dog *in-vivo*	pig *in-vivo*	cat *in-vivo*	dog *in-vivo*	various *in-vitro*	calf *in-vitro*	various *in-vitro*	rabbit *in-vitro*	cow *in-vitro*
Temp, °C	37	37	37		37	25	37	37	37
Ref	10	9	18	62	21(1,3,7)	36*	1,3	59	61
Freq, Hz									
100	1.5×10^5								
1k	5.0×10^4								
10k	2.0×10^4								
1M		300							
10M		40	47	46.9†					
30M			17	23.4†					
50M					11–13		6.8–7.7		
100M		14	14						
200M					4.5–7.5				
300M			14			20–47			
400M				16.3†	4.0–7.0				
1G			13	14.6†	5.3–7.5		4.3–7.3		
2.45G				11.2					
8.5G					3.5–4.5		4.4–5.4		
40G								3.6⊗	11
90G									10

* Values for various adipose tissues at 300 MHz are given. ϵ lies in the range 2 to 18.
† Measurements at 13.6, 27.1, 433 and 918 MHz. ⊗ Measured at 35 GHz.

Table 6.8b Low–water–content tissues: fat, bone–marrow, conductivity, mS cm^{-1}

Tissue	fat	fatty tissue	fat tissue	fat	fat	bone marrow	bone marrow	fat	fat
Animal	dog *in–vivo*	pig *in–vivo*	cat *in–vivo*	dog *in–vivo*	various *in–vitro*	calf *in–vitro*	various *in–vitro*	rabbit *in–vitro*	cow *in–vitro*
Temp, °C	37	37	37		37	25	37	37	37
Ref	10	9	18	62	21(1,3,7)	36*	1,3	59	61
Freq, Hz									
1k	0.2–0.7								
1M		1.7							
10M		1.9	2.1	2.0†					
30M			2.1	2.0†					
50M					0.4–0.59		0.2–0.36		
100M		2.1	2.1						
200M					0.29–0.95				
300M			2.7			2.1–6.7			
400M				2.6†	0.36–1.11				
1G			3.6	3.3†	0.83–1.49		0.43–1.0		
2.45G				6.3					
8.5G					2.7–4.17		1.67–4.76		
40G								27⊗	80
90G									100

* Values for various adipose tissues at 300 MHz are given. σ lies in the range 0.06 to 2.0 mS cm^{-1}.
† Measurements given at 13.7, 27.1, 433 and 918 MHz. ⊗ Measured at 35 GHz.

Table 6.9a **Kidney, relative permittivity**

Animal	cow	cat	cat	human	dog	pig	human	dog	rat
Tissue state	*in-vitro* 1.5hr pm	*in-vitro*	*in-vivo*	*in-vitro* 1-2hr pm	*in-vitro*	*in-vivo*	*in-vitro*	*in-vivo*	*in-vivo*
Temp, °C	24	35	35-36.5	36.8	37	37	27	37	32
Ref	38	45	26,52	51	28	19	7	18	53
Freq. Hz									
10k	$2.8x10^4$	$3.0x10^4$	$4.7x10^4$	$8.14x10^4$					
100k	$8.0x10^3$	$9.6x10^3$	$1.4x10^4$	$1.12x10^4$	$1.09,1.25 x10^4$				
1M	1300	2000	2000	2450	2400,2700	1300			
10M	280	350	190-204	469	431,499	190			
22M			118-119		268,283	90			
50M			70-73		128,136				
100M	80	60	56-62	83.5	89,95	35		78.5	72.5
300M			47-48				57-60		56.6
500M			44-45					57	54.1
700M			43-44				53-56		
1G			43					53	52
3G								49	50.6
10G								34	39.7

Table 6.9b Kidney, conductivity, mS cm^{-1}

Animal	cow	cat	cat	human	dog	pig	human	dog	rat
Tissue state	*in-vitro* 1.5hr pm	*in-vitro*	*in-vivo*	*in-vitro* 1-2hr pm	*in-vitro*	*in-vivo*	*in-vitro*	*in-vivo*	*in-vivo*
Temp, °C	24	35	35-36.5	36.8	37	37	27	37	32
Ref	38	45	26,52	51	28	19	7	18	53
Freq. Hz									
10k	0.6	1.3	1.5	2.44					
100k	1.0	2.1	2.6	3.28	2.4,2.5				
1M	2.0	3.4	4.1	4.93	3.7,3.9	3.6			
10M	3.6	5.7	5.0-5.7	7.71	6.4,6.8	4.9			
22M			5.3-6.2		7.7,8.3	5.5			
30M			5.5-6.5						
50M			5.6-6.6		8.8,9.6				
100M	6.7	8.2	6.6-7.2	11.68	9.4,10.5	7.0		7.8	8.0
300M			7.2-7.4				10.0		8.4
500M			7.4-7.7					8.3	9.1
700M			8.4-9.0				11.2		
1G			9.5-9.7					12.0	10.5
3G								30	22.5
10G								100	97.4

Table 6.10 Liver, relative permittivity and conductivity 100 Hz – 100 MHz

Animal	dog	rabbit	cow	cat	cat	human	dog	pig	cow,pig
Tissue state	*in–vivo*	*in–vitro*	*in–vitro* 1.5 hr pm	*in–vivo*	*in–vitro*	*in–vitro* 1–2 hr pm	*in–vitro*	*in–vivo*	*in–vitro*
Temp, °C	37	37	24	35–36.5	34	36.8	37	37	37
Ref	10	28,44	38	26,27,52	45	51	28	19	1
Freq. Hz	**Relative permittivity**								
100	$8.5x10^5$								
1k	$1.45x10^5$	$1.0x10^5$							
10k	$5.5x10^4$	$5.0x10^4$	$2.0x10^4$	$4.0x10^4$	$2.4x10^4$	$2.101x10^4$			
100k		$13.7x10^3$	$6x10^3$	$10.6x10^3$	$10x10^3$	$6.94x10^3$	$9.76x10^3$		
1M		1970	1000	1800	2300	1940	1970	1300	
10M		300	190	251–265	350	409	338	150	
22M		175		148–162			190	80	
30M				111					136,138
50M		110		85–88	80		110		88–93
100M		79	60	65–68		73.9	77	38	76–79
	Conductivity, mS cm^{-1}								
100	1.25								
1k	1.3	0.5							
10k	1.46	0.65	1.2	1.3	0.7	1.23			
100k		1.6	1.4	2.2	1.1	1.41	1.5		
1M		3.0	2.3	3.7	2.6	2.27	2.7	3.0	
10M		4.6	3.2	4.2–4.6	4.8	4.06	4.7	4.3	
22M		5.5		4.7–5.0			5.7	4.8	
30M				4.9–5.3					4.7–5.4
50M		6.3		5.3–5.6			6.7		5.1–5.8
100M		7.0	4.8	6.0–7.1	7.3	7.02	7.2	5.8	5.6–6.5

Table 6.10 cont. Liver, relative permittivity and conductivity above 100 MHz

Animal	cat	human	human	rat	dog
	in-vivo	*in-vitro*	*in-vitro*	*in-vivo*	*in-vitro*
Temp, °C	35–36.5	27	37	31	20
Ref	26,27,52	7	3	53	60
Freq, Hz	Relative Permittivity				
300M	52–54	48–56		52.2	
500M	49–50			49.6	57.5
700M	47–49	46–54		48.9	56.3*
1G	47–49		46–47	47.9	54.6
3G	45		42–43	46.0	49.1
8.5G	40		37–38	36.2	38.3*
12G				33.3	32.1*
	Conductivity, mS cm^{-1}				
300M	6.9–7.7	6.5–8.3		7.2	
500M	7.3–7.9			8.1	6.4
700M	8.5–9.2	7.7–10.0		8.7	8.25*
1G	9.5–10.3		9.4–10.2	9.6	8.6
3G	21.5		20.0–20.4	21.2	23.5
8.5G	60		58.8–66.7	71.2	125*
12G				115.8	161*

* Measurements at 750 MHz, 9 GHz and 11.8 GHz.

Table 6.11 Lung, relative permittivity and conductivity

Animal	dog	cat	cat	rat	pig	pig	rat	human
Tissue		inflated	deflated		inflated	deflated	inflated	deflated
state	*in-vivo*	*in-vivo*	*in-vivo*	*in-vitro*	*in-vivo*	*in-vivo*	*in-vivo*	*in-vitro*
Temp, °C	37	34	34	37	37	37		27
Ref	10	48	48	16	19	19	62	7
Freq, Hz	**Relative permittivity**							
1k	8.5×10^4							
10k	2.5×10^4	2.5×10^4	5.0×10^4					
100k		0.45×10^4	1.0×10^4					
1M		800	1500	800	200	500		
10M		105	250	160	50	120	3.2*	
25M					30	60	2.1*	
100M		30	50	42	12	22		
300M							1.5*	36
2.45G							1.45	
	Conductivity, mS cm^{-1}							
1k	0.96							
10k	1.05	0.6	1.4					
100k		1.0	2.2					
1M		1.7	3.2	1.15	0.7	1.5		
10M		2.4	4.8	1.94	1.1	2.5	0.07*	
25M					1.3	3.1	0.08*	
100M		3.2	6.4	3.1	1.5	3.6		
300M							0.11*	5.9
2.45G							0.17	

* Measured at 13.6 MHz, 27.1 MHz and 433 MHz.

Table 6.12 Muscle: cardiac, relative permittivity and conductivity

Animal	dog	human	pig	dog	human	rat
Tissue state	*in-vivo*	*in-vitro* pm+1-2 hr	*in-vivo*	*in-vitro*	*in-vitro*	*in-vitro*
Temp, °C	37	36.8	37	20	27	
Ref	10	51	19	60	7	40
Freq, Hz		**Relative permittivity**				
100	$8.2x10^5$					
1k	$3.2x10^5$					
10k	$10.0x10^4$	$4.08x10^4$				
100k		5500				
1M		1245	1500			
2.5M			600			
10M		312	90			
100M		71.9	25	61.2		
300M					55–62	
1G				53.0		
9.4G				39.6*		39.6
		Conductivity, mS cm^{-1}				
100	1.08					
1k	1.18					
10k	1.67	4.61				
100k		4.81				
1M		5.28	4.9			
3M			5.6			
10M		6.86	6.6			
100M		8.90	8.3	4.6		
300M					8.3–9.5	
1G				11.8		
9.4G				13.1*		9.76

* Measured at 9.0 GHz.

Table 6.13 Muscle: skeletal, relative permittivity and conductivity 1 kHz – 100 MHz

Animal	dog	dog	dog	cat	cat,rat	dog	rabbit	pig
Tissue state	*in–vitro* parallel	*in–vitro* perp.	*in–vivo*	*in–vitro*	*in–vivo*	*in–vitro*	*in–vitro*	*in–vivo*
Temp, °C	37	37	37	34	31–36	37	37	37
Ref	29	29	10	45	26,27,52	28	28	19
Freq, Hz	Relative permittivity							
1k	2.2×10^5	1.2×10^5	1.3×10^5					
10k	8.0×10^4	7.0×10^4	5.5×10^4	8.8×10^4	5.0×10^4			
100k	1.4×10^4	2.7×10^4		1.58×10^4	2.6×10^4	$1.44–1.58 \times 10^4$	$2.48–2.73 \times 10^4$	
1M	800	3600		1900	1900	1900–2190	2400–2530	1900
10M				130	170–190	162–181	187–204	90
25M				85	104	96–110	125–135	50
50M					83	73–81	100–102	
100M				60	67–72	64–70	89–90	39
	Conductivity, mS cm^{-1}							
100	5.24	0.76	1.14					
1k			1.20					
10k	6.0	1.1	1.3	2.5	2.5			
100k	7.0	3.4		4.0	4.5	3.8–4.4	5.6,5.9	
1M	8.1	6.1		6.3	7.4	5.8–6.3	8.3,8.5	5.8
10M				7.5	8.2–8.6	6.9–7.5	9.2,9.6	6.7
25M					9.0	7.2–7.8	9.4,9.9	7.0
50M					9.0–9.2	7.5–8.0	9.6,10.2	
100M				8.3	9.0–9.9	7.5–8.2	10.5	7.7

Table 6.13 cont. **Muscle: skeletal, relative permittivity and conductivity >100 MHz**

Animal	human *in-vitro*	cat,rat *in-vivo*	rat,dog *in-vivo*	dog *in-vitro*	dog *in-vitro*	rabbit *in-vitro*	rat *in-vivo*
Temp. °C	27	31–36	31,34	20	37	37	
Ref	7	26,27,52	18	60	21	59	61
Freq. Hz	Relative permittivity						
300M	55–62	60					
500M		60–64	63	61.5			
700M	55–66	57–59		58.3*			
1G		56–59	61	52.0	54		
3G		53–55	56	52.9	50		
8.5G		42–46		40.8*	42		
10G		44	41		40		
17G					34		
40G						19.1†	av ~25
90G							av ~20
	Conductivity, mS cm^{-1}						
300M	9.1–9.5	9.0–9.9					
500M		10.8–11.6	11.1	10.1			
700M	10.8–11.5	12.3–12.7		12.6*			
1G		13.0–14.5	13.1	11.3	14.9		
3G		27–29	28.4	25	32		
8.5G		63–84	83.3	133*	100		
10G			128		116		
17G					222		
40G						399†	~750
90G							~600

* Measurements at 750 MHz and 9 GHz. † Measurement at 35 GHz.

Table 6.14 Skin, relative permittivity and conductivity

Animal	human	human	human	dog	rat	human	human	human	rabbit
Tissue	Stratum corneum	*in-vivo* various	*in-vitro*	*in-vitro*	*in-vivo, in-vitro*	*in-vivo*	various	*in-vivo*	*in-vitro*
Temp, °C			37	20	32		37	37	32.5
Ref	15	49	2,4,9	60	40,46	62	6	25	59
Freq, Hz		**Relative permittivity**							
10k	1100								
100k	1005								
1M	450								
100M		48.7–65.7	65	38.3					
500M		32.7–47.1	46.5	45.4	11.8	40			
1G		30.2–43.5	44	41.1	8.8	37†			
3G			43.5	39.2	11.6	34†	40–42		
9.4G			35.5	28.7⊕	13.7			24*	
18G			23#					13	
35G									17.3
		Conductivity, mS cm^{-1}							
10k	0.001								
100k	0.01								
1M	0.2								
100M		3.3–5.3	7.2–8.3	2.9					
500M		4.3–7.3	8.0	5.5		5.0			
1G		4.8–10.0	11.0	6.5		7.0†			
3G			30.6	18.4	2.2	10.0†	20.5–21.9		
9.4G			83.9	84.6⊕	2.4			53*	
18G			171#					125	
35G									374

* Measured at 8.5 GHz. † Measured at 918 and 2450 MHz. ⊕ Measured at 9.0 GHz. # Measured at 24 GHz.

Table 6.15 Spleen, relative permittivity and conductivity

Animal	cow	cow	cat,rat	cat	human	pig	dog	cow,pig	dog
Tissue state	*in-vitro* 2.2hr pm	*in-vitro* 31hr pm	*in-vivo*	*in-vitro*	*in-vitro* 1-2hr pm	*in-vivo*	*in-vitro*	*in-vitro*	*in-vitro*
Temp, °C	24	24	34-36.5	32.5	36.8	37	37	37	37
Ref	38	38	26,27,52	45	51	19	28	1	21
Freq. Hz	Relative permittivity								
10k	$2.2x10^4$		$2.0x10^4$	$1.8x10^4$	$5.087x10^4$				
100k	5800	3500	4500	4500	9200		3260		
1M	1300	1500	2000	2000	1940	1800	1450		
10M	300	400	352-410	420	451	310	321		
50M			102-114				110	135-140	
100M		90	71-76	65	76.3	43		100-101	
300M			54-56						
1G			50-55						52
3G			50-52						47
10G			40.7						38
17G									34
	Conductivity, mS cm^{-1}								
10k	0.95		1.5	1.2	1.42				
100k	1.2	2.1	2.5	1.6	1.53		6.2		
1M	1.7	3.6	4.1	2.4	2.35	2.0	6.3		
10M	3.4	4.8	5.0-5.8	5.4	4.59	4.8	8.5		
50M			6.7-7.0				9.9	7.5	
100M		8.1	7.3-7.6	7.9	10.53	7.5	10.5	10.0	
300M			7.9-8.3						
1G			10.9-12.1						12
3G			25-26						28
10G			101						110
17G									208

Table 6.16 Various tissues, relative permittivity and conductivity

Tissue	artery	pancreas	large bowel	gut muscle,	stomach, intestine	breast milk	breast	breast through skin
Animal	cow *in–vitro*	dog *in–vitro*	pig *in–vivo*	cat *in–vivo*	rat *in–vitro*	human	human *in–vitro*	human *in–vivo*
Temp, °C	37	37	37	35		40	37	
Ref	23	28	19	52	40	43	58	62
Freq, Hz	**Relative permittivity**							
10k							800	
100k		1.0×10^4					100	
300k		5800						
1M		2300	500			176	25	
10M		320	90	459–462		80	12	
100M		85	25	77–78		73.4	6	
300M				60				43*
1G				55–56				38*
3G	43							32*
9.4G					62			
	Conductivity, mS cm^{-1}							
10k							2.8	
100k		3.0					2.8	
300k		3.5						
1M		4.3				2.2	3.0	
10M		6.0		7.0–7.2		6.4	3.2	
100M		8.5		9.4–9.7		8.8	4.0	
300M				10.1–10.2				4.0*
1G				13.2–13.3				6.0*
3G	13							10.0*
9.4G					14.5			

* Measured at 500 MHz, 918 MHz and 2450 MHz.

Reference codes used in Tables 6.4 to 6.16, 6.19 and 6.21.

1	Osswald, 1937	32	Rogers *et al*, 1983
2	England, 1950	33	Kosterich *et al*, 1983
3	Herrick *et al*, 1950	34	Reddy & Saha, 1983
4	Cook, 1951	35	Thurai *et al*, 1984
5	Cook, 1952	36	Smith & Foster, 1985
6	Roberts & Cook, 1952	37	Steel & Sheppard, 1985
7	Schwan & Li, 1953	38	Surowiec *et al*, 1985
8	Schwan & Piersol, 1954	39	Thurai *et al*, 1985
9	Schwan, 1957	40	Karolkar *et al*, 1985
10	Schwan & Kay, 1957	41	Gabriel & Grant, 1985
11	Pauly & Schwan, 1964	42	Burdette *et al*, 1986
12	Kinnen *et al*, 1964	43	Laogun, 1986
13	Geddes & Sadler, 1973	44	Smith *et al*, 1986
14	Lin, 1975	45	Surowiec *et al*, 1986a
15	Yamamoto & Yamamoto, 1976	46	Zywietz & Knochel, 1986
16	Bottomley & Andrew, 1978	47	Surowiec *et al*, 1986b
17	Foster *et al*, 1979	48	Surowiec *et al*, 1987a
18	Burdette *et al*, 1980	49	Grant *et al*, 1988
19	Hahn *et al*, 1980	50	Zore *et al*, 1967
20	Schepps & Foster, 1980	51	Surowiec *et al*, 1987b
21	Schwan & Foster, 1980	52	Stuchley *et al*, 1981
22	Stuchly & Stuchly, 1980a	53	Kraszewski *et al*, 1982
23	Brady *et al*, 1981	54	Fricke & Curtis, 1935
24	Dawkins *et al*, 1981	55	Ray & Behari, 1986
25	Hey-Shipton *et al*, 1982	56	Bengtsson *et al*, 1963
26	Stuchly *et al*, 1982a	57	Mudgett *et al*, 1979
27	Stuchly *et al*, 1982b	58	Surowiec *et al*, 1988
28	Stoy *et al*, 1982	59	Steel & Sheppard, 1988
29	Epstein & Foster, 1983	60	Deming Xu *et al*, 1987
30	Gabriel *et al*, 1983	61	Edrich & Hardee, 1976
31	Nightingale *et al*, 1983	62	Burdette, 1982

to give good estimates of dielectric values at intermediate frequencies. Commonly used therapeutic frequencies are 27.12, 433, 915 and 2450 MHz. Other frequencies used for medical, industrial and scientific purposes include 13.56 MHz, 40.68 MHz and 5800 MHz. Some authors, for instance Pethig (1984) and Pethig and Kell (1987) have reported estimates of dielectric properties of some tissues at these frequencies. Wherever possible, details of tissues state are included. If measurements were reported at more than one temperature, those made closest to body temperature have been selected for conciseness. The data included have been drawn from published tables, calculations from empirically-fitted functions or from published graphs as appropriate. The values should be used with some care, and the following discussion is included to identify the sources of variability in the measurements.

6.1.9 *Factors affecting dielectric properties*

6.1.9.1 Frequency

For biological materials the variation of ϵ' and σ with frequency typically shows three major dispersions (α,β and γ) and several of minor importance (Schwan, 1957; Grant *et al*, 1978). The major dispersions for muscle tissue are centred near 0.1 kHz (α), 100 kHz (β) and 20 GHz (γ) and are characteristic for most tissues with high water content. The origin of the α dispersion is still not completely understood, but cell membrane capacitance and ionic movement around the outer surface of the cell may be a predominant factor. The β dispersion results from Maxwell-Wagner type relaxations resulting from the capacitative charging of cell membranes. Dielectric relaxation of free water gives rise to the γ dispersion. Bound water exhibits a relaxation, designated δ, at around 100 MHz and a further relaxation corresponding to the rotation of macromolecules, β_1, is seen in the dielectric dispersion curves from aqueous solutions of biological molecules. However, these latter peaks are substantially masked in tissue owing to the much larger α and β dispersions, and cannot be resolved in the values included here, which are primarily for whole tissue.

The conductivity increases and permittivity falls monotonically with frequency for all tissues as shown in Figures 6.2 and 6.3. α-dispersion is seen in the data from Schwan (1957) for liver, muscle and liver measured *in-vivo*. As remarked earlier, the increase of conductivity with frequency up to about 1 kHz is, in general, small. Low-frequency permittivity is commonly in excess of 10^6 relative units. At these frequencies the effective conductivity of the cells may be considered to be negligible, and the conduction is entirely through the extra-cellular space.

A broad β-dispersion is seen for all tissues, with relaxation frequencies around 100 kHz. At this frequency the conductive pathway includes the cells but excludes the proteins and associated water of hydration. A further dispersion occurs at the upper frequency limit of the data presented, where at around 10 GHz σ is increasing very rapidly, and ϵ' decreasing. At these frequencies the dielectric properties of tissues are dominated by those of

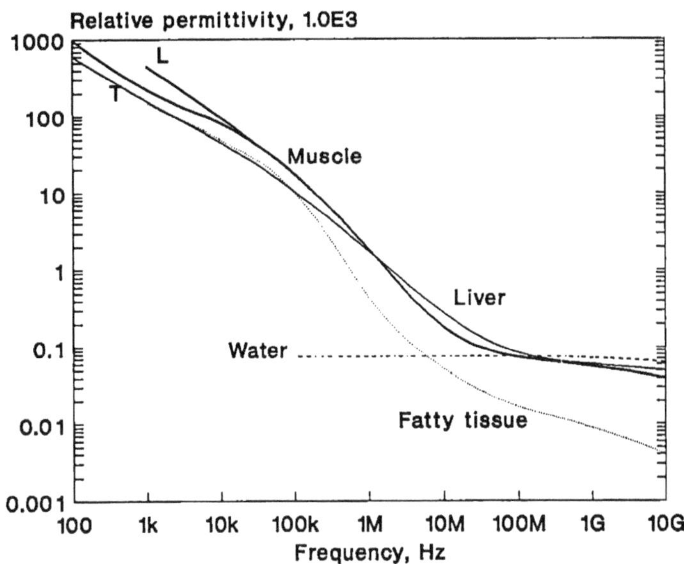

Figure 6.2 Dispersion of relative permittivity of muscle, liver and fatty tissue over the frequency range 100 Hz to 10 GHz.

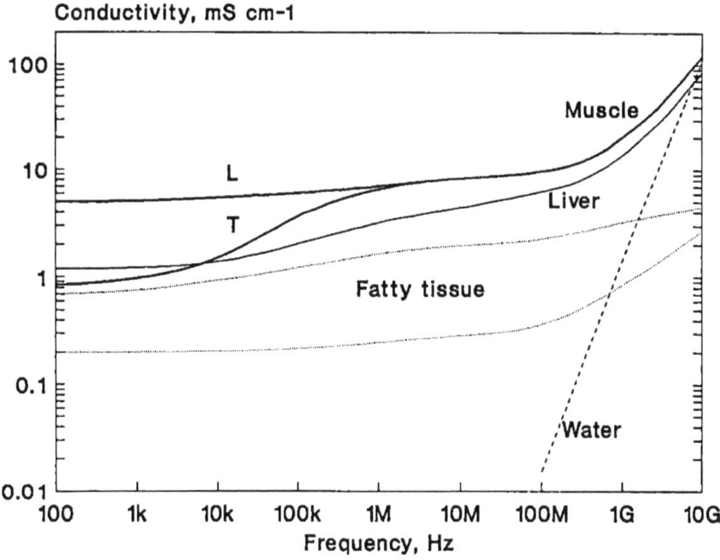

Figure 6.3 Dispersion of conductivity for muscle, liver, fatty tissue over the frequency range 100 Hz to 10 GHz.

water with a relaxation frequency of about 25 MHz. Prediction of tissue properties should be possible at these frequencies using the Debye equations for pure water with proper allowance being made for the fraction of tissue which is occupied by low permittivity protein (Schwan and Foster, 1980).

6.1.9.2 Temperature

Schwan and Foster (1980) discuss the temperature coefficients expected both for well separated relaxations, and for a broader distribution of relaxation times. The permittivity temperature coefficient may be expected to vary through the dispersion range from about −0.3% °C^{-1}, through a maximum of +2% °C^{-1} and back to −0.3% °C^{-1}. For conductivity, for a single relaxation, temperature coefficients of about 2% °C^{-1} at frequencies well above f_R fall to about 1% °C^{-1} in the dispersion range. The difference is somewhat smaller for multiple relaxations.

Temperature coefficients for ϵ' and σ in the range 20°C to 40°C can be calculated from some experimental reports when observations were recorded at more than one temperature. Tables 6.17 and 6.18 give values reported by Schwan (1957) derived from earlier work, together with other more recent data. For most tissues at most frequencies both ϵ' and σ increase with temperature, the coefficient being typically about +2% °C^{-1} for σ and 1.5% °C^{-1} for ϵ' up to about 50 MHz. At frequencies above 400 MHz $\Delta\epsilon'/\epsilon'$ becomes negative. $\Delta\sigma/\sigma$ also becomes negative at frequencies in excess of 1 GHz. Tissues with high fat content can show high positive conductivity temperature coefficients.

Table 6.17 Relative permittivity temperature coefficient, temperature range 20–40°C (% °C^{-1})

Freq,MHz	0.1	1	10	50	100	200	900	3000	10000
Muscle	1.4	2.2	0.8	0.3	0.3		−0.2	−0.1*	
Liver	1.2	1.5	1.4	0.3	0.3	0.2	−0.2		
Spleen	0.2	0.6	1.2	1.0					
Kidney	0.8	1.4	1.7	0.5	0.3	0.2	−0.4		
Brain	1.3	1.2	1.0	1.1					
Pancreas	2.3	2.7	2.0						
Eye,lens			2.0		0.0		0.0		2.0
–,retina			3.0		0.0		0.0		2.0
Fat						1.3			
Blood				0.3				−0.3	0.6
Serum, 0.9% saline				−0.4		−0.4	−0.4		
Milk,human		−2.2	−1.2	−0.6					

* Temperature range 40–140°C, 2800 MHz.

Table 6.18 Conductivity temperature coefficient, 20–40°C (% °C^{-1})

Freq,MHz	0.1	1	10	50	100	200	900	3000	10000
Muscle	2.1	2.1	2.1	2.5	2.0	1.5	1.0	0.3*	
Liver	0.6	1.3	1.3	2.0	1.6	1.8	1.4		
Spleen		2.8	2.6	2.3	2.7				
Kidney	0.9	0.8	1.2	1.6	1.7	2.0	1.3		
Brain	1.8	1.7	1.6	1.4	1.3		1.0		−1.0
Pancreas	1.7	2.5	2.6		2.6				
Eye,lens			2.0		2.0		1.5		−0.5
–,retina			0.5		1.0		0.5		−0.5
Fat				1.7–4.3		4.9	4.2		
Blood				2.7				−1.3	−1.8
Serum, 0.9% saline				2.0		1.7	1.3		
Milk,human		−3.6	−2.2	−1.5	−1.5				

* Temperature range 40–140°C, 2800 MHz.

The dielectric properties of beef and pork have been reported over the wider range of temperatures relevant in the food processing industry. Over the temperature range 40°C to 140°C Ohlsson and Bengtsson (1975) report temperature coefficients for beef similar to those given in Tables 6.17 and 6.18 in both magnitude and sign for frequencies of 450, 900 and 2800 MHz. The values for 2800 MHz have been included in the tables.

At temperatures below 0°C values of both ϵ' and σ are lower than the corresponding values in the unfrozen state, both for muscle and fat (Bengtsson *et al*, 1963). The change in dielectric properties in the temperature range −25°C to 10°C during thawing and freezing takes place over an extended temperature range and not abruptly at the freezing point of the tissue. This is particularly noticeable for changes in conductivity, and may be attributed to the percentage of water content in the tissue which may be considered to be 'frozen' at any particular temperature.

Some values of dielectric properties of tissues at temperatures below 0°C are given in Table 6.19.

6.1.9.3 Changes following death

The electrical characteristics of tissue alter following death, and these changes are particularly noticable at low frequencies. The changes may be categorised as those that take place immediately following death, which may be related to a redistribution of the blood volume, and those which take place due to cellular changes in the hours following death. Surowiec *et al* (1986a) have measured the former changes, showing that for all frequencies in the range 10 kHz to 10 MHz conductivity *in–vivo* is higher than that *in–vitro*. Muscle, liver, kidney and spleen were all investigated, the changes being greatest in

Table 6.19 Dielectric properties of frozen tissue

	Tissue	muscle	muscle	fat	blood
	Animal	cow	cow 72.5% water	cow	dog
	Ref*	56	57	56	22
Freq,Hz	Temp,°C	**Relative permittivity**			
10k	-17				400
100k	-17				45
1M	-17				10.5
10M	-17				8
	-10	11.7–13.5		4.4	
	-25	~8			
100M	-10	7.6–8.9		1.8	
	-25	~5			
300M	-20		5.4		
	-40		3.9		
915M	-20		4.8		
	-40		3.6		
2.45G	-20		4.4		
	-40		3.5		
		Conductivity, $\mu S\ cm^{-1}$			
10M	-10	71–87		9	
100M	-10	147–165		20	
300M	-20		162		
	-40		57		
915M	-20		275		
	-40		107		
2.45G	-20		695		
	-40		177		

* Reference codes are given immediately following Table 6.16.

spleen and liver. Permittivity below 300 kHz fell after death, whilst above 1 MHz there was a slight increase. Van Oosteram *et al* (1979) have demonstrated a comparable decrease in the conductivity of canine heart muscle *in-vivo* following induced local ischaemia, σ reducing by about a factor of three after 2.7 hours, at 1 kHz. At 2.45 GHz, Burdette *et al* (1986) have shown slight increases in σ and ϵ' for canine brain (both white and grey matter) following death. However, the subsequent time progression differed showing both σ and ϵ' continuing to increase for white matter following death, while for grey matter both values returned approximately to *in-vivo* values after about 30 minutes. Surowiec *et al* (1985, 1986b) examined tissue properties from 20 kHz to 100 MHz at times from 1.25 to 24 hours following death. Changes within the first 10 hours were observed for kidney, spleen, and both grey and white matter in the brain, with the conductivity increasing and ϵ' decreasing, especially at frequencies below 1 MHz.

Significant liver tissue changes were not seen within the first 10 hours. In the frequency range 200 MHz to 8 GHz both σ and ϵ' of frog heart muscle fell following death (Schwartz and Mealing, 1985). However, in the same frequency range, Kraszewski *et al* (1982) observed no change, within experimental uncertainties, during periods up to 4 hours following death, in a range of tissues from both cats and rats.

In summary, for many tissues at frequencies below 1 MHz, both ϵ' and σ show an immediate decrease at death. Subsequently ϵ' continues to decrease whilst conductivity increases. At frequencies greater than 1 MHz small changes may occur following death, but the patterns with tissue type and frequency are variable and both increases and decreases in ϵ' and σ have been reported.

6.1.9.4 Animal species

The tables include data from human, dog, cat, pig, and cow, rat, mouse and rabbit tissues. Other values have been published from non-mammalian species. There is insufficient evidence to identify consistent variation between species, and reports suggest that other sources of variability (local tissue inhomogeneities, *post mortem* changes, age of animal) probably mask any possible species-specific variations. Stuchly *et al* (1982b) failed to identify any significant differences *in-vivo* between the properties of skeletal muscle, brain cortex, spleen and liver of cats and of rats. Frog tissue properties for skeletal and heart muscle skin and blood, not included in the tables also compare well in the frequency range 200 MHz to 80 GHz both *in-vivo* and *in-vitro* (Schwartz and Mealing, 1985). Brain tissue from five different species could not be separated at 3 GHz (Lin, 1975). On the other hand Surowiec *et al* (1987b) report measurements demonstrating that dielectric constants for human kidney and spleen are higher than corresponding values for feline tissues in the frequency range 10 kHz to 1 MHz and at around 100 MHz. No significant differences for liver tissue were observed. In view of the known dependence of electrical properties on water content and the variation of water content of specific organs with species (see Chapter 9) it may be inferred that species-dependent differences, if and when they occur, are most probably primarily related to variations in tissue water content.

One tissue showing clear species-dependent variation is skin. The skin impedance of nude mouse skin has been measured to be about ten times lower than human skin (Yamamoto and Yamamoto, 1981) at 10 Hz and 100 Hz. Likewise values given for rat skin in Table 6.15 at microwave frequencies are lower than those for human skin.

6.1.9.5 Water content

Much of the variability in tissue dielectric properties may be attributed to its water content. The large variations reported for adipose tissue and bone marrow (Tables 6.8 and 6.20) are caused by the wide range of water content in these tissues. A single value cannot be chosen for the conductivity or permittivity in this case, but a range of values must be used in calculations unless the exact composition of the tissues is known. The interpretation of

Table 6.20 Permittivity and conductivity related to tissue water content at 25°C and 300 MHz

Volume fraction of water	Permittivity relative to water	Conductivity mS cm^{-1}
20%	0.18	1.5
40%	0.33	3.0
60%	0.51	6.1
80%	0.74	8.3

Source: Smith and Foster (1985).

the dielectric properties has been in terms of tissue water content, and the regional variations in brain properties (Table 6.6) variation of brain properties with age (Thurai *et al*, 1984, 1985) and tumour properties (Table 6.21) can be interpreted in this way.

The state of hydration of bone has considerable effect on its electrical properties (Marino *et al*, 1967). For fully saturated bone the electrical properties are primarily determined by the immersion fluid (Kosterich *et al*, 1984). Other factors affecting the measurement of bone impedance, such as pH and length of exposure to air, have been discussed by Saha *et al* (1984). Similar dependence on tissue water content has been demonstrated for bovine tendon by Lim and Shanus (1971).

A discussion of the investigation of bound water in tissue has been given by Grant (1982) using the variation in high frequency dielectric properties from those of pure water.

6.1.9.6 Local tissue variability

Detailed measurements on specific regions of brain tissue, particularly white matter, grey matter, dura and pia mater (Table 6.6) demonstrate the local variation which may occur over small distances on a tissue which may well be considered otherwise as having constant electrical properties. Stuchley *et al* (1981) remark on the spatial variability of grey matter in the brain, and in particular on the range of loss tangent and conductivity values. It is possible that other tissues with local regional variations, for instance kidney, heart or arterial wall, may well have spatially variable electrical properties yet to be resolved.

De Mercato and García-Sánchez (1988) have noted differences in the dielectric properties of wet bone taken from bovine tibia, depending upon position. The conductivity and permittivity were both lowest in bone samples taken from the diaphysis, and higher in samples from the epiphyses. The frequency range in this study was from 1 kHz to 1 MHz.

Table 6.21 Tissue pathology, relative permittivity and conductivity

Animal	rabbit	human	rat	dog	mouse	dog	mouse	human	rat
Tumour	VX-2	breast Ca.	hepatoma D23	fibro-sarcoma	KHT, RIF/1	several visceral	several	breast tumours	R1H
	in-vitro	*in-vitro*	*in-vitro*	*in-vivo*	*in-vitro*	*in-vitro*		*in-vivo*	*in-vivo*
Temp, °C	25	37	37	37*	37	37			32
Ref	44	58	16	19	32	20	62	62	46
Freq, Hz	**Relative permittivity**								
1k	$2.0–6.5x10^4$								
10k	$1.0–2.5x10^4$	$5–20x10^3$							
100k	3500–8000	$1–5x10^3$							
1M	1000–1500	200–1100	2500	389					
10M	200–250	40–300	420	61	89–91	150–400	202–269†		
100M		12–80	88	12.7	56–59	70–80			
500M							34.7–65†	36–56†	61.6
1G					35–38	45–60	31.4–65†	34–50	59.9
2.45G							27–65	31–44	52.9
10G						40–45			
	Conductivity, mS cm^{-1}								
1k	3.0–4.0								
10k	3.2–4.3	0.8–6.0							
100k	3.5–4.7	0.85–6.1							
1M	4.7–5.2	0.9–6.3	4.3	1.4					
10M	5.5–8.0	1.1–8.2	7.1	1.7		6.0–11.0	3.3–8.6†		
100M		1.3–14.0	10.0	2.0	9.0–10.1	3.5–13.7			
500M							6.6–12.1†	3.5–8.0	10.8
1G					15.0–17.0	10.0–15.0	7.8–13.4†	6.0–10.0†	13.3
2.45G							11.8–26.3	9.0–14.0	24.7
10G						110–128			

* Values also given following heating to 43°C. † Measurements at 13.6, 433 and 918 MHz. Reference codes listed after Table 6.16.

6.1.9.7 Tissue pathology

Measurements from a variety of pathological tissues have been reported both from animal and human tissues. Some values for pathological tissues have been included in Table 6.21.

6.1.10 *Dielectric properties of some materials other than tissue*

6.1.10.1 Water

Values of dielectric properties for pure water mixtures are given in Tables 6.22 and 6.23. The variation with frequency and the temperature coefficients are given in Table 6.22, with more detail of the temperature dependence of the low frequency permittivity of water being given in Table 6.23. The variation of both ϵ' and σ with frequency is also shown in Figure 6.2 and Figure 6.3. Water has a region of dispersion at 20°C, centred around 17 GHz, rising to about 25 GHz at body temperature. For frequencies above 1 MHz ϵ' and ϵ'' (and hence σ) may also be calculated from the Debye equations using ϵ_0, ϵ_∞ and τ. Literature values for these quantities have been thoroughly reviewed by Kaatze and Uhlendorf (1981) who give best estimates as $\epsilon_0 = 78.38 \pm 0.03$, $\epsilon_\infty = 5.44 \pm 0.06$, and $\tau = 8.30 \pm 0.01$ ps, at 25°C. Temperature dependence is also reviewed. The static relative permittivity as a function of temperature in the range 0–60°C is given by:

$$\epsilon_0 = 88.15 - 41.4\theta + 13.1\theta^2 - 4.6\theta^3 \qquad (6.12)$$

where θ = centigrade temperature/100
The relaxation time τ is given by:

Temp °C	0	10	20	30	40	50	60
τ, ps	17.7	12.6	9.2	7.1	5.7	4.8	3.9

The infinite frequency permittivity decreases from about 5.0 in the infrared to 1.8 in the optical region (Kaye and Laby, 1973). The complex dielectric spectrum for water at frequencies to 10 THz has been reported by Asfar and Hasted (1977), showing serious discrepancies between predictions based on microwave data and the far infrared measurements.

6.1.10.2 Aqueous solutions

Dielectric properties of a variety of aqueous solutions are also given in the literature. Of most interest are the properties of physiological saline (0.9% NaCl). Geddes and Baker (1967) give a conductivity at 37.5°C of 19.8 mS cm^{-1} measured at 1 kHz, whilst average high frequency values at 27°C given by Schwan and Li (1953) are $\epsilon' = 76.7$ and $\sigma = 17.5$ mS cm^{-1} in the

Table 6.22 Water, relative permittivity and conductivity at 25°C

Freq GHz	Relative permittivity	Conductivity mS cm^{-1}	Temp coeff, % °C^{-1}	
			ϵ'	σ
0.1	78.3	0.024		
0.3	78.3	0.199		
0.7	78.3	1.11		
1.05	78.2	2.61		
3.0	76.5	19.0	−0.3	−3.1
4.63	73.6	43.0	−0.1	−2.9
9.39	63.1	149.4	0.5	−2.3
23.77	34.7	471.4	2.4	0.0
35.0	22.4	586.1	2.5	0.9

Sources: Schwan *et al* (1976); Cook (1952); Steel and Sheppard (1988).

Table 6.23 Relative permittivity of water; temperature variation

Temp °C	Permittivity
0	87.740
10	83.832
20	80.103
30	76.546
40	73.151
50	69.910
60	66.815
70	63.857
80	61.027
90	58.319
100	55.720

Source: Parsons (1954).

range 200 MHz to 900 MHz. Values at other concentrations of NaCl are given, for instance, by Cooper (1946), Lane and Saxton (1952) and Chambers *et al* (1956). Cooper gives values of σ at percentage concentrations of NaCl by weight from 1% to 4% which vary from 14 to 55 mS cm^{-1} at 20 °C. No variation in the frequency range 0.95 to 13 MHz was observed. The temperature dependence of the dielectric properties of saline solutions has been reviewed by Foster and Schwan (1986). Analytic expressions allowing the calculation of the dielectric properties of specific concentrations are given by Stogryn (1971). Further values for saline, and for alcohol–water mixtures, have been given by Mudgett (1986).

Table 6.24 Tissue substitute materials, dielectric properties at room temperature

Tissue mimic	Freq MHz	ϵ'	σ mS/cm	Reference	Composition
Solid mixtures					
Bone,fat	30	19	0.28	Stuchly & Stuchly 1980a*	79% Laminac 4110, 20.72% Al powder, 0.28% acetylene black, 3.5cc/kg MEK peroxide
	100 900	13.6 7.4	0.08 1.6	Hartsgrove *et al* 1987	35% epoxy, 35% hardener, 28% KCl solution
	915 2450	5.6 4.5	0.68 1.14	Guy 1971†; Durney *et al* 1980	85.2% Laminac 4110, 14.5% Al powder,0.24% acetylene black, 3.75cc/kg MEK peroxide
Brain	915 2450	34.4 33.6	7.7 12.4	Guy 1971†; Durney *et al* 1980; Chung-Kwang *et al* 1984	62.61% water, 0.5823% NaCl, 29.8% polyethylene powder,7.01% Super Stuff(TX150)
	2450	42	25.9	Stuchly & Stuchly 1980a*	59% water, 40% gelatine, 1% NaCl
Muscle	10 50 100	65.8 44.1 44.8	11.7 12.0 12.0	Stuchly & Stuchly 1980a*	75.5% water, 0.9% NaCl, 15.2% polyethylene powder, 8.4% Super Stuff(TX150)
	30	110	6.5	Stuchly & Stuchly 1980a*	76.57% water, 0.153% NaCl, 13.78% Al powder, 9.495% Super Stuff(TX150)
	915 2450	50.6 49.6	13.5 22.5	Guy 1971†; Durney *et al* 1980	75.44% water, 0.907% NaCl, 15.2% polyethylene powder, 8.45% Super Stuff(TX150)
	2450	50	21.8	Stuchly & Stuchly 1980a	69% water, 30% gelatine, 1% NaCl
Skin	2450	43		Stuchly & Stuchly 1980a	60% water, 1% NaCl, 39% cellulose paper.

cont.

Table 6.24 cont. Tissue substitute materials

Tissue	F,MHz	ϵ'	mS/cm	Reference	Composition
Various	4–50	$\epsilon'=79.25-1.009P$ $\sigma=4.144\times10^{-3}\times(85-P)^{1.235}\times(146.6N+0.059)$		Kato *et al* 1986 Kato & Ishida 1987	4% agar,0–44.5% PVC powder(P),sodium azide powder, 0–8x10^{-3}%(N), water
Liquid mixtures					
Bone	100	10.8	0.35	Hartsgrove *et al* 1987	TWEEN 57%, n–amyl alcohol 28.5%, paraffin oil 9.5%, water 4.5%, NaCl 0.5%
	900	7.2	1.2		
Brain	100	63.0	4.7		Water 10.4%, NaCl 2.5%, sugar 56%, HEC 1%
	900	41.2	12.2		
Lung	100	37.0	3.4		Brain recipe 47%, microspheres 53%
	900	28.0	6.6		
Muscle	100	70.5	6.8		Water 52.4%, NaCl 1.4%, sugar 45%, HEC 1%
	900	54.7	13.8		
Tissue	915	71.8	26.6	Hand *et al* 1979	Krebs–Ringer solution
	2450	70.8	33.5		
	915	67.6	14.8		Krebs soln. with 200g/l dextran
	2450	64.3	24.8		
	915	62.0	13.5		Krebs soln. with 400g/l dextran
	2450	57.8	24.3		
Body	915	50		Kopecky 1980	20% ethanol, 80% physiological saline, 37.5°C
	2450	50			
Tissue	35GHz	15.7	446	Steel & Sheppard 1988	Ethanediol 14.04%, water 85.96%
Fat	35GHz	3.82	23.4		Ethanol

* Referenced in this review paper. † The original formulation (Guy, 1971) is modified slightly in Durney *et al* (1981); the latter source is used.

6.1.10.3 Tissue substitute materials

Values of dielectric properties of some mixtures reported as being suitable tissue substitute materials are given in Table 6.24. Mixture recipes are also given. The effect of varying the salinity of the mixture on its conductivity measured at 100 kHz is given by Guy *et al* (1976), and over the frequency range 13.6 to 2450 MHz by Chung-Kwang Chou *et al* (1984). These latter authors also report several alternative recipes for this frequency range, and give the temperature dependence of their dielectric properties. The thermal properties are reported in Chapter 2. Sources of the materials given are reported by Durney *et al* (1980). Cheung and Coopman (1976) have also reported the dielectric properties of several different examples of mixtures using similar component materials suitable for bone or muscle substitutes at 8.5 and 10 GHz. Some data for gelatin mixtures are given by Fricke and Jacobson (1939), Masuzawa and Sterling (1968) and Masszi (1972).

6.1.10.4 Gases

Values of relative permittivity of some gases are given in Table 6.25.

Table 6.25 Relative permittivity of gases and vapours at 1 atm

Material	Temp °C	$10^4(\epsilon-1)$
Dry air	20	5.36
Nitrogen	20	5.47
Oxygen	20	4.94
Carbon dioxide	20	9.21
Nitrous oxide	25	10.3
Water vapour	100	60
Water vapour (10 mm Hg)	20	1.21

6.2 Electromagnetic radiation reflection and absorption

6.2.1 *Terminology and definitions*

The **absorption coefficient,** μ, for plane electromagnetic waves in a homogenous substance is defined by the relationship:

$$E = E_0{}^{-\mu d} \tag{6.13}$$

where E is the electrical field at a distance d from the source of field strength E_0. If the values of conductivity and permittivity are known then μ can be calculated using the following expression:

$$\mu^2 = \left[\frac{2\pi}{\lambda}\right]^2 \frac{\epsilon'}{2}\left[\left\{1 + \left[\frac{60\lambda\sigma}{\epsilon'}\right]^2\right\}^{\frac{1}{2}} - 1\right] \tag{6.14}$$

where λ is the wavelength of the radiation in air. The **penetration depth** is $1/\mu$, where the field is reduced 1/e of its initial value. At this depth, the power density is reduced by a factor $1/e^2$.

The **refractive index** n of tissue is given by:

$$n^2 = \frac{\epsilon'}{2}\left[\left\{1 + \left[\frac{60\lambda\sigma}{\epsilon'}\right]^2\right\}^{\frac{1}{2}} + 1\right] \tag{6.15}$$

(see, for instance, Schwan and Piersol, 1954).

6.2.2 *Calculated values*

These standard equations (Equations 6.13, 6.14) have been used to calculate penetration depths for two 'typical' tissues with high and low water content for frequencies between 1 MHz and 10 GHz (Johnson and Guy, 1972). Some values from these calculations are given in Table 6.26. Calculated penetration depths are shown also in Figure 6.4.

In addition, values for reflection coefficients R between these 'typical' tissues were calculated using:

$$R^2 = r\, e^{j\varphi} = \frac{(\epsilon_1^*)^{\frac{1}{2}} - (\epsilon_2^*)^{\frac{1}{2}}}{(\epsilon_1^*)^{\frac{1}{2}} + (\epsilon_2^*)^{\frac{1}{2}}} \tag{6.16}$$

where ϵ_1^* and ϵ_2^* are the complex permittivities of the two media. Some results of these calculations are given in Table 6.27. These data may be used to give the basis for calculations of absorbed power in tissues.

Table 6.26 Wavelengths and penetration depths for 'typical' tissue with low and high water content

Freq MHz	λ,air cm	λ,tissue cm	ε'	σ mS cm^{-1}	Penetration depth
	High water content				
1	30000	436	2000	4.0	91.3
10	3000	118	160	5.2	21.6
27.12	1106	68.1	113	6.1	14.3
100	300	27	72	8.9	6.7
300	100	11.9	54	13.7	3.9
915	32.8	4.46	51	16.0	3.04
2450	12.2	1.76	47	22.1	1.70
8000	3.75	0.58	40	76.5	0.41
	Low water content				
27.12		214	20	1.1–4.3	159
100		106	7.5	1.9–7.6	60.4
300		41	5.7	3.2–10.7	32.1
915		13.7	5.6	5.6–14.7	17.7
2450		5.21	5.5	9.6–21.3	11.2
8000		1.73	4.7	25.5–43.1	4.61

Source: Johnson and Guy (1972).

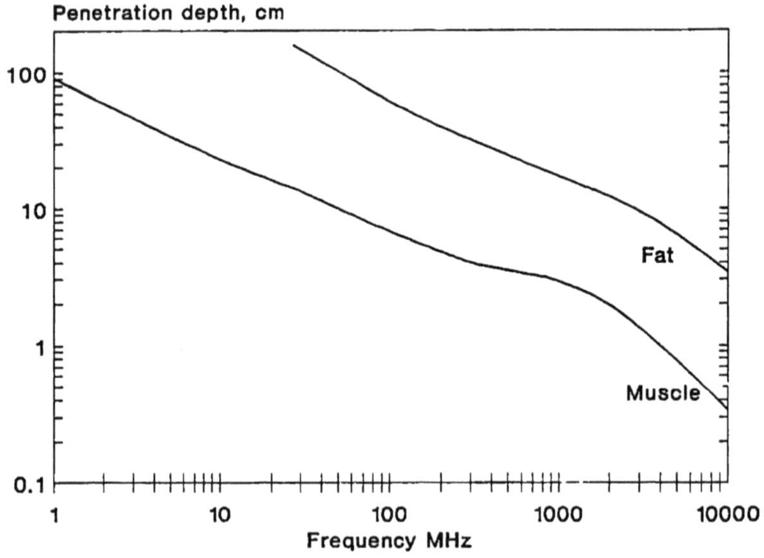

Figure 6.4 Calculated penetration depth for tissues with high and low water content over the frequency range 1 MHz to 10 GHz. Johnson & Guy (1972).

Table 6.27 Calculated reflection coefficients for 'typical' interfaces

Freq	Air–muscle		Muscle–fat		Air–fat	
MHz	r	θ	r	θ	r	θ
1	0.982	+179				
10	0.956	+178				
27.12	0.925	+177	0.651	−11	0.660	+174
100	0.881	+175	0.650	−8	0.511	+168
300	0.825	+175	0.592	−8	0.438	+169
915	0.772	+177	0.519	−4	0.417	+173
2450	0.754	+177	0.500	−4	0.406	+176
8000	0.744	+176	0.513	−7	0.371	+173

Source: Johnson and Guy (1972).

Penetration depths for some specific tissues are given by Schwan (1984), Schwan and Piersol (1954), Bottomley and Andrew (1978), by Röschmann (1987) for 10–220 MHz and by Edrich and Hardee (1976) for 40–90 GHz.

6.3 Other electrical properties

6.3.1 *Electromechanical properties: piezoelectricity*

In piezoelectric materials, electric charge is generated as a result of an applied stress. Conversely, the material will alter its dimensions (if allowed to do so) under the effect of an applied electric field. The **piezoelectric coefficients**, d_{ij}, characterize the effect in terms of the charge generated for unit applied stress under short circuit conditions. Subscript i defines the electrical direction (i= 1,2,3) and j defines the mechanical direction (j=1,...,6): j=1,2,3 apply to tensile stresses and j=4,5,6 to shear stresses. A practical unit used is picocoulomb newton^{-1} (pC N^{-1}).

Studies by Fukada and Yasuda (1957) established that dry bone showed both direct and inverse piezoelectric effects. Electromechanical effects were subsequently also shown to occur in dentine (Braden *et al*, 1966), and in soft tissues such as tendon and arterial wall (Fukada, 1968a,b) and skin (Shamos and Lavine, 1967). It is now generally agreed that collagen is responsible for the piezoelectric effect, having the primary requirements that it is an ordered structure and lacks a centre of symmetry. Recent work indicates that two separate mechanisms may be responsible for the overall electromechanical response of tissue, and their relative importance depends upon state of hydration of the tissue. For dried tissues a piezoelectric effect predominates. On the other hand in normally hydrated tissue a streaming potential has been proposed (Anderson and Eriksson, 1968). The electrical potentials

Table 6.28 Piezoelectric coefficients of dehydrated tissue at room temperature

Tissue	d_{14} pC N^{-1}	d_{15} pC N^{-1}	d_{33} pC N^{-1}	d_{31} pC N^{-1}
Aorta,natural	−0.033			
–,elongated	−0.17			
Bone,horse femur	−0.22	0.043	0.003	0.003
Intestine	−0.007			
Tendon,achilles, horse	−1.89	0.53	0.066	0.013
–,–,cow	−2.66	1.39	0.086	0.066
Trachea	0.007			

Source: Fukada (1968a,b). The z axis is longitudinal for bone, tendon or vessel. For vessels, x is radial and y is circumferential.

generated in bone *in–vivo* are currently considered to originate primarily from the streaming potential rather than the piezoelectric property of the bone itself.

Piezoelectric charge constants for dried tissue have been reported by Fukada (1968a,b) and are given in Table 6.28. However the assumption that bone is a simple linear piezoelectric material is probably not valid (Bruce–Martin, 1979), giving rise to inconsistencies in values given in the literature. Unlike a classic piezoelectric, the generated potential is dependent on the strain gradient, and also the modulus of elasticity is frequency–dependent. Further discussion of these problems is given by Hastings and Mahmud (1988).

Singh *et al* (1985) have reported some piezoelectric properties of dried bovine bone and collagen in both poled and unpoled states. Curie temperatures are gived as 120°C for bone and 95°C for collagen. The piezoelectric charge constants at 25°C are reported as 2.63 pC N^{-1} for bone and 0.84 pC N^{-1} for collagen.

6.3.2 *Pyroelectric properties*

Pyroelectric coefficients for dried bone and tendon have also been reported (Lang, 1966), varying from $(2.5\pm1.8)\times10^{-13}$ coulomb cm^{-2} °C^{-1} for phalangeal bone to $(4.1\pm2.0)\times10^{-13}$ coulomb cm^{-2} °C^{-1} for hoof tendon. Liboff and Furst (1974) have inferred that the pyroelectric response of dried bone originates from its collagen content since there is a change in polarity in the pyroelectric current in bovine tendon at 157°C, close to the temperature for denaturation of collagen. Pyroelectric coefficients for human tooth material lie in the range 1.5×10^{-13} to 25×10^{-13} C cm^{-2} °C^{-1} (Athenstaedt, 1971).

6.3.3 *Photoelectric properties*

Associated photoelectric effects have been reported in dried human cortical bone (Becker and Brown, 1965).

References

Almasi J.J. and Schmitt O.H., 1970, Systematic and random variations of ECG electrode system impedance, Ann NY Acad Sci, 170, 509–519.

Anderson J.C. and Eriksson C., 1968, Electrical properties of wet collagen, Nature, 218, 166–168.

Asfar M.N. and Hasted J.B., 1977, Measurements of the optical constants of liquid H_2O and D_2O between 6 and 450 cm^{-1}, J Opt Soc Am, 67, 902–904.

Athenstaedt H., 1971, Pyroelectric and piezoelectric behaviour of human dental hard tissues, Archs Oral Biol, 16, 495–501.

Becker R.O. and Brown F.M., 1965, Photoelectric effects in human bone, Nature, 206, 1325–1328.

Bengtsson N.E., Melin J., Remi K. and Söderlind S., 1963, Measurements of the dielectric properties of frozen and defrosted meat and fish in the frequency range 10–200 MHz, J Sci Food Agric, 14, 592–604.

Bolte A., 1961, Zur Ableitung und Bewertung fetaler Herzaktionspotentiale bei schwangeren Frauen, Arch Gynäkol, 194, 594–610.

Bottomley P.A. and Andrew E.R., 1978, RF magnetic field penetration, phase shift and power dissipation in biological tissue: implications for NMR imaging, Phys Med Biol, 23, 630–643.

Braden M., Bairstow A.G., Beider I. and Ritter B.G., 1966, Electrical and piezo-electrical properties of dental hard tissues, Nature, 212, 1565–1566.

Brady M.M., Symons S.A. and Stuchly S.S., 1981, Dielectric behaviour of selected animal tissues *in vitro* at frequencies from 2 to 4 GHz, IEEE Trans Biomed Eng, 28, 305–307.

Bruce-Martin R., 1979, Analysis of bone and other piezoelectric textures. In *Electrical Properties of Bone and Cartilage,* C.T.Brighton, J.Black and S.R.Pollack, (eds), Grune and Stratton, New York, pp.141–151.

Burger H.C. and van Milaan J.B., 1943, Measurement of the specific resistance of the human body to direct current, Acta Med Scand, 114, 584–607.

Burger H.C. and van Dongen R., 1961, Specific electrical resistance of body tissues, Phys Med Biol, 5, 431–447.

Burdette E.C., 1982, Electromagnetic and acoustic properties of tissues. In *Physical Aspects of Hyperthermia,* G.H.Nussbaum (ed.), AAPM Medical Physics Monographs No. 8, pp.105–150.

Burdette E.C., Cain F.L., and Seals J., 1980, *In vivo* probe measurement technique for determining dielectic properties at VHF through microwave frequencies, IEEE Trans Microwave Theory and Techniques, 28, 414–27.

Burdette E.C., Friederich P.G., Seaman R.L. and Larsen L.E., 1986, *In situ* permittivity of canine brain: regional variations and postmortem changes, IEEE Trans Microwave Theory and Techniques, 34, 38–50.

Campbell S.D., Kraning K.K., Schibli E.G. and Momii S.T., 1977, Hydration characteristics and electrical resistivity of stratum corneum using a noninvasive four-point microelectrode method, J Invest Dermatol, 69, 290–295.

Chakkalakal D.A., Johnson M.W., Harper R.A. and Katz J.L., 1980, Dielectric properties of fluid-saturated bone, IEEE Trans Biomed Eng, 27, 95–100.

Chambers J., Stokes J., and Stokes R., 1956, Conductance of concentrated aqueous sodium and potassium chloride solutions at 25°C, J Phys Chem, 60, 985–986.

Cheung A.Y. and Coopman D.W., 1976, Experimental development of simulated materials for dosimetry studies of hazardous microwave radiation, IEEE Trans Microwave Theory and Techniques, 24, 669–673.

Chung-Kwang Chou, Gang-Wu Chen, Guy A.W. and Luk K.H., 1984, Formulas for preparing phantom muscle tissue at various radiofrequencies, Bioelectromagnetics, 5, 435–441.

Clerc L., 1976, Directional differences of impulse spread in trabecular muscle from mammalian heart, J Physiol, 255, 335–346.

Cole K.S. and Cole R.H., 1941, Dispersion and absorption in dielectrics. 1. Alternating current characteristics, J Chem Phys, 9, 341–351.

Cook H.F., 1951, Dielectric behaviour of human blood at microwave frequencies, Nature, 168, 247–248.

Cook H.F., 1952, A comparison of the dielectric behaviour of pure water and human blood at microwave frequencies, Br J Appl Phys, 3, 249–255.

Cooper R., 1946, The electical properties of salt-water solutions over the frequency range 1–4000 Mc/s, J Inst Elec Engineers, 93 III, 69–75.

Crile G.W., Hosmer H.R. and Rowland A.F., 1922, The electrical conductivity of animal tissues under normal and pathological conditions, Am J Physiol, 60, 59–106.

Dawkins A.W.J., Gabriel C., Sheppard R.J. and Grant E.H., 1981, Electrical properties of lens material at microwave frequencies, Phys Med Biol, 26, 1–9.

De Mercato G. and García-Sánchez F.J., 1988, Dielectric properties of fluid-saturated bone; a comparison between diaphysis and epiphysis, Med and Biol Eng and Comput, 26, 313–316.

Deming Xu, Liping Liu and Zhiyan Jiang, 1987, Measurement of the dielectric properties of biological substances using an improved open-ended coaxial line resonator method, IEEE Trans Microwave Theory and Techniques, 35, 1424–1428.

Durney C.H., Iskander M.F., Massoudi H., Allen S.J. and Mitchell J.C., 1980, Radiofrequency radiation dosimetry handbook (3rd edition) SAM-TR-80-32 USAF School of Aerospace Medicine.

Edrich J. and Hardee P.C., 1976, Complex permittivity and penetration depth of muscle and fat tissues between 40 and 90 GHz, IEEE Trans Microwave Theory and Techniques, 25, 273–275.

Edelburgh R., 1977, Relation of the electrical properties of skin to structure and physiologic state, J Invest Derm, 69, 324–327.

England T.S., 1950, Dielectric properties of the human body for wave-lengths in the 1–10 cm range, Nature, 166, 480–481.

Epstein B.R. and Foster K.R., 1983, Anisotropy in the dielectric properties of skeletal muscle, Med Biol Eng Comput, 21, 51–55.

Foster K.R., Schepps J.L., Stoy R.D. and Schwan H.P., 1979, Dielectric properties of brain tissue between 0.01 and 10 GHz, Phys Med Biol, 24, 1177–1187.

Foster K.R. and Schwan H.P., 1986, Dielectric properties of tissues. In *CRC Handbook of Biological Effects of Electromagnetic Fields,* C.Polk and E.Pastow (eds), CRC Press, Boca Raton, Florida, pp.27–96.

Freygang W.H. and Landau W.M., 1955, Some relations between resistivity and electrical activity in the cerebral cortex of the cat, J Cell Comp Physiol, 45, 377–392.

Fricke H., 1924, A mathematical treatment of the electrical conductivity and capacity of disperse systems. 1. The electical conductivity of a suspension of homogenous spheroids, Phys Rev, 24, 575–584.

Fricke H., 1925, The electric capacity of suspensions with special reference to blood, J Gen Physiol, 9, 137–152.

Fricke H. and Curtis H.J., 1935, The electrical impedance of hemolized suspensions of mammalian erythrocytes, J Gen Physiol, 18, 821–836.

Fricke H. and Jacobson L.E., 1939, A dielectric study of the gelatin–water system: anomalous dispersion in bound (oriented) water, J Phys Chem, 43, 781–795.

Fukada E., 1968a, Piezoelectricity in polymers and biological materials, Ultrasonics, 6, 229–234.

Fukada E., 1968b, Mechanical deformation and electrical polarization in biological substances, Biorheology, 5, 199–208.

Fukada E. and Yasuda I., 1957, On the piezoelectric effect of bone, J Phys Soc Japan, 12, 1158–1162.

Gabriel C., Sheppard R.J. and Grant E.H., 1983, Delectric properties of ocular tissues at 37°C, Phys Med Biol, 28, 43–49.

Gabriel C. and Grant E.H., 1985, Dielectric properties of ocular tissues in the supercooled and frozen states, Phys Med Biol, 30, 975–983.

Galeotti G., 1902, Uber die elektrische leitfahigkeit der tierschen Gewebe, Z Biol, 43, 289–340.

Geddes L.A. and Baker L.E., 1967, The specific resistance of biologic material – a compendium of data for the biomedical engineer and physiologist, Med Biol Eng, 5, 271–293.

Geddes L.A. and da Costa C.P., 1973, The specific resistance of canine blood at body temperature, IEEE Trans Biomed Eng, 20, 51–53.

Geddes L.A. and Sadler C., 1973, The specific resistance of blood at body temperature, Med Biol Eng, 11, 336–339.

Goodwin R.S. and Sapirstein L.A., 1957, Measurement of the cardiac output of dogs by a conductivity method after single intravenous injections of autogenous plasma, Circ Res, 5, 531–538.

Grant E.H.; 1982, The dielectric method of investigating bound water in biological material: an appraisal of the technique, Bioelectromagnetics, 3, 17–24.

Grant E.H., Sheppard R.J. and South G.P., 1978, *Dielectric Behaviour of Biological Molecules in Solution*, Oxford University Press, Oxford.

Grant J.P., Clarke R.N., Symm G.T. and Spyrou N.M., 1988, *In vivo* dielectric properties of human skin from 50 MHz to 2.0 GHz, Phys Med Biol, 33, 607–612.

Guy A.W., 1971, Analyses of electromagnetic fields induced in biological tissues by thermographic studies of equivalent phantom models, IEEE Trans Microwave Theory and Techniques, 19, 205–214.

Guy A.W., Webb M.D. and Sorensen C.C, 1976, Determination of power absorption in man exposed to high frequency electromagnetic fields by thermographic measurements on scale models, IEEE Trans Biomed Eng, 23, 361–371.

Hahn G.M., Kenahan A., Martinez A., Pounds D. and Prionas S., 1980, Some heat transfer problems associated with heating by ultrasound, microwaves or radio frequency, Ann NY Acad Sci, 335, 327–346.

Hand J.W., Robinson J.E., Szwarnowski S., Sheppard R.J. and Grant E.H., 1979, A physiologically compatible tissue–equivalent liquid bolus for microwave heating of tissues, Phys Med Biol, 24, 426–431.

Hartsgrove G., Kraszewski A. and Surowiec A., 1987, Simulated biological materials for electromagnetic radiation absorption studies, Bioelectromagnetics, 8, 29–36.

Hastings G.W. and Mahmud F.A., 1988, Electrical effects in bone, J Biomed Eng, 10, 515–521.

Hermann L., 1871, Ueber eine Wirking galvanischer Strome auf Muskeln und Nerven, Pflug Arch ges Physiol, 5, 223–275.

Herrick J.F., Jelatis D.G. and Lee G.M., 1950, Dielectric properties of tissues important in microwave diathermy, Fed Proc, 9, 60 (Abstract only; cited by Schwan and Li, 1953).

Hey–Shipton G.L., Mathews P.A. and McStay J., 1982, The complex permittivity of human tissue at microwave frequencies, Phys Med Biol, 27, 1067–1071.

Johnson C.C. and Guy A.W., 1972, Non–ionising electromagnetic wave effects in biological materials and systems, Proc IEEE, 60, 692–718.

Kaatze U. and Uhlendorf V., 1981, The dielectric properties of water at microwave frequencies, Zeit für Physik Chemie Neue Folge, 126, 151–165.

Karolkar B.D., Behari J., and Prim A., 1985, Biological tissues characterisation at microwave frequencies, IEEE Trans Microwave Theory and Techniques, 33, 64–66.

Kato H. and Ishida T., 1987, Development of an agar phantom adaptable for simulation of various tissues in the range 5–40 MHz, Phys Med Biol, 32, 221–226.

Kato H., Hiraoka M. and Ishida T., 1986, An agar phantom for hyperthermia, Med Phys, 13, 396–398.

Kaye G.W.C. and Laby T.H., 1973, *Tables of physical and chemical constants*, 14th edition, Longman, London, p.116.

Kinnen E., Kubicek W., Hill P. and Turton G., 1964, Thoracic cage impedance measurements. (Tissue resistivity *in vivo* and transthoracic impedance at 100 kc/s) Tech. Doc. Rep. SAM–TDR 64–5. School of Aerospace Medicine Brooks, AFB Texas.

Kopecky W.J., 1980, Using liquid dielectrics to obtain spatial thermal distributions, Med Phys, 7, 566–570.

Kosterich J.D., Foster K.R. and Pollack S.R., 1983, Dielectric permittivity and electrical conductivity of fluid saturated bone, IEEE Trans Biomed Eng, 30, 81–86.

Kosterich J.D., Foster K.R. and Pollack S.R., 1984, Dielectric properties of fluid saturated bone – the effect of variation in conductivity of the immersion fluid, IEEE Trans Biomed Eng, 31, 369–373.

Kraszewski A., Stuchly M.A., Stuchly S.S. and Smith A.M., 1982, *In vivo* and *in vitro* dielectric properties of animal tissues at radio frequencies, Bioelectromagnetics, 3, 421–432.

Lane J.A. and Saxton J.A., 1952, Dielectric dispersion in pure polar liquids at very high radio frequencies III. The effect of electrolytes in solution, Proc R Soc, A213, 531–545.

Lang S.B., 1966, Pyroelectric effect in bone and tendon, Nature, 212, 704–705.

Laogun A.A., 1986, Effect of temperature on the R.F. dielectric properties of human breast milk, Phys Med Biol, 31, 893–900.

Lawler J.C., Davis M.J. and Griffiths E.C., 1960, Electrical characteristics of the skin, J Invest Dermatol, 34, 301–308.

Liboff A.R. and Furst M., 1974, Pyroelectric effect in collagenous structures, Ann NY Acad Sci, 238, 26–35.

Lim J.J. and Shanus M.H., 1971, An investigation of the bound water in tendon by dielectric measurement, Biophys J, 11, 648–663.

Lin J.C., 1975, Microwave properties of fresh mammalian brain tissue at body temperature, IEEE Trans Microwave Theory and Techniques, 22, 74–76.

Little V.I. and Smith V., 1955, The ionic conductivity of dilute potassium chloride solutions at centimetre wavelengths, Proc Phys Soc B, 68, 65–74.

Marino A.A., Becker R.O. and Bachman C.H., 1967, Dielectric determination of bound water of bone, Phys Med Biol, 12, 367–378.

Masszi G., 1972, Dielectric relaxation and water structure in gelatine solutions, Acta Biochim Biophys Acad Sci Hung, 7, 349–357.

Masuzawa M. and Sterling C., 1968. Gel-water relationships: dielectric dispersion, Biopolymers, 6, 1453–1459.

Maxwell J.C., *Treatise on Electricity and Magnetism*. Oxford University Press, Oxford.

Mazzoleni A.P., Sisken B.F. and Kahler R.L., 1986, Conductivity values of tissue culture medium from 20°C to 40°C, Bioelectromagnetics, 7, 95–99.

McDougall E.I., 1964, The proteins of the thoracic-duct and intestinal-duct lymph of sheep, Biochem J, 90, 160–162.

Mohapatra S.N. and Hill D.W., 1975, The changes in blood resistivity with haematocrit and temperature, Europ J Intensive Care Med, 1, 153–162.

Mudgett R.E., 1986, Electrical properties of foods. In *Engineering Properties of Foods*, M.A. Rao and S.S.H. Rizvi (eds), Marcel Dekker Inc., New York, pp.329–390.

Mudgett R.E., Mudgett D.R., Goldblith S.A., Wang D.I.C. and Westphal W.B., 1979, Dielectric properties of frozen meats, J Microwave Power, 14, 209-216.

Mumford J.M., 1967, Resistivity of human enamel and dentine, Archs Oral Biol, 12, 925-927.

Nicholson P.W., 1965, Specific impedance of cerebral white matter, Exp Neurol, 13, 386-401.

Nightingale N.R.V., Goodridge V.D., Sheppard R.J. and Christie J.L., 1983, The dielectric properties of the cerebellum, cerebrum and brain stem of mouse brain at radiowave and microwave frequencies, Phys Med Biol, 28, 897-903.

Ohlsson T. and Bengtsson N.E., 1975, Dielectric food data for microwave sterilisation processing, J Microwave Power, 10, 93-108.

Oostendorp T.F., van Oosterom A. and Jongsma H.W., 1989, Electrical properties of tissues involved in the conduction of foetal ECG, Med & Biol Eng & Comput, 27, 322-324.

Osswald K., 1937, Messung der Leitfahigkeit und Dielektrizitatkonstante biologischer Gewebe und Flussigkeiten bei kurzen Wellen, Hochfrequentz Tech Elektroakustik, 49, 40-49.

Parsons R., 1954, *Handbook of Electrochemical Constants*, Butterworths, London.

Pauly H. and Schwan H.P., 1964, The dielectric properties of the bovine eye lens, IEEE Trans Bio-Med Eng, 11, 103-109.

Pethig R., 1984, Dielectric properties of biological materials: biophysical and medical applications. IEEE Trans Electrical Insulation, 19, 453-474.

Pethig R. and Kell D.B., 1987, The passive electrical properties of biological systems: their significance in physiology, biophysics, and biotechnology, Phys Med Biol, 32, 933-970.

Philipson M., 1920, Sur la resistance electrique des cellules et des tissues, C Hebd Seanc Mem de la Soc de Biol, 83, 1399-1402.

Rajewsky B., 1938, Ultrakurzwellan, Ergebuisse der biophysikalischon Forschung, I. Georg Thieme, Leipzig, Germany.

Ranck J.B., 1963, Specific impedance of rabbit cerebral cortex, Exp Neurol, 7, 144-152.

Ranck J.B. and Be Merit S.L., 1965, The specific impedance of the dorsal columns of cat; an anisotropic medium, Exp Neurol, 11, 451-463.

Ray S. and Behari J., 1986, Electical conduction in bone in frequency range 0.4 - 1.3 GHz, Biomat Med Dev Art Org, 14, 153-165.

Reddy G.N. and Saha S., 1984, Electrical and dielectric properties of wet bone as a function of frequency, IEEE Trans Biomed Eng, 31, 296-303.

Reinish G.B. and Nowick A.S., 1979, A model for delectric behaviour of wet bone. In *Electrical properties of Bone and Cartilage*, C.T. Brighton, J.B. Black and S.R. Pollock (eds), Grune and Stratton, New York, pp.13-29.

Roberts J.E. and Cook H.F., 1952, Microwaves in medical and biological research, Brit J Appl Phys, 3, 33-40.

Röschmann P., 1987, Radiofrequency penetration and absorption in the human body: limitations to high-field whole-body nuclear magnetic resonance imaging, Med Phys, 14, 922-931.

Rogers J.A., Sheppard R.J., Grant E.H., Bleehan N.M. and Honess D.J., 1983, The dielectric properties of normal and tumour mouse tissue between 50 MHz and 10 GHz, Br J Radiol, 56, 335–338.

Rosenthal R.L. and Thobias C.W., 1948, Measurement of electrical resistance of human blood use in coagulations studies and cell volume determination, J Lab Clin Med, 33, 1110–1122.

Rush S., Abildskov J.A. and McFee R., 1963, Resistivity of body tissues at low frequencies, Circ Res, 12, 40–50.

Saha S., Reddy G.N. and Albright J.A., 1984, Factors affecting the measurement of bone impedance, Med Biol Eng Comput, 22, 123–129.

Schepps J.L. and Foster K.R., 1980, The UHF and microwave dielectric properties of normal and tumour tissues: variations in dielectric properites with tissue water content, Phys Med Biol, 25, 1149–1159.

Schwan H.P., 1955, Electrical properties of body tissues and impedance plethysmography, IRE Trans Med Electron, PGME-3, 32–45.

Schwan H.P., 1957, Electrical properties of tissue and cell suspensions, Adv in Biol Med Phys, 5, J.A. Lawrence and C.A. Tobias, (eds), Academic Press, New York, pp.147–209.

Schwan H.P., 1984, Electrical and acoustical properties of biological materials and biomedical applications, IEEE Trans Biomed Eng, 31, 872–878.

Schwan H.P. and Foster K.R., 1980, RF-Field interactions with biological systems: electrical properties and biophysical mechanisms, Proc IEEE, 68, 104–113.

Schwan H.P. and Kay C.F., 1956, Specific resistance of body tissues, Circ Res, 4, 664–670.

Schwan H.P. and Kay C.F., 1957, The conductivity of living tissues, Ann NY Acad Sci, 65, 1007–1013.

Schwan H.P. and Li K., 1953, Capacity and conductivity of body tissues at ultra high frequencies, Proc IRE, 41, 1735–1740.

Schwan H.P. and Piersol G.M., 1954, The absorption of electromagnetic energy in body tissues, Amer J Phys Med, 33, 371–404.

Schwan H.P., Sheppard R.J. and Grant E.H., 1976, Complex permittivity of water at 25°C, J Chem Phys, 64, 2257–2258.

Schwartz J.L. and Mealing G.A.R., 1985, Dielectric properties of frog tissues *in-vivo* and *in-vitro*, Phys Med Biol, 30, 117–124.

Shamos M.H. and Lavine L.S., 1967, Piezoelectricity as a fundamental property of biological tissues, Nature, 213, 267–269.

Sigman E., Kolin A., Katz L.N. and Jochim K., 1937, Effect of motion on the electrical conductivity of the blood, Am J Physiol, 118, 708–719.

Singh S., Behari J. and Ranu H.S., 1985, Piezoelectric properties of bone materials, Proc 7th Int Conf Med Phys, Espoo, Finland, 123–124.

Smith S.R. and Foster K.R., 1985, Dielectric properties of low-water-content tissues, Phys Med Biol, 30, 965–973.

Smith S.R., Foster K.R. and Wolf G.L., 1986, Dielectric properties of VX-2 carcinoma versus normal liver tissue, IEEE Trans Biomed Eng, 33, 522–524.

Spector W.S., 1956, *Handbook of Biological Data*, W. B. Saunders Co.

Steel M.C. and Sheppard R.J., 1985, Dielectric properties of mammalian brain tissue between 1 and 18 GHz, Phys Med Biol, 30, 621–630.

Steel M.C. and Sheppard R.J., 1988, The dielectric properties of rabbit tissue, pure water and various liquids suitable for tissue phantoms at 35 GHz, Phys Med Biol, 33, 467–472.

Stogryn A., 1971, Equations for calculating the dielectric constant of saline water, IEEE Trans Microwave Theory & Techniques, 19, 733–736.

Stoy R.D., Foster K.R. and Schwan H.P., 1982, Dielectric properties of mammalian tissues from 0.1 to 100 MHz: a summary of recent data, Phys Med Biol, 27, 501–513.

Stuchly M.A. and Stuchly S.S., 1980a, Dielectric properties of biological substances – tabulated, J Microwave Power, 15, 19–26.

Stuchly M.A. and Stuchly S.S., 1980b, Coaxial line reflection methods for measuring dielectric properties of biological substances at radio and microwave frequencies – a review, IEEE Trans Instrum Meas, 29, 176–183.

Stuchly M.A., Athey T.W., Stuchly S.S., Samoras G.M. and Taylor G., 1981, Dielectric properties of animal tissues *in vivo* at frequencies 10 MHz–1 GHz, Bioelectromagnetics, 2, 93–103.

Stuchly M.A., Athey T.W., Samoras G.M. and Taylor G.E., 1982a, Measurement of radiofrequency permittivity of biological tissues with an open-ended coaxial line: Part II – experimental results, IEEE Trans Microwave Theory & Techniques, 30, 87–92.

Stuchly M.A., Kraszewski A., Stuchly S.S. and Smith A.W., 1982b, Dielectric properties of animal tissues *in-vivo* at radio and microwave frequencies: comparison between species, Phys Med Biol, 27, 927–936.

Surowiec A., Stuchly S.S. and Swarup A., 1985, Radiofrequency dielectric properties of animal tissues as a function of time following death, Phys Med Biol, 30, 1131–1141.

Surowiec A., Stuchly S.S., Keaney M., and Swarup A., 1986a, *In-vivo* and *in-vitro* dielectric properties of feline tissues at low radiofrequencies, Phys Med Biol, 31, 901–909.

Surowiec A., Stuchly S.S. and Swarup A., 1986b, Postmortem changes of the dielectic properties of bovine brain tissues at low radiofrequencies, Bioelectromagnetics, 7, 31–43.

Surowiec A.J., Stuchley S. S., Keaney M. and Swarup A., 1987a, Dielectic polarisation of animal lung at radio frequencies, IEEE Trans Biomed Eng, 34, 62–67.

Surowiec A., Stuchley S.S., Eidus L. and Swarup A., 1987b, *In-vitro* dielectic properties of human tissues at radiofrequencies, Phys Med Biol, 32, 615–621.

Surowiec A.J., Stuchly S.S., Barr J.R. and Swarup A., 1988, Dielectric properties of breast carcinoma and the surrounding tissues, IEEE Trans Biomed Eng, 35, 257–263.

Thurai M., Goodridge V.D., Sheppard R.J. and Grant E.H., 1984, Variation with age of the dielectric properties of mouse brain cerebrum, Phys Med Biol, 29, 1133–1136.

Thurai M., Steel M.C., Sheppard R.J. and Grant E.H., 1985, Dielectric properties of developing rabbit brain at 37°C, Bioelectromagnetics, 6, 235–242

Troutman E.D. and Newbower R.S., 1983, A practical analysis of the electrical conductivity of blood, IEEE Trans Biomed Eng, 30, 141–153.

Van Harreveld A., Murphy T. and Nobel K.W., 1963, Specific impedance of rabbit's cortical tissue, Am J Physiol, 205, 203–207.

Van Oosterom A., de Boer R.W. and van Dam R.Th., 1979, Intramural resistivity of cardiac tissue, Med & Biol Eng & Comput, 17, 337–343.

Vellick S. and Gavin M., 1940, The electrical conductance of suspension of ellipsoids and its relationship to the study of avian erythrocytes, J Gen Physiol, 23, 753–771.

Weidmann S., 1970, Electrical constants of trabecular muscle from mammalian heart, J Physiol, 210, 1041–1054.

Witsoe D.A. and Kinnen E., 1967, Electrical resistivity of lung at 100 kHz, Med Biol Engng, 5, 239–247.

Yamakoshi K., Shimazu H., Togawa T., Fukuoka M. and Ito H., 1980, Non–invasive measurement of hematocrit by electrical admittance plethysmography technique, IEEE Trans Biomed Eng, 27, 156–161

Yamamoto T. and Yamamoto Y., 1976, Dielectric constant and resistivity of epidermal stratum corneum, Med Biol Eng, 14, 494–499.

Yamamoto T. and Yamamoto Y., 1978, Dispersion and correlation of the parameters for skin impedance, Med & Biol Eng & Comput, 16, 592–594.

Yamamoto T. and Yamamoto Y., 1981, Non–linear electrical properties of skin in the low frequency range, Med & Biol Eng & Comput, 19, 302–310.

Zheng E., Shao S. and Webster J.G., 1984, Impedance of skeletal muscle from 1 Hz to 1 MHz, IEEE Trans Biomed Eng, 31, 477–481.

Zore V.A., Kimel'fel'd O.D., Suzdaleva V.V., Kobyzeva L.P. and Genkina Ye.S., 1967, Complex dielectic permiability in the frequency range 100–500 Mc/s of human blood serum in normal conditions and in certain diseases, Biophys, 12, 142–145.

Zywietz F. and Knochel R., 1986, Dielectric properties of Co–γ–irradiated and microwave–heated rat tumour and skin measured *in–vivo* between 0.2 and 2.4 GHz, Phys Med Biol, 31, 1021–1029.

Chapter 7

Ionising Radiation and Tissue

This chapter considers the interaction between tissue and all forms of ionising radiation, both particles and photons. The interaction properties of x–radiation and gamma radiation for photon energies between 10 keV and 100 MeV are presented, together with the properties of the energetic subatomic particles of major importance, namely electrons, positrons, protons and neutrons. The energy ranges for the particulate radiations are; electrons and positrons, 10 keV to 1 GeV; protons 1 to 500 MeV; neutrons 25 meV to 100 MeV.

7.1 Photons: x–radiation and gamma radiation

7.1.1 *Terminology and definitions*

For photons the **mass attenuation coefficient**, μ/ρ, is given by the equation:

$$\frac{\mu}{\rho} = \frac{1}{\rho N}\frac{dN}{dl} \qquad (7.1)$$

where dN/N is the fraction of photons experiencing interactions while traversing a distance dl in a material of density ρ. The unit is metre^2 per kilogram ($m^2\ kg^{-1}$).

For x–radiation and gamma radiation the total mass attenuation coefficient is composed of the sum of four components:

$$\frac{\mu}{\rho} = \frac{\tau}{\rho} + \frac{\sigma_c}{\rho} + \frac{\sigma_{coh}}{\rho} + \frac{\kappa}{\rho} \qquad (7.2)$$

where the component mass attenuation coefficients refer to the four processes contributing to the total loss, namely **photoelectric absorption** τ/ρ, **Compton scattering** σ_c/ρ, **coherent scattering** σ_{coh}/ρ, and **pair production** κ/ρ.

For a mixture such as tissue, the mass attenuation coefficient can be closely approximated from a simple mixture rule, adding the weighted sum of the mass attenuation coefficients of the constituent elements:

$$\frac{\mu}{\rho} = \sum_i w_i \frac{\mu_i}{\rho_i} \tag{7.3}$$

where w_i is the proportion by weight of the *i*th constituent with attenuation coefficient μ_i and density ρ_i. For a compound with the formula $(A_1)_{b1}(A_2)_{b2}...(A_n)_{bn}$, the weight fraction is given by

$$w_i = b_i A_i \left[\sum_{j=1}^{n} b_j A_j\right]^{-1} \tag{7.4}$$

where A_i and b_i are the number and atomic mass of the ith element.

Limitations to the mixture rule are that it ignores changes in the atomic wave function resulting from the molecular or crystalline environment of an atom. However, errors from this source are no more than a few per cent for photon energies above 10 keV (Hubbell, 1969).

The mass attenuation coefficient is also proportional to the total photon interaction cross-section per atom, σ_a, and to the total cross-section per electron σ_e. The appropriate relationships are:

$$\frac{\mu}{\rho} = \sigma_a \frac{N_A}{A} = \sigma_e \frac{N_A Z}{A} \tag{7.5}$$

where N_A is the Avagadro constant, 6.022×10^{23} mol^{-1}, A is the atomic mass of the material having Z electrons per atom. The quantity $N_A Z/A$ is the mass electron density, which has an approximately constant value, $N_A/2$, for all elements except hydrogen.

The **effective atomic number**, Z_{eff}, can be defined for each of the partial processes, photoelectric absorption, coherent and incoherent scatter and pair production:

$$Z_{eff} = \left[\sum_i \alpha_i Z_i^{\,m}\right]^{1/m} \tag{7.6}$$

where α_i is the fractional electron number and Z_i the atomic number of the constituent element. m is a parameter to fit the cross-section per electron as a function of Z for the particular interaction process (McCullough, 1975, White, 1977a). It is possible also to calculate the total effective atomic number for tissues from the total photon interaction cross-sections per electron (Yang *et al*, 1987), giving values which are less dependent on Z. Other definitions for the effective atomic number may give considerably different values (Thirumala Rao *et al*, 1985; Parthasaradhi *et al*, 1989).

The **mass energy transfer coefficient**, μ_{tr}/ρ is given by

$$\frac{\mu_{tr}}{\rho} = \frac{1}{\rho EN}\frac{dE_{tr}}{dl} \qquad (7.7)$$

where E is the energy of each particle, N is the number of particles, and dE_{tr}/EN is the fraction of incident energy that is transferred to kinetic enegy of charged particles in traversing a distance dl.

The **mass energy absorption coefficient**, μ_{en}/ρ, quantifies that part of the mass energy transfer from the incident radiation which is deposited in the material, omitting that fraction, g, which is lost to brehmsstrahlung in the material.

$$\frac{\mu_{en}}{\rho} = \frac{\mu_{tr}}{\rho}(1 - g) \qquad (7.8)$$

The unit for both μ_{en}/ρ and μ_{tr}/ρ is $m^2\ kg^{-1}$.

μ is termed the **total linear attenuation coefficient**, with unit m^{-1}. For use in the imaging of linear attenuation coefficient by means of computed x-ray tomography (CT), the attenuation is measured by **CT number**:

$$\text{CT number} = \frac{\mu_t - \mu_w}{\mu_w} \times 1000 \qquad (7.9)$$

μ_t and μ_t are the linear attenuation coefficients of tissue and water, evaluated at the same energy. Since the energy dependence of μ_t and μ_w are not necessarily identical, there is a small but finite variation of CT number with energy over the range of interest, using tube potentials between about 90 to 140 kV (Zatz, 1976). Furthermore the use of polychromatic radiation in x-ray CT scanners must give rise to some variation beween estimates of μ from CT measurements and those from direct measurement (McCullough, 1975, Bydder and Kreel, 1980). However, these differences appear to be minor in practice for many soft tissues (Phelps *et al*, 1975). In early reports the attenuation derived from CT scan measurements was quoted in Hounsfield units. The relationship is (Hounsfield unit) = (CT number)/2.

7.1.2 *Measurement and calculation of interaction coefficients*

The detailed understanding of the interaction of ionising radiation and matter has resulted in the ability to predict with considerable accuracy many of the quantities defined in Section 7.1.1. It has therefore been possible to compile for this chapter predicted values of quantities where other chapters have presented largely or solely experimentally measured values. The predictive expressions require knowlege of both density and the elemental composition of tissue. Values of the percentage composition of tissues are given in Chapter 9, and of mass density and electron density in Table 5.1. Using such information it has been possible to compile extensive tables of predicted tissue properties covering a wide range of tissue type, form of radiation and energy (see, for instance, Kim, 1974a,b). Values primarily drawn from the latest available authorative compilation (ICRP, 1989) are included in the tabulations of data in this chapter.

The measurement methods which have been used to validate the predicted values fall into two main categories. The first uses the gamma emissions from selected radionuclides to make energy–specific measurements through slabs of tissue of known thickness. Phelps *et al* (1975) used five separate radioactive sources to cover the energy range 17.7 to 136.3 keV, namely ^{241}Am (17.7, 21.1, 26.4 and 59.5 keV), ^{125}I (27.4, 31.1 and 35.5 keV), ^{153}Gd (41.4, 47.2, 97.4, and 103.2 keV), ^{170}Tm (52.0, 59.6, and 84.3 keV) and ^{57}Co (121.9 and 136.4 keV). Others have extended the energy range upwards ^{203}Hg (279 keV) and ^{137}Cs (662 keV) (Rao and Gregg, 1975; Joyet *et al*, 1974). The use of fluorescent x–rays allowed measurements to be made at energies as low as 9.88 keV using germanium k_α radiation (White *et al,* 1980). The detectors used have varied, with the earlier use of NaI(Tl) scintillation crystals (Rao and Gregg, 1975; Joyet *et al*, 1974) being superceded by the use of solid state germanium detectors (Phelps *et al*, 1975; White *et al*, 1980) giving higher spectral resolution. When density of the tissue sample was also measured then mass attenuation coefficients could also be calculated (Joyet *et al*, 1974; White *et al*, 1980).

The second alternative source of photons is the x–ray tube. The method used by Spiers (1946) to generate a narrow band photon beam was by the use of a pair of balanced filters composed of adjacent elements in the periodic table, silver and cadmium. The transmission is only due to the energies between the elemental absorption edges. Spiers used an ionisation chamber as a detector. More recently use was made of a high–purity germanium spectroscopy system, where the photon source was the polychromatic x–radiation from an x–ray tube operated at 120 kV (Johns and Yaffe, 1987).

The remaining experimental results derive from many measurements made using a variety of commercial x–ray computed tomography systems. Selected examples from these measurements are included in the tabulated data.

7.1.3 *Values of attenuation and absorption coefficients for tissue*

Calculated values of mass attenuation coefficient, μ/ρ, and mass energy absorption coefficient μ_{en}/ρ for photons for a selection of twenty 'standard' tissues over the energy range 10 keV to 100 MeV are shown in Tables 7.1 and 7.2. The cross–sections for photoelectric, coherent, Compton and pair production for four of these standard tissues, muscle, adipose (fatty), cortical bone and cartilage are given in Tables 7.3 and 7.4. The values for muscle can be taken to closely characterise the behaviour of most soft tissues. These values have been drawn from a more detailed tabulation given in ICRU, 1989, using average recommended compositions. The data for the separate cross–sections and mass attenuation coefficients were calculated using the XCOM computer code (Berger and Hubbell, 1987). Mass energy absorption coefficients from 0.01 to 20 MeV were calculated using the elemental data of Hubbell (1982). Calculations were extended to 100 MeV by normalisation to the elemental data of Storm and Israel (1970).

The estimated uncertainties associated with these predictions vary depending on the radiation and energy considered. For photons in the energy range dominated by Compton scattering, the estimated uncertainty is 1%. Contributions from low–Z elements in tissue at energies below 30 keV, may give rise to uncertainties of 5–10%.

For comparison, Table 7.5 shows mass attenuation coefficients for several tissues measured *in–vitro*, at energies from 9.9 to 662 keV. Measured linear attenuation coefficients over the same energy range are shown in Tables 7.6 to 7.8 for a range of human tissues, with values for breast tissue (Table 7.7) and some pathological tissues (Table 7.8) presented separately.

CT numbers calculated at a photon energy of 100 keV from mass attenuation coefficients and measured density values (Table 5.1) are shown in Table 7.9. Table 7.10 gives measured values of CT number from a variety of sources, both *in–vivo* and *in–vitro*.

Calculated effective atomic numbers for several tissues, from Yang *et al* (1987) are given in Table 7.11.

7.1.4 *Factors affecting photon attenuation*

7.1.4.1 Photon energy

All the contributing processes, scattering, photoelectric and pair production, are dependent on the photon energy, and therefore μ/ρ is energy dependent. The pattern of relative contribution from the four processes for skeletal muscle as a typical soft tissue is shown in Figure 7.1, and for cortical bone in Figure 7.2. Absorption edges associated with ionisation do not appear since the ionisation energies of the low atomic weight components of tissue are too low to be of significance.

Table 7.1a Mass attenuation coefficients, μ/ρ, for photons in the range 10 to 80 keV ($m^2\ kg^{-1}\ x10^{-2}$)

Photon energy,keV	10	15	20	30	40	50	60	80
Blood	55.2	17.4	8.43	3.85	2.71	2.28	2.06	1.83
Bone,cortical	285	90.3	40.0	13.3	6.66	4.24	3.15	2.23
Bone,trabecular	121	38.5	17.6	6.63	3.88	2.86	2.38	1.95
Bone,red marrow	43.5	14.1	7.04	3.46	2.55	2.20	2.01	1.81
Bone,yellow marrow	31.0	10.3	5.48	3.01	2.37	2.11	1.97	1.80
Brain	54.1	17.1	8.28	3.81	2.70	2.27	2.06	1.83
Breast	43.0	13.8	6.89	3.40	2.53	2.19	2.01	1.81
Cartilage	63.7	20.0	9.48	4.15	2.83	2.33	2.08	1.83
Cell nucleus	60.5	19.0	9.07	4.04	2.79	2.32	2.08	1.84
Eye,lens	48.3	15.3	7.51	3.57	2.59	2.21	2.01	1.80
Fatty tissue	32.7	10.8	5.68	3.06	2.40	2.12	1.97	1.80
Intestine	51.7	16.3	7.94	3.71	2.66	2.25	2.04	1.82
Kidney	53.6	16.9	8.20	3.78	2.69	2.26	2.05	1.82
Liver	54.0	17.1	8.26	3.80	2.69	2.27	2.05	1.82
Lung,inflated	54.6	17.2	8.32	3.82	2.70	2.27	2.05	1.83
Lymph	54.6	17.2	8.29	3.81	2.70	2.28	2.06	1.83
Muscle,cardiac	54.8	17.3	8.38	3.84	2.71	2.28	2.06	1.83
Muscle,skeletal	53.6	16.9	8.21	3.78	2.69	2.26	2.05	1.82
Ovary	54.1	17.1	8.25	3.80	2.69	2.27	2.05	1.83
Pancreas	50.9	16.1	7.87	3.69	2.65	2.25	2.04	1.82
Skin	49.5	15.7	7.68	3.63	2.62	2.23	2.03	1.81
Spleen	54.3	17.1	8.29	3.81	2.70	2.27	2.05	1.83
Testicle	53.6	16.9	8.18	3.78	2.69	2.27	2.05	1.83
Thyroid gland	53.3	16.9	8.19	3.79	2.87	2.37	2.11	1.86

Table 7.1b Mass attenuation coefficients, μ/ρ, for photons in the range 0.1 to 1 MeV (m^2 kg^{-1} $x10^{-2}$)

Photon energy,MeV	0.1	0.2	0.3	0.4	0.5	0.6	0.8	1.0
Blood	1.70	1.36	1.18	1.05	0.960	0.887	0.779	0.701
Bone,cortical	1.86	1.31	1.11	0.991	0.902	0.833	0.731	0.657
Bone,trabecular	1.75	1.35	1.16	1.04	0.945	0.874	0.767	0.689
Bone,red marrow	1.69	1.36	1.18	1.05	0.962	0.890	0.781	0.702
Bone,yellow marrow	1.69	1.37	1.19	1.06	0.970	0.897	0.788	0.708
Brain	1.70	1.36	1.18	1.06	0.964	0.891	0.783	0.704
Breast	1.69	1.36	1.18	1.06	0.963	0.890	0.782	0.703
Cartilage	1.69	1.35	1.17	1.05	0.954	0.882	0.775	0.697
Cell nucleus	1.71	1.36	1.18	1.06	0.963	0.890	0.782	0.703
Eye,lens	1.68	1.35	1.17	1.05	0.955	0.883	0.775	0.697
Fatty tissue	1.69	1.37	1.19	1.06	0.970	0.897	0.787	0.708
Intestine	1.70	1.36	1.18	1.06	0.963	0.891	0.782	0.703
Kidney	1.69	1.36	1.18	1.05	0.961	0.888	0.780	0.701
Liver	1.69	1.36	1.18	1.05	0.960	0.887	0.779	0.701
Lung,inflated	1.70	1.36	1.18	1.05	0.961	0.888	0.780	0.701
Lymph	1.70	1.37	1.18	1.06	0.965	0.892	0.784	0.705
Muscle,cardiac	1.70	1.36	1.18	1.05	0.961	0.888	0.780	0.701
Muscle,skeletal	1.69	1.36	1.18	1.05	0.960	0.887	0.779	0.701
Ovary	1.70	1.36	1.18	1.05	0.962	0.890	0.781	0.703
Pancreas	1.70	1.36	1.18	1.06	0.963	0.891	0.782	0.703
Skin	1.69	1.35	1.17	1.05	0.958	0.886	0.778	0.699
Spleen	1.70	1.36	1.18	1.05	0.961	0.888	0.780	0.701
Testicle	1.70	1.36	1.18	1.06	0.963	0.891	0.782	0.703
Thyroid gland	1.71	1.36	1.18	1.05	0.962	0.889	0.781	0.702

Table 7.1c Mass attenuation coefficients, μ/ρ, for photons in the range 2 to 100 MeV (m^2 kg^{-1} $x10^{-2}$)

Photon energy,MeV	2	4	6	10	20	40	60	100
Blood	0.490	0.337	0.274	0.220	0.179	0.166	0.166	0.170
Bone,cortical	0.461	0.326	0.273	0.231	0.207	0.207	0.214	0.226
Bone,trabecular	0.482	0.333	0.273	0.221	0.183	0.172	0.173	0.179
Bone,red marrow	0.490	0.336	0.272	0.216	0.173	0.158	0.156	0.160
Bone,yellow marrow	0.494	0.338	0.272	0.214	0.169	0.151	0.149	0.151
Brain	0.492	0.338	0.275	0.220	0.179	0.165	0.165	0.169
Breast	0.491	0.337	0.273	0.217	0.175	0.159	0.158	0.162
Cartilage	0.487	0.336	0.274	0.220	0.181	0.168	0.169	0.174
Cell nucleus	0.491	0.338	0.276	0.221	0.181	0.168	0.168	0.173
Eye,lens	0.487	0.335	0.272	0.217	0.177	0.163	0.162	0.167
Fatty tissue	0.494	0.338	0.272	0.214	0.170	0.152	0.150	0.153
Intestine	0.491	0.338	0.275	0.220	0.179	0.165	0.165	0.170
Kidney	0.490	0.337	0.274	0.219	0.179	0.165	0.165	0.170
Liver	0.490	0.337	0.274	0.219	0.179	0.165	0.165	0.170
Lung,inflated	0.490	0.337	0.275	0.220	0.179	0.166	0.166	0.171
Lymph	0.492	0.339	0.276	0.221	0.181	0.167	0.167	0.172
Muscle,cardiac	0.490	0.337	0.274	0.220	0.179	0.166	0.165	0.170
Muscle,skeletal	0.490	0.337	0.274	0.219	0.179	0.165	0.165	0.169
Ovary	0.491	0.338	0.275	0.220	0.180	0.166	0.166	0.171
Pancreas	0.491	0.338	0.275	0.219	0.178	0.164	0.164	0.168
Skin	0.489	0.336	0.273	0.218	0.177	0.163	0.163	0.167
Spleen	0.490	0.337	0.274	0.220	0.179	0.166	0.166	0.170
Testicle	0.491	0.338	0.275	0.220	0.180	0.166	0.166	0.170
Thyroid gland	0.490	0.338	0.275	0.220	0.179	0.166	0.165	0.170

Table 7.2a Mass energy absorption coefficients, μ_{en}/ρ, for photons in the range 10 to 80 keV ($m^2\ kg^{-1}\ x10^{-2}$)

Photon energy,keV	10	15	20	30	40	50	60	80
Blood	49.9	14.1	5.70	1.63	0.730	0.441	0.329	0.263
Bone,cortical	264	82.8	35.5	10.6	4.46	2.31	1.38	0.684
Bone,trabecular	111	33.9	14.4	4.28	1.84	1.00	0.647	0.391
Bone,red marrow	38.7	10.9	4.42	1.29	0.594	0.376	0.294	0.250
Bone,yellow marrow	26.9	7.41	2.97	0.875	0.429	0.295	0.249	0.233
Brain	49.1	13.8	5.57	1.59	0.713	0.433	0.325	0.262
Breast	38.4	10.7	4.28	1.23	0.567	0.361	0.285	0.246
Cartilage	58.5	16.6	6.72	1.92	0.843	0.496	0.359	0.274
Cell nucleus	55.3	15.6	6.31	1.80	0.796	0.473	0.347	0.270
Eye,lens	43.6	12.1	4.86	1.39	0.629	0.390	0.300	0.251
Fatty tissue	28.5	7.87	3.16	0.925	0.448	0.304	0.254	0.235
Intestine	46.8	13.0	5.24	1.49	0.671	0.411	0.313	0.257
Kidney	48.6	13.6	5.50	1.57	0.703	0.427	0.322	0.260
Liver	49.0	13.8	5.55	1.59	0.710	0.431	0.323	0.261
Lung,inflated	49.6	13.9	5.60	1.60	0.714	0.432	0.324	0.261
Lymph	49.6	13.8	5.56	1.58	0.705	0.428	0.322	0.261
Muscle,cardiac	49.5	14.0	5.65	1.62	0.725	0.439	0.328	0.263
Muscle,skeletal	48.6	13.6	5.50	1.57	0.705	0.428	0.322	0.260
Ovary	49.1	13.7	5.53	1.58	0.705	0.428	0.322	0.260
Pancreas	46.1	12.9	5.18	1.48	0.667	0.410	0.312	0.257
Skin	44.7	12.5	5.02	1.44	0.649	0.400	0.306	0.254
Spleen	49.3	13.8	5.58	1.59	0.712	0.431	0.324	0.261
Testicle	48.6	13.6	5.47	1.56	0.698	0.425	0.320	0.260
Thyroid gland	48.3	13.5	5.47	1.57	0.770	0.479	0.358	0.280

Table 7.2b Mass energy absorption coefficients, μ_{en}/ρ, for photons in the range 0.1 to 1 MeV ($m^2\ kg^{-1}\ x10^{-2}$)

Photon energy,MeV	0.1	0.2	0.3	0.4	0.5	0.6	0.8	1.0
Blood	0.255	0.294	0.316	0.325	0.327	0.325	0.318	0.307
Bone,cortical	0.456	0.300	0.303	0.307	0.307	0.305	0.297	0.287
Bone,trabecular	0.317	0.297	0.313	0.320	0.322	0.320	0.312	0.302
Bone,red marrow	0.249	0.294	0.317	0.326	0.328	0.326	0.318	0.308
Bone,yellow marrow	0.242	0.296	0.320	0.328	0.330	0.329	0.321	0.311
Brain	0.255	0.295	0.318	0.326	0.328	0.327	0.319	0.309
Breast	0.247	0.294	0.317	0.326	0.328	0.327	0.319	0.308
Cartilage	0.260	0.293	0.315	0.323	0.325	0.323	0.316	0.305
Cell nucleus	0.259	0.296	0.318	0.326	0.328	0.326	0.319	0.308
Eye,lens	0.248	0.292	0.315	0.323	0.325	0.324	0.316	0.306
Fatty tissue	0.243	0.296	0.319	0.328	0.330	0.329	0.321	0.311
Intestine	0.253	0.295	0.317	0.326	0.328	0.327	0.319	0.308
Kidney	0.254	0.294	0.317	0.325	0.327	0.326	0.318	0.308
Liver	0.254	0.294	0.316	0.325	0.327	0.325	0.318	0.307
Lung,inflated	0.254	0.294	0.317	0.325	0.327	0.326	0.318	0.308
Lymph	0.255	0.296	0.318	0.327	0.329	0.327	0.319	0.309
Muscle,cardiac	0.255	0.295	0.317	0.325	0.327	0.326	0.318	0.308
Muscle,skeletal	0.254	0.294	0.316	0.325	0.327	0.325	0.318	0.307
Ovary	0.254	0.295	0.317	0.326	0.328	0.326	0.318	0.308
Pancreas	0.252	0.295	0.317	0.326	0.328	0.327	0.319	0.308
Skin	0.250	0.293	0.316	0.324	0.326	0.325	0.317	0.307
Spleen	0.254	0.294	0.317	0.325	0.327	0.326	0.318	0.308
Testicle	0.254	0.295	0.318	0.326	0.328	0.327	0.319	0.308
Thyroid gland	0.266	0.297	0.318	0.326	0.328	0.326	0.318	0.308

Table 7.2c Mass energy absorption coefficients, μ_{en}/ρ, for photons in the range 2 to 100 MeV ($m^2\ kg^{-1}\ x10^{-2}$)

Photon energy,MeV	2	4	6	10	20	40	60	100
Blood	0.258	0.204	0.179	0.155	0.137	0.128	0.124	0.120
Bone,cortical	0.242	0.197	0.178	0.164	0.156	0.153	0.149	0.143
Bone,trabecular	0.254	0.202	0.177	0.155	0.139	0.132	0.127	0.123
Bone,red marrow	0.259	0.204	0.177	0.152	0.132	0.123	0.118	0.115
Bone,yellow marrow	0.261	0.205	0.177	0.150	0.129	0.119	0.114	0.110
Brain	0.259	0.205	0.179	0.155	0.137	0.128	0.124	0.119
Breast	0.259	0.204	0.178	0.153	0.133	0.124	0.120	0.116
Cartilage	0.256	0.204	0.178	0.155	0.138	0.130	0.126	0.121
Cell nucleus	0.259	0.205	0.179	0.156	0.138	0.130	0.125	0.121
Eye,lens	0.257	0.203	0.177	0.153	0.135	0.126	0.122	0.118
Fatty tissue	0.261	0.205	0.177	0.151	0.129	0.120	0.115	0.111
Intestine	0.259	0.205	0.179	0.155	0.137	0.128	0.124	0.119
Kidney	0.258	0.204	0.179	0.155	0.137	0.128	0.124	0.119
Liver	0.258	0.204	0.178	0.155	0.137	0.128	0.124	0.119
Lung,inflated	0.258	0.205	0.179	0.155	0.137	0.129	0.124	0.120
Lymph	0.259	0.206	0.180	0.156	0.138	0.129	0.125	0.121
Muscle,cardiac	0.258	0.204	0.179	0.155	0.137	0.128	0.124	0.120
Muscle,skeletal	0.258	0.204	0.178	0.155	0.136	0.128	0.124	0.119
Ovary	0.259	0.205	0.179	0.155	0.137	0.129	0.124	0.120
Pancreas	0.259	0.205	0.179	0.155	0.136	0.127	0.123	0.119
Skin	0.258	0.204	0.178	0.154	0.135	0.127	0.122	0.118
Spleen	0.258	0.204	0.179	0.155	0.137	0.128	0.124	0.120
Testicle	0.259	0.205	0.179	0.155	0.137	0.129	0.124	0.120
Thyroid gland	0.258	0.205	0.179	0.155	0.137	0.128	0.124	0.120

Table 7.3 Photon mass attenuation coefficient components for muscle and fatty tissue ($m^2\ kg^{-1}$)

Photon energy	Skeletal muscle				Fatty tissue			
MeV	τ/ρ	σ_{coh}/ρ	σ_c/ρ	κ/ρ	τ/ρ	σ_{coh}/ρ	σ_c/ρ	κ/ρ
0.01	4.98×10^{-1}	2.23×10^{-2}	1.53×10^{-2}	0	2.93×10^{-1}	1.74×10^{-2}	1.59×10^{-2}	0
0.02	5.59×10^{-2}	8.61×10^{-3}	1.76×10^{-2}	0	3.18×10^{-2}	6.83×10^{-3}	1.81×10^{-2}	0
0.03	1.51×10^{-2}	4.55×10^{-3}	1.81×10^{-2}	0	8.49×10^{-3}	3.57×10^{-3}	1.86×10^{-2}	0
0.04	5.94×10^{-3}	2.79×10^{-3}	1.81×10^{-2}	0	3.30×10^{-3}	2.18×10^{-3}	1.85×10^{-2}	0
0.06	1.58×10^{-3}	1.35×10^{-3}	1.75×10^{-2}	0	8.64×10^{-4}	1.05×10^{-3}	1.78×10^{-2}	0
0.08	6.14×10^{-4}	7.92×10^{-4}	1.68×10^{-2}	0	3.33×10^{-4}	6.13×10^{-4}	1.71×10^{-2}	0
0.10	2.95×10^{-4}	5.19×10^{-4}	1.61×10^{-2}	0	1.59×10^{-4}	4.00×10^{-4}	1.63×10^{-2}	0
0.20	3.13×10^{-5}	1.35×10^{-4}	1.34×10^{-2}	0	1.66×10^{-5}	1.03×10^{-4}	1.36×10^{-2}	0
0.30	8.88×10^{-6}	6.03×10^{-5}	1.17×10^{-2}	0	4.69×10^{-6}	4.62×10^{-5}	1.18×10^{-2}	0
0.40	3.81×10^{-6}	3.40×10^{-5}	1.05×10^{-2}	0	2.01×10^{-6}	2.60×10^{-5}	1.06×10^{-2}	0
0.60	1.28×10^{-6}	1.52×10^{-5}	8.86×10^{-3}	0	6.74×10^{-7}	1.16×10^{-5}	8.95×10^{-3}	0
0.80	6.49×10^{-7}	8.53×10^{-6}	7.78×10^{-3}	0	3.40×10^{-7}	6.52×10^{-6}	7.87×10^{-3}	0
1.00	4.03×10^{-7}	5.46×10^{-6}	7.00×10^{-3}	0	2.11×10^{-7}	4.18×10^{-6}	7.07×10^{-3}	0
2.00	1.16×10^{-7}	1.37×10^{-6}	4.86×10^{-3}	3.82×10^{-5}	6.09×10^{-8}	1.04×10^{-6}	4.91×10^{-3}	3.26×10^{-5}
3.00	2.48×10^{-8}	6.07×10^{-7}	3.82×10^{-3}	1.10×10^{-4}	3.41×10^{-8}	4.64×10^{-7}	3.86×10^{-3}	9.47×10^{-5}
4.00	4.44×10^{-8}	3.42×10^{-7}	3.19×10^{-3}	1.82×10^{-4}	2.34×10^{-8}	2.61×10^{-7}	3.22×10^{-3}	1.57×10^{-4}
6.00	2.70×10^{-8}	1.52×10^{-7}	2.43×10^{-3}	3.08×10^{-4}	1.43×10^{-8}	1.16×10^{-7}	2.46×10^{-3}	2.67×10^{-4}
8.00	1.93×10^{-8}	8.54×10^{-8}	1.99×10^{-3}	4.12×10^{-4}	1.02×10^{-8}	6.53×10^{-8}	2.01×10^{-3}	3.57×10^{-4}
10.00	1.50×10^{-8}	5.47×10^{-8}	1.69×10^{-3}	4.98×10^{-4}	7.96×10^{-9}	4.18×10^{-8}	1.71×10^{-3}	4.33×10^{-4}
20.00	7.09×10^{-9}	1.37×10^{-8}	1.01×10^{-3}	7.80×10^{-4}	3.77×10^{-9}	1.04×10^{-8}	1.02×10^{-3}	6.81×10^{-4}
30.00	4.64×10^{-9}	6.07×10^{-9}	7.33×10^{-4}	9.50×10^{-4}	2.46×10^{-9}	4.64×10^{-9}	7.40×10^{-4}	8.32×10^{-4}
40.00	3.44×10^{-9}	3.42×10^{-9}	5.82×10^{-4}	1.07×10^{-3}	1.83×10^{-9}	2.61×10^{-9}	5.88×10^{-4}	9.36×10^{-4}
60.00	2.27×10^{-9}	1.52×10^{-9}	4.19×10^{-4}	1.23×10^{-3}	1.21×10^{-9}	1.16×10^{-9}	4.23×10^{-4}	1.08×10^{-3}
80.00	1.70×10^{-9}	0.85×10^{-9}	3.30×10^{-4}	1.34×10^{-3}	0.90×10^{-9}	0.65×10^{-9}	3.34×10^{-4}	1.18×10^{-3}
100.00	1.35×10^{-9}	0.55×10^{-9}	2.74×10^{-4}	1.42×10^{-3}	0.72×10^{-9}	0.42×10^{-9}	2.77×10^{-4}	1.25×10^{-3}

Table 7.4 Photon mass attenuation coefficient components for bone and cartilage (m^2 kg^{-1})

Photon energy	Cortical bone				Cartilage			
MeV	τ/ρ	σ_{coh}/ρ	σ_c/ρ	κ/ρ	τ/ρ	σ_{coh}/ρ	σ_c/ρ	κ/ρ
0.01	2.80	4.18×10^{-2}	1.29×10^{-2}	0	5.98×10^{-1}	2.39×10^{-2}	1.51×10^{-2}	0
0.02	3.68×10^{-1}	1.70×10^{-2}	1.55×10^{-2}	0	6.82×10^{-2}	9.21×10^{-3}	1.74×10^{-2}	0
0.03	1.08×10^{-1}	9.08×10^{-3}	1.63×10^{-2}	0	1.86×10^{-2}	4.88×10^{-3}	1.80×10^{-2}	0
0.04	4.44×10^{-2}	5.68×10^{-3}	1.64×10^{-2}	0	7.35×10^{-3}	3.00×10^{-3}	1.80×10^{-2}	0
0.06	1.26×10^{-2}	2.84×10^{-3}	1.61×10^{-2}	0	1.96×10^{-3}	1.45×10^{-3}	1.74×10^{-2}	0
0.08	5.07×10^{-3}	1.69×10^{-3}	1.55×10^{-2}	0	7.68×10^{-4}	8.54×10^{-4}	1.67×10^{-2}	0
0.10	2.51×10^{-3}	1.12×10^{-3}	1.49×10^{-2}	0	3.70×10^{-4}	5.60×10^{-4}	1.60×10^{-2}	0
0.20	2.84×10^{-4}	2.98×10^{-4}	1.25×10^{-2}	0	3.94×10^{-5}	1.46×10^{-4}	1.33×10^{-2}	0
0.30	8.29×10^{-5}	1.35×10^{-4}	1.09×10^{-2}	0	1.12×10^{-5}	6.53×10^{-5}	1.16×10^{-2}	0
0.40	3.62×10^{-5}	7.65×10^{-5}	9.80×10^{-3}	0	4.83×10^{-6}	3.69×10^{-5}	1.04×10^{-2}	0
0.60	1.24×10^{-5}	3.42×10^{-5}	8.29×10^{-3}	0	1.63×10^{-6}	1.64×10^{-5}	8.80×10^{-3}	0
0.80	6.27×10^{-6}	1.93×10^{-5}	7.28×10^{-3}	0	8.23×10^{-7}	9.24×10^{-6}	7.74×10^{-3}	0
1.00	3.90×10^{-6}	1.23×10^{-5}	6.55×10^{-3}	0	5.12×10^{-7}	5.92×10^{-6}	6.96×10^{-3}	0
2.00	1.12×10^{-6}	3.09×10^{-6}	4.54×10^{-3}	5.87×10^{-5}	1.48×10^{-7}	1.48×10^{-6}	4.83×10^{-3}	3.98×10^{-5}
3.00	6.14×10^{-7}	1.37×10^{-6}	3.58×10^{-3}	1.68×10^{-4}	8.21×10^{-8}	6.58×10^{-7}	3.80×10^{-3}	1.15×10^{-4}
4.00	4.16×10^{-7}	7.73×10^{-7}	2.98×10^{-3}	2.74×10^{-4}	5.62×10^{-8}	3.70×10^{-7}	3.17×10^{-3}	1.90×10^{-4}
6.00	2.50×10^{-7}	3.43×10^{-7}	2.28×10^{-3}	4.57×10^{-4}	3.41×10^{-8}	1.64×10^{-7}	2.42×10^{-3}	3.20×10^{-4}
8.00	1.78×10^{-7}	1.93×10^{-7}	1.86×10^{-3}	6.05×10^{-4}	2.44×10^{-8}	9.25×10^{-8}	1.98×10^{-3}	4.27×10^{-4}
10.00	1.38×10^{-7}	1.24×10^{-7}	1.59×10^{-3}	7.28×10^{-4}	1.90×10^{-8}	5.92×10^{-8}	1.68×10^{-3}	5.16×10^{-4}
20.00	6.45×10^{-8}	3.09×10^{-8}	9.42×10^{-4}	1.13×10^{-3}	8.96×10^{-9}	1.48×10^{-8}	1.00×10^{-3}	8.08×10^{-4}
30.00	4.20×10^{-8}	1.37×10^{-8}	6.86×10^{-4}	1.36×10^{-3}	5.85×10^{-9}	6.58×10^{-9}	7.28×10^{-4}	9.84×10^{-4}
40.00	3.12×10^{-8}	7.73×10^{-9}	5.45×10^{-4}	1.53×10^{-3}	4.35×10^{-9}	3.70×10^{-9}	5.79×10^{-4}	1.11×10^{-3}
60.00	2.05×10^{-8}	3.43×10^{-9}	3.92×10^{-4}	1.75×10^{-3}	2.87×10^{-9}	1.64×10^{-9}	4.16×10^{-4}	1.27×10^{-3}
80.00	1.53×10^{-8}	1.93×10^{-9}	3.09×10^{-4}	1.90×10^{-3}	2.14×10^{-9}	0.93×10^{-9}	3.28×10^{-4}	1.38×10^{-3}
00.00	1.22×10^{-8}	1.24×10^{-9}	2.57×10^{-4}	2.00×10^{-3}	1.71×10^{-9}	0.59×10^{-9}	2.72×10^{-4}	1.47×10^{-3}

Table 7.5 Measured mass photon attenuation coefficients, μ/ρ ($m^2\ kg^{-1}\ x10^{-2}$)

Tissue keV	9.9	17.4	28.5	39.9	59.3	59.6	662
Artery (aorta)			4.05			2.06	0.853
Bone,cortical	305	60.9		6.98	3.33		
Bone,crushed							
femur, body			14.6–16.7			3.46–3.61	0.827–0.899
–, head			10.0–10.5			2.67–2.87	0.875–0.895
–, trochanter			10.6–12.2			2.81–3.02	0.873–0.904
rib			13.5			3.03,3.25	0.875,0.893
skull			14.7,15.4			3.29,3.31	0.826,0.882
Brain			4.01			2.04	0.851
Breast	45.5	9.65		2.51	2.00		
Cartilage			...			2.12	0.876
Fat,adipose	31.4	6.54	2.88,3.09	2.33	1.96	1.95	0.850
Kidney			4.10			2.04	0.845
Liver			4.00			2.03	0.845
Lung	57.0	10.31	4.03	2.51	1.94	2.03	0.837
Muscle	57.3	9.85	3.74,4.02	2.47	1.99	2.03	0.846
Skin			3.63,3.86			2.01	0.840
Testis			4.01			2.03	0.848
Thyroid	48.5	9.66	4.04	2.35	...	2.05	0.836
Vein (vena cava)			3.88			2.03	0.857

Sources: Joyet *et al*, 1974 (28.5, 59.6 and 662 keV), White *et al*, 1980 (9.9, 17.4, 39.9 and 59.3 keV), and Spiers, 1946 (28.5 keV).

Table 7.6a Measured linear attenuation coefficients for photons, in the range 9.9 to 41.4 keV (cm^{-1})

Ref*	Tissue keV	9.9	17.4	17.7	21.1	27.4	31.1	39.9	41.4
R	Blood					0.454–0.463			
P	clot			1.159	0.762	0.459	0.376		0.272
P	erythrocytes			1.165	0.766	0.465	0.384		0.280
P	plasma			1.115	0.736	0.438	0.360		0.264
W	Bone, cortical	57.7	11.5					1.320	
	Brain								
P	grey matter			1.122	0.731	0.445	0.368		0.267
P	white matter			1.092	0.711	0.434	0.358		0.263
P	Cerebrospinal fluid			1.116	0.724	0.438	0.361		0.260
R						0.446–0.448			
	Fat								
R	subcutaneous					0.328			
P	subcutaneous,monkey			0.665	0.462	0.317	0.276		0.219
W	adipose tissue	2.91	0.606					0.216	
R	Liver					0.441–0.461			
P	Liver,monkey			1.175	0.773	0.460	0.377		0.277
W	Lung	6.15	1.114					0.271	
R	Muscle,cardiac					0.448			
W	Muscle,skeletal	6.09	1.046					0.262	
P	Muscle,monkey			1.153	0.759	0.455	0.375		0.274
P	Pancreas,monkey			1.142	0.755	0.454	0.375		0.274
R	Pancreas					0.436			
R	Spleen					0.458			
W	Thyroid	5.33	1.063					0.258	

* Sources: Phelps *et al*, 1975 (P); Rao & Gregg, 1975 (R); White *et al*, 1980 (W).

Table 7.6b Measured linear attenuation coefficients for photons in the range 52 to 662 keV (cm^{-1})

Ref*	Tissue keV	52.0	59.5†	84.3	103.2	121.9	136.3	279	662
R	Blood		0.214–0.222			0.168–0.170		0.127–0.130	0.091
P	clot	0.237	0.221	0.190	0.176	0.161	0.155		
P	erythrocytes	0.238	0.222	0.194	0.181	0.170	0.164		
P	plasma	0.222	0.208	0.182	0.169	0.162	0.156		
W	Bone, cortical		0.629						
R	Brain		0.218		0.171			0.128	0.0907
R	cerebellum		0.214						0.0893
P	grey matter	0.226	0.210	0.184	0.171	0.163	0.158		
R	–		0.213						0.0884
P	white matter	0.225	0.208	0.182	0.171	0.163	0.159		
R	–		0.215						0.0882
P	Cerebrospinal fluid	0.222	0.207	0.181	0.167	0.160	0.155		
	Fat								
R	subcutaneous		0.190			0.153		0.120	0.0825
P	–,monkey	0.195	0.185	0.166	0.162	0.152	0.147		
W	adipose tissue		0.182						
R	Liver		0.208–0.221		0.165–0.172			0.126–0.128	0.0900–0.0930
P	Liver,monkey	0.233	0.216	0.190	0.178	0.167	0.162		
W	Lung		0.210						
R	Muscle:cardiac		0.212			0.165			0.0878
W	Muscle:skeletal		0.212						
P	–,monkey	0.233	0.214	0.189	0.177	0.169	0.164		
P	Pancreas,monkey	0.232	0.215	0.189	0.176	0.167	0.161		
R	–		0.214–0.215			0.165		0.126–0.127	0.0896
R	Spleen		0.216			0.168		0.127	0.0896

* Sources: Phelps *et al*, 1975 (P); Rao & Gregg, 1975 (R); White *et al*, 1980 (W). † White *et al* 59.3 keV.

Table 7.7 Linear attenuation coefficients for photons in breast tissue (cm^{-1})

Energy	Johns & Yaffe, 1987			Rao & Gregg	White *et al*
keV	fat	non–fat	tumour	1975	1980
9.9					4.646
17.4					0.984
18	0.538–0.585 (0.558)	1.014–1.045 (1.028)	1.061–1.137 (1.085)		
20	0.441–0.476 (0.456)	0.791–0.816 (0.802)	0.826–0.884 (0.844)		
25	0.314–0.333 (0.322)	0.499–0.516 (0.506)	0.519–0.552 (0.529)		
27				0.317	
30	0.259–0.271 (0.264)	0.372–0.386 (0.378)	0.387–0.408 (0.392)		
40	0.212–0.218 (0.215)	0.269–0.280 (0.273)	0.276–0.291 (0.281)		0.256
50	0.191–0.196 (0.194)	0.228–0.239 (0.233)	0.233–0.245 (0.238)		
59.5				0.189	0.204
80	0.165–0.169 (0.167)	0.186–0.194 (0.189)	0.188–0.198 (0.192)		
110	0.150–0.154 (0.152)	0.167–0.175 (0.170)	0.169–0.178 (0.173)		
122				0.152	
279				0.111	
662				0.0812	

Table 7.8 Linear attenuation coefficients for photons in some pathological tissues (cm^{-1})

Tissue keV	17.7	21.1	27.4	41.4	59.6	121.9	279	662
Brain								
astrocytoma	1.11	0.724	0.439–0.453	0.265	0.208–0.220	0.164	0.123–0.126	0.0887–0.0901
edema	1.07–1.08	0.719–0.721	0.429–0.431	0.261–0.262	0.206–0.208	0.162		
glioblastoma	1.12–1.16	0.729–0.765	0.442–0.454	0.261–0.269	0.208–0.219	0.161–0.164		
meningioma	1.13–1.14	0.745, 0.750	0.448–0.460	0.268, 0.270	0.212–0.214	0.162, 0.163		0.0901
metastasis (ca breast)	1.46	0.947	0.521	0.288	0.220	0.168		
other	1.11–1.28	0.724–0.833	0.440–0.482	0.258–0.279	0.206–0.218	0.161–0.171		0.0882
Liver tumours			0.445–0.462		0.209–0.219	0.165–0.174	0.125, 0.128	0.0891, 0.0891
Pancreas tumour			0.436		0.211	0.168	0.126	0.0888

Sources: Rao and Gregg (1975); Phelps *et al* (1975).

Table 7.9 Estimated CT numbers at 100 keV

Tissue	CT Number
Blood	46 to 58
Bone, cortical	>1000
–,trabecular,bone matrix	~1000
with marrow	106
–,red marrow	–20 to 35
–,yellow marrow	–84 to 15
Brain	24 to 35
Breast	33
Cartilage	79 to 91
Eye, lens	16 to 100 (with age)
Fatty tissue (adipose)	–95 (min)
Intestine	34 to 41
Kidney	38
Liver	28 to 63
Lung, fully inflated	~ –750
Lung, deflated	34 to 86
Lymph	24
Muscle, cardiac	54
Muscle, skeletal	29 to 44
Ovary	41 to 43
Pancreas	34 to 44
Skin	80 to 176
Spleen	48
Testis	38
Thyroid gland	36 to 66
Water	0
Air	~ –1000

CT numbers have been calculated using mass attenuation coefficient from Table 7.1 and density from Table 5.1. Single values imply that only a single average density is available for this tissue.

7.1.4.2 Atomic number

All four principal photon interaction processes are dependent on atomic number (Walter, 1926; Spiers, 1946). The variations with atomic number, Z, follow approximately simple power laws. The power law may describe either the variation of cross–section per atom with Z, or the variation of mass attenuation with Z. For cross–sections (perhaps the less practically useful of the two) for photoelectric absorption, cross–sections per atom at energies above aborption edges and below 150 keV the exponent is 4.1 to 4.9 (Weber and van den Berge, 1969; Cho *et al*, 1975; McCullough, 1975; White, 1977a). For Compton cross–sections per atom the exponent for Z is 1.0, for coherent scatter it is 2.4 to 2.8 (Weber and van den Berge, 1969; White,

1977a) and for pair production at 40 MeV it is 1.7 to 2.0 (White, 1977a).

The power–law dependence on Z for mass attenuation coefficient have exponents which are reduced by about 1.0 from those for corresponding cross–sections. These exponents are therefore approximately 3.5, 0 ,1.5 and 1.0 for photoelectric, Compton, coherent scatter and pair productions respectively. The Compton mass attenuation coefficient is noticably influenced by its hydrogen content. The relative contributions to μ/ρ and μ_{en}/ρ in muscle from the principle elements present are shown in Figures 7.3 and 7.4.

Table 7.10 Measured CT numbers at 140kVp

Tissue	CT number(±SD)	Reference
Normal tissues		
Bile	16±8	Mategrano *et al* 1977
Blood	42±18	-
–,*in–vitro*,27.3% Hb	86.6	Zatz 1976
–,–,15.6% Hb	57.4	-
–,–,45% Hct	56	New & Aronow 1976
–,–,plasma	26	-
Bone marrow	142±48	Mategrano *et al* 1977
Brain	31.3±6.2	Pullan *et al* 1978
Fat	–90±18	Mategrano *et al* 1977
–,subcutaneous,*in–vitro*	–121.4	Zatz 1976
–,omental,*in–vitro*	–115.6	-
–,orbital	–101.4,–122.6	-
Kidney	32±10	Mategrano *et al* 1977
–,*in–vitro*	47.4,62.0	Zatz 1976
Liver	60±14	Mategrano *et al* 1977
-	64.8±22.2	Pullan *et al* 1978
Muscle	44±14	Mategrano *et al* 1977
Pancreas	40±14	-
Spleen	51.2±20.5	Pullan *et al* 1978
–,*in–vitro*	52.4,80.4	Zatz 1976
Urine	–5.1 to 19.3	Bydder & Kreel 1980
–,*in–vitro*	–3.9 to 94.3	-
Pathological tissues		
Abcess	–6.0 to 39.4	Bydder & Kreel 1980
Ascites	–5.5 to 17.9	-
Brain,atrophied	31.3±6.4	Pullan *et al* 1978
Cyst	–0.4 to 33.8	Bydder & Kreel 1980
–,–,hemorrhagic	21.6,26.4	-
–,dermoid	–30	Rutherford *et al* 1976
Liver,cirrhotic	46.0±21.8	Pullan *et al* 1978
Meningioma	190(max)	Rutherford *et al* 1976

In–vivo, unless otherwise noted.

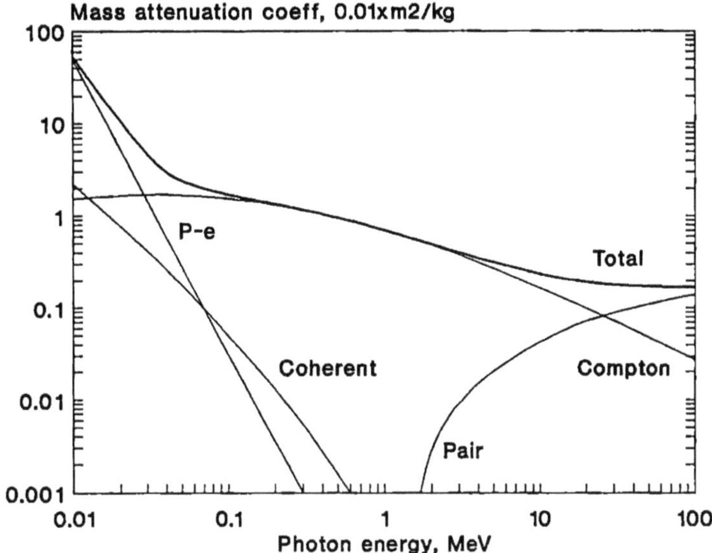

Figure 7.1 The variation of total mass attenuation coefficient with photon energy for skeletal muscle. Contributions from photoelectric absorption, coherent scatter, Compton scatter and pair production are also shown.

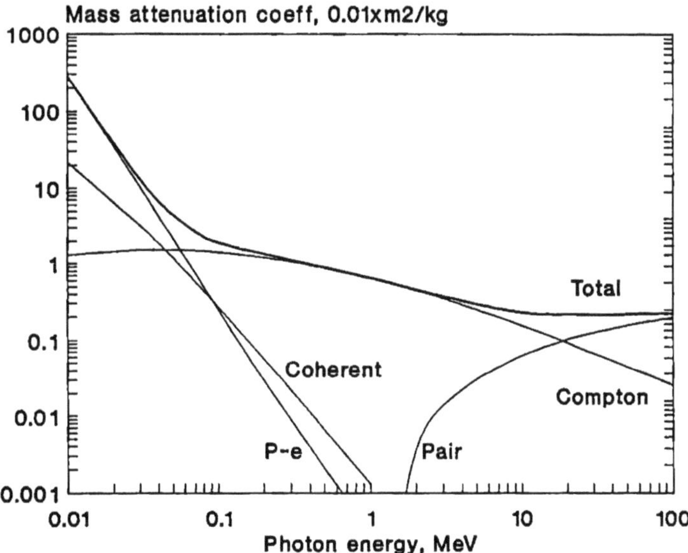

Figure 7.2 The variation of total mass attenuation coefficient with photon energy for cortical bone. Contributions from photoelectric absorption, coherent scatter, Compton scatter and pair production are also shown.

Table 7.11 Partial effective atomic numbers at 60 keV

Tissue	Z(coh)	Z(pe)	Z(Comp)
Blood	7.00	7.75	5.25
Bone	9.82	12.72	6.73
Brain	6.94	7.77	5.09
Fat	5.48	5.60	4.15
Kidney	6.94	7.65	5.21
Liver	6.92	7.76	5.14
Lung	6.98	7.63	5.20
Muscle:cardiac	6.92	7.67	5.18
Muscle:skeletal	6.98	7.75	5.23
Ovary	6.97	7.48	5.19
Pancreas	6.91	7.65	5.15
Spleen	6.91	7.46	5.18
Water	7.04	7.55	5.20

Source: Yang *et al* (1987).

7.1.4.3 Temperature

It is usually assumed that photon attenuation is not dependent on temperature over the relevant temperature range. However, Bydder and Kreel (1979) have reported that care needs to be taken when comparing *in-vivo* with *in-vitro* measurements of CT number because of small but significant differences in attenuation coefficient with temperature. These occur because of the thermal expansion of both the water reference and the tissue (see Section 5.1.1), giving rise to a smaller electron density and hence lower attenuation at body temperature than at room temperature.

7.1.4.4 Tissue variability

The calculated values given in the tables are for a average representation of each tissue based on reviews of the literature values for tissue composition (see Chapter 9). The extent to which the may be taken to represent any particular tissue sample depends upon the degree of variability in tissue composition from the selected standard tissue. For instance in a critical review of the literature (ICRP, 1975) it was stated that the water content of adipose tissue may vary from 10.9 to 21.0% by mass, whilst lipid content is in the range 62.0 to 91.0%. Bone exhibits similar variations. Body tissue content is also affected by age, sex, state of health and even elevation above sea level. For these reasons it is important when using the data from calculations to take note of the variability shown both in the tables of body composition (Tables 9.1) and that in the tables of experimentally determined coefficients (Tables 7.6, 7.7, 7.8).

For blood *in-vitro* a linear dependence between CT number and haematocrit was noted by New and Aronow (1976). A range of blood cell

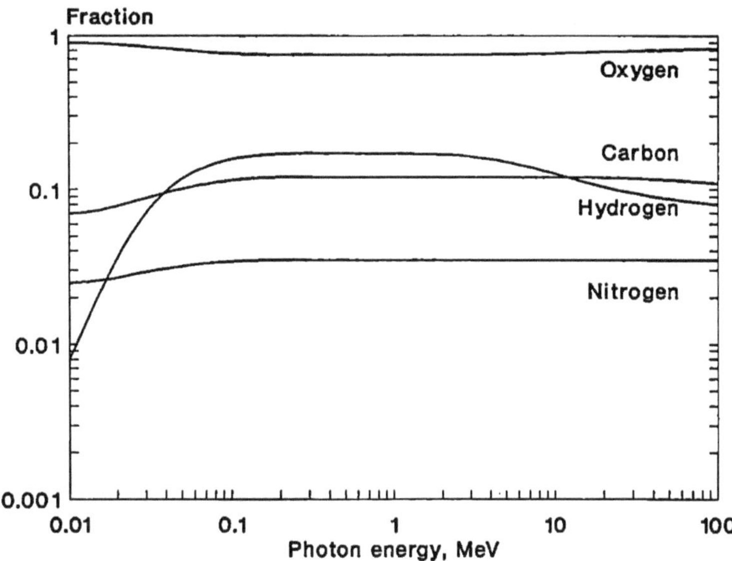

Figure 7.3 Relative contributions to the mass attenuation coefficient in muscle from interaction with different elements. From ICRU (1989).

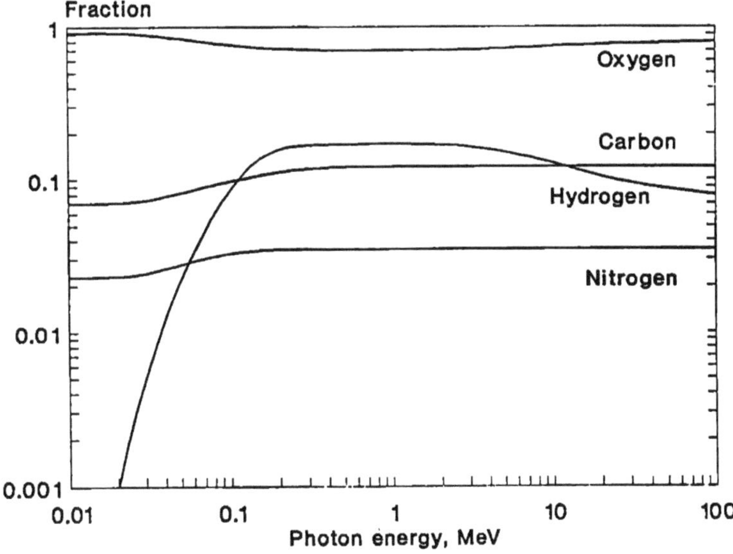

Figure 7.4 Relative contributions to the mass absorption coefficient in muscle from interaction with different elements. From ICRU (1989).

concentrations was studied from pure plasma (CT number = 26) to 60% haematocrit (CT number = 68).

In a study of the CT numbers of various abdominal fluids, Bydder and Kreel (1980) noted a correlation between the attenuation in ascetic fluid *in-vitro* and the protein concentration. However, this agreement was lost in *in-vivo* measurements: an effect thought to be due to spectral errors.

7.1.4.5 Tissue fixation and storage

It is generally assumed that if due care is taken with tissue handling, to prevent dehydration or obvious compression, then measurements on fresh autopsy specimens can be taken to represent well the properties of tissue *in-vivo*, provided due regard is taken of temperature variations. Storage of body fluids under refrigeration gives rise to no subsequent alteration in attenuation.

Fixation of tissue specimens has however been shown to alter their attenuation. Rao and Gregg (1975) noted that formalin-fixed specimens of normal and pathological brain and lung tissues exhibited lower linear attenuation coefficients than unfixed tissue. The difference for brain tissue was about 5% and was explained on the basis that formalin has a lower mean atomic number and lower density than brain.

7.1.5 *Photon attenuation in some materials other than tissue*

7.1.5.1 Water

Values of photon attenuation and absorption coefficients for water have been included in Table 7.12 with detail including values for the separate components of mass attenuation coefficient.

7.1.5.2 Tissue substitute materials

Probably more mixtures have been devised with the intention of being tissue sustitutes for photon interactions than for any other physical property of tissue. These materials have been used to fabricate phantoms for radiation protection, to allow depth-dose measurement and to calibrate radiation detection systems. Tissue equivalent materials are also required in the construction of radiation detectors, and for bolus materials used to provide extra scattering or build-up during radiotherapy. The properties of a wide range of materials have been reviewed (ICRU, 1989) from which a summary has been compiled. The selection includes four common tissue substitute materials (A150, acrylic, paraffin wax and polystyrene) together with a selection of mixtures and materials designed for equivalence to bone, soft tissue, fatty tissue and lung. Details of compositions and densities are given in Table 7.13. Calculated values for mass attenuation coefficient and mass energy absorption coefficient are given in Tables 7.14 and 7.15. Other

Table 7.12 Photon mass attenuation coefficients ($m^2\ kg^{-1}$) electron mass stopping powers S/ρ ($MeV\ m^2\ kg^{-1}$) and electron mass scattering powers T/ρ ($radian^2\ m^2\ kg^{-1}$) for water

Energy	Photon mass attenuation coefficients						For electrons	
MeV	τ/ρ	σ_{coh}/ρ	σ_c/ρ	κ/ρ	μ/ρ	μ_{en}/ρ	S/ρ	T/ρ
0.01	4.91×10^{-1}	2.31×10^{-2}	1.55×10^{-2}	0	5.33×10^{-1}	4.84×10^{-1}	2.26	686
0.02	5.44×10^{-2}	8.86×10^{-3}	1.77×10^{-2}	0	8.10×10^{-2}	5.37×10^{-2}	1.32	205
0.03	1.46×10^{-2}	4.69×10^{-3}	1.83×10^{-2}	0	3.76×10^{-2}	1.52×10^{-2}	0.965	101
0.04	5.68×10^{-3}	2.87×10^{-3}	1.83×10^{-2}	0	2.68×10^{-2}	6.80×10^{-3}	0.778	60.9
0.06	1.49×10^{-3}	1.39×10^{-3}	1.77×10^{-2}	0	2.06×10^{-2}	3.15×10^{-3}	0.580	30.2
0.08	5.77×10^{-4}	8.16×10^{-4}	1.70×10^{-2}	0	1.84×10^{-2}	2.58×10^{-3}	0.476	18.4
0.10	2.76×10^{-4}	5.35×10^{-4}	1.63×10^{-2}	0	1.71×10^{-2}	2.54×10^{-3}	0.412	12.6
0.20	2.89×10^{-5}	1.39×10^{-4}	1.35×10^{-2}	0	1.37×10^{-2}	2.97×10^{-3}	0.280	4.04
0.30	8.16×10^{-6}	6.21×10^{-5}	1.18×10^{-2}	0	1.19×10^{-2}	3.19×10^{-3}	0.236	2.13
0.40	3.49×10^{-6}	3.51×10^{-5}	1.06×10^{-2}	0	1.06×10^{-2}	3.28×10^{-3}	0.215	1.37
0.60	1.17×10^{-6}	1.56×10^{-5}	8.94×10^{-3}	0	8.96×10^{-3}	3.28×10^{-3}	0.197	0.742
0.80	5.92×10^{-7}	8.79×10^{-6}	7.86×10^{-3}	0	7.87×10^{-3}	3.21×10^{-3}	0.190	0.482
1.00	3.68×10^{-7}	5.63×10^{-6}	7.07×10^{-3}	0	7.07×10^{-3}	3.10×10^{-3}	0.186	0.345
2.00	1.06×10^{-7}	1.41×10^{-6}	4.90×10^{-3}	3.91×10^{-5}	4.94×10^{-3}	2.60×10^{-3}	0.185	0.119
3.00	5.94×10^{-8}	6.25×10^{-7}	3.86×10^{-3}	1.13×10^{-4}	3.97×10^{-3}	2.28×10^{-3}	0.189	6.23×10^{-2}
4.00	4.07×10^{-8}	3.52×10^{-7}	3.22×10^{-3}	1.87×10^{-4}	3.40×10^{-3}	2.06×10^{-3}	0.193	3.88×10^{-2}
6.00	2.48×10^{-8}	1.56×10^{-7}	2.45×10^{-3}	3.16×10^{-4}	2.77×10^{-3}	1.80×10^{-3}	0.201	1.96×10^{-2}
8.00	1.78×10^{-8}	8.80×10^{-8}	2.01×10^{-3}	4.21×10^{-4}	2.43×10^{-3}	1.66×10^{-3}	0.208	1.19×10^{-2}
10.00	1.39×10^{-8}	5.63×10^{-8}	1.71×10^{-3}	5.09×10^{-4}	2.22×10^{-3}	1.57×10^{-3}	0.215	8.03×10^{-3}
20.00	6.55×10^{-9}	1.41×10^{-8}	1.02×10^{-3}	7.97×10^{-4}	1.81×10^{-3}	1.39×10^{-3}	0.245	2.31×10^{-3}
30.00	4.29×10^{-9}	6.25×10^{-9}	7.40×10^{-4}	9.71×10^{-4}	1.71×10^{-3}	1.32×10^{-3}	0.274	1.10×10^{-3}
40.00	3.19×10^{-9}	3.52×10^{-9}	5.88×10^{-4}	1.09×10^{-3}	1.68×10^{-3}	1.30×10^{-3}	0.301	6.47×10^{-4}
60.00	2.10×10^{-9}	1.56×10^{-9}	4.23×10^{-4}	1.26×10^{-3}	1.68×10^{-3}	1.26×10^{-3}	0.356	3.02×10^{-4}
80.00	1.57×10^{-9}	0.88×10^{-9}	3.33×10^{-4}	1.37×10^{-3}	1.70×10^{-3}	1.23×10^{-3}	0.410	1.71×10^{-4}
100.00	1.25×10^{-9}	0.56×10^{-9}	2.77×10^{-4}	1.45×10^{-3}	1.73×10^{-3}	1.21×10^{-3}	0.464	1.10×10^{-4}

Table 7.13 **Tissue substitute materials for photons and electrons**

Substitute material	Description	ρ $kg.m^{-3}$	n_o $m^{-3}x10^{-26}$	Reference
For muscle and similar tissues				
A150	Polythene, nylon, carbon, calcium fluoride	1120	3700	Smathers *et al* 1977
Acrylic	$(C_5H_8O_2)_n$	1170	3800	
Alderson muscle	Isocyanate rubber, microspheres and antimony trioxide (Rando)	1000	3270	Alderson *et al* 1962
Lincolnshire bolus	Sucrose, magnesium carbonate	1050,bulk	3340	Lindsay & Stern 1953
Mix D	Paraffin wax/polyethyene, magnesium oxide, and titanium dioxide	990	3370	Jones & Raine 1949
Polystyrene	$(C_8H_8)_n$	1050	3400	
Temex	Rubber with oil, carbon, sulphur, stearic acid, titanium dioxide	1010	3310	Stacy *et al* 1961
For fatty tissues				
AP6	Epoxy CB4 with PTFE, polythene, microspheres	920	2990	White *et al* 1977
Ethoxyethanol	$C_4H_{10}O_2$	930	3110	White 1976
For lipid				
Paraffin wax	C_nH_{2n+1}, n>15	930	3210	
Polyethylene	$(C_2H_4)_n$	920	3160	
For lung				
Alderson lung	Epoxy, microspheres, antimony trioxide	320	1020	Alderson *et al* 1962
LN 10/75	Foamed epoxy CB4, microspheres, polythene, calcium carbonate	310	1000	White *et al* 1977, 1986
For cortical bone				
B110	As A150, different proportions	1790	5500	Spokas & White 1982
Plaster of Paris	$CaSO_4.2H_2O$	2320	7140	Spiers 1946
SB5	Epoxy CB4, calcium carbonate	1870	5770	White *et al* 1987

Table 7.14 Mass attenuation coefficients, μ/ρ, for photons in tissue substitute materials ($m^2\ kg^{-1} x10^{-2}$)

Photon energy,MeV	0.01	0.03	0.1	0.3	1	3	10	100
For muscle and similar tissues								
A150	41.5	3.47	1.68	1.17	0.699	0.391	0.212	0.150
Acrylic	33.6	3.03	1.64	1.15	0.687	0.384	0.210	0.154
Alderson muscle	43.4	3.47	1.68	1.16	0.691	0.387	0.212	0.155
Lincolnshire bolus	48.0	3.50	1.62	1.13	0.673	0.378	0.212	0.165
Mix D	44.8	3.66	1.73	1.21	0.719	0.402	0.216	0.150
Polystyrene	22.2	2.64	1.62	1.15	0.684	0.382	0.206	0.144
Temex	44.1	3.68	1.68	1.17	0.696	0.389	0.210	0.148
For fatty tissues								
AP6	30.8	2.95	1.64	1.15	0.688	0.385	0.209	0.151
Ethoxyethanol	34.2	3.11	1.69	1.18	0.707	0.395	0.215	0.155
For lipid								
Paraffin wax	20.7	2.71	1.73	1.22	0.730	0.407	0.215	0.142
Polyethylene	20.9	2.71	1.72	1.22	0.726	0.405	0.215	0.142
For lung								
Alderson lung	32.2	3.00	1.63	1.13	0.672	0.376	0.206	0.152
LN 10/75	52.6	3.74	1.66	1.15	0.688	0.386	0.214	0.162
For cortical bone								
B110	279	13.2	1.84	1.11	0.652	0.371	0.227	0.216
Plaster of Paris	344	15.6	1.91	1.11	0.652	0.374	0.238	0.247
SB5	282	13.3	1.85	1.11	0.654	0.372	0.228	0.219

Table 7.15 Mass energy absorption coefficients, μ_{en}/ρ, for photons in tissue substitute materials ($m^2\ kg^{-1} x10^{-2}$)

Photon energy,MeV	0.01	0.03	0.1	0.3	1	3	10	100
For muscle and similar tissues								
A150	36.4	1.34	0.251	0.315	0.307	0.224	0.148	0.110
Acrylic	29.5	0.940	0.236	0.310	0.301	0.221	0.148	0.112
Alderson muscle	38.8	1.34	0.258	0.313	0.303	0.222	0.149	0.112
Lincolnshire bolus	43.3	1.38	0.241	0.304	0.295	0.217	0.149	0.117
Mix D	39.1	1.47	0.261	0.325	0.316	0.231	0.152	0.110
Polystyrene	18.5	0.606	0.229	0.309	0.300	0.219	0.145	0.107
Temex	35.6	1.50	0.256	0.313	0.304	0.222	0.147	0.108
For fatty tissues								
AP6	26.8	0.865	0.235	0.310	0.302	0.221	0.147	0.110
Ethoxyethanol	30.0	0.962	0.243	0.319	0.310	0.227	0.151	0.112
For lipid								
Paraffin wax	17.0	0.573	0.243	0.329	0.321	0.234	0.151	0.106
Polyethylene	17.2	0.576	0.242	0.328	0.319	0.233	0.150	0.106
For lung								
Alderson lung	28.1	0.944	0.254	0.304	0.295	0.216	0.145	0.110
LN 10/75	47.7	1.58	0.250	0.311	0.302	0.221	0.150	0.115
For cortical bone								
B110	257	10.5	0.456	0.301	0.285	0.212	0.160	0.138
Plaster of Paris	320	12.7	0.504	0.303	0.285	0.213	0.169	0.152
SB5	259	10.6	0.460	0.302	0.286	0.213	0.161	0.139

compositions have also been reported. The following are readily accessible and reference may be made to the original publications (Frigerio, 1962; Goodman, 1969; Hermann *et al*, 1986; Rossi and Failla, 1956; White, 1977b). Tissue substitute gas mixtures intended primarily for dosimetry purposes have been reported by Awschalom and Attix (1980).

7.2 Electrons and positrons

7.2.1 *Terminology and definitions*

The **total mass stopping power,** S/ρ, of a material for charged particles is defined as the rate of total energy loss with distance per unit density:

$$\frac{S}{\rho} = \frac{1}{\rho}\frac{dE}{dl} \tag{7.10}$$

The unit is joule metre^2 kilogram^{-1} (J m^2 kg^{-1}), or eV m^2 kg^{-1}.

Electrons in the energy interval 10 keV to 1000 MeV lose energy by two main energy-dependent processes, electronic collisions and brehmsstrahlung production (Bichsel, 1968; ICRU, 1984). The total mass stopping power can therefore be separated into two components:

$$\frac{S}{\rho} = \left[\frac{S}{\rho}\right]_{col} + \left[\frac{S}{\rho}\right]_{rad} \tag{7.11}$$

where the **mass collision stopping power,** $(S/\rho)_{col}$, includes all energy losses in particle collisions which directly produce secondary electrons and atomic excitations. Losses associated with the production of Cerenkov radiation are also included. The second component, $(S/\rho)_{rad}$, the **mass radiative stopping power** includes all energy losses which lead to brehmsstralung radiation.

Calculations of stopping powers are based on various formulae of the Bethe theory (Bethe and Ashkin, 1953; Uehling, 1954), and their use is well summarised in ICRU (1984). A key parameter in the Bethe stopping-power theory is the mean excitation energy of the medium.

The scattering of electrons passing through a certain thickness of body tissue may be described in terms of the **mass scattering power,** T/ρ (ICRU, 1984). The multiple scattering is described by the mean-square angle of scattering, $(\overline{\theta^2})$. T/ρ expresses the rate of change of $(\overline{\theta^2})$ with distance l through matter of density ρ:

$$\frac{T}{\rho} = \frac{1}{\rho}\frac{d(\overline{\theta^2})}{dl} \tag{7.12}$$

The unit is radian^2 metre^2 kilogram^{-1}. The mass scattering power must be evaluated with care for thin absorbers (thickness<1 mg cm^{-2} of unit density material) where single large angle scattering events may be important.

It is possible to estimate the **electron range** in tissue using the continuous-slowing-down approximation (csda). In this approximation energy-loss fluctuations are neglected and electrons are assumed to lose energy continuously along their track with a mean energy loss per unit pathlength given by the stopping power. The csda range in g cm^{-2} is evaluated from the expression

$$r_0(T_0 \rightarrow T_f) = \rho \int_{T_f}^{T_0} [S_{col}(T) + S_{rad}(T)]^{-1} dT \qquad (7.13)$$

where $r_0(T_0 \rightarrow T_f)$ represents the average pathlength travelled by an electron as it slows down from an initial energy, T_0, to a final energy, T_f.

Minor differences beween the behaviour of **positrons** and electrons result from the possibility that a positron may lose its entire energy at a single collision whereas an electron can at most lose only half its energy. The use of different cross-sections in the calculations of losses from positrons is discussed in ICRU (1984).

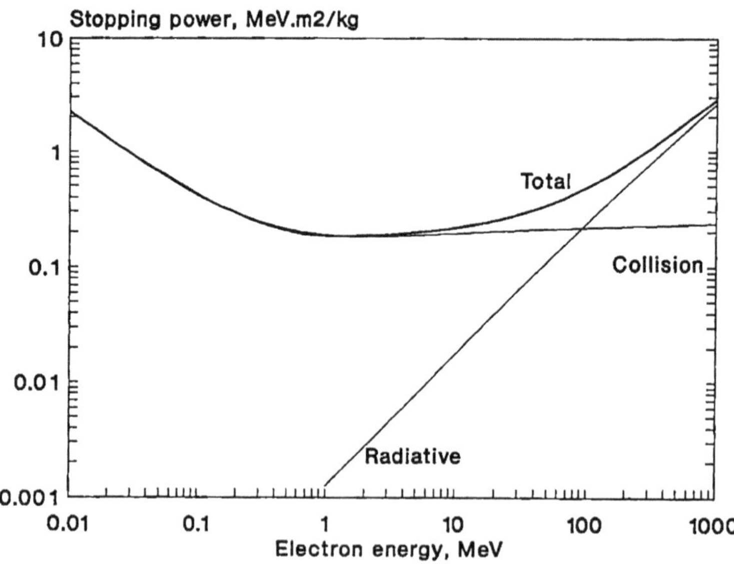

Figure 7.5 Calculated stopping powers for electrons, 0.01 to 1000 MeV

7.2.2 *Calculated values for electrons and positrons for tissue*

Calculated values for total mass stopping power, S/ρ, are given in Table 7.16, and mass scattering power, T/ρ, in Table 7.17 for a range of tissues, from ICRU (1989). The mass scattering powers were calculated using expressions recommended by Rossi (1952) and Brahme (1971) and given in ICRU (1984). The uncertainties for collision stopping powers are 2–3% for low Z elements below 0.1 MeV, reducing to 1–2% above 0.1 MeV. For three 'standard' tissues, skeletal muscle, cortical bone and adipose tissue, more detail is given in Tables 7.18 7.19 and 7.20, including collision and radiative stopping powers, csda ranges and radiation yields for both electrons and positrons for energies up to 1000 MeV. These data are from ICRU (1984). The variation of electron stopping power with energy for muscle, being typical of most soft tissue, is shown in graphical form in Figure 7.5.

7.2.3 *Electron interactions with some materials other than tissue*

7.2.3.1 Water

Values for water are given in Table 7.15 including values of S/ρ and T/ρ, and also in Table 7.21 presenting collision and radiative stopping powers, csda range and radiation yield.

7.2.3.2 Tissue substitute materials

The mass scattering power and total mass stopping power in the energy range 0.01–100 MeV for several common tissue substitute materials are included in Tables 7.22 and 7.23.

Table 7.16a Total mass stopping power, S/ρ, for electrons in the energy range 10 to 300 keV (MeV m^2 kg^{-1})

Electron energy, keV	10	15	20	30	40	60	100	300
Blood	2.24	1.63	1.31	0.956	0.771	0.575	0.408	0.234
Bone,cortical	1.93	1.42	1.14	0.837	0.677	0.506	0.361	0.209
Bone,trabecular	2.18	1.59	1.27	0.934	0.753	0.562	0.399	0.229
Bone,red marrow	2.28	1.66	1.33	0.972	0.783	0.583	0.414	0.237
Bone,yellow marrow	2.33	1.70	1.36	0.993	0.799	0.595	0.422	0.241
Brain	2.25	1.64	1.32	0.963	0.776	0.579	0.411	0.235
Breast	2.27	1.66	1.33	0.970	0.782	0.582	0.413	0.237
Cartilage	2.21	1.61	1.29	0.945	0.762	0.568	0.404	0.231
Cell nucleus	2.24	1.64	1.31	0.959	0.773	0.576	0.409	0.234
Eye,lens	2.23	1.63	1.30	0.953	0.768	0.573	0.407	0.233
Fatty tissue	2.33	1.69	1.35	0.990	0.797	0.593	0.421	0.240
Intestine	2.25	1.64	1.31	0.962	0.775	0.578	0.410	0.235
Kidney	2.24	1.64	1.31	0.958	0.772	0.576	0.409	0.234
Liver	2.24	1.63	1.31	0.957	0.771	0.575	0.408	0.234
Lung,inflated	2.24	1.63	1.31	0.957	0.771	0.575	0.408	0.234
Lymph	2.25	1.64	1.31	0.961	0.774	0.577	0.410	0.235
Muscle,cardiac	2.24	1.63	1.31	0.958	0.772	0.576	0.409	0.234
Muscle,skeletal	2.24	1.63	1.31	0.958	0.772	0.575	0.409	0.234
Ovary	2.24	1.64	1.31	0.959	0.773	0.576	0.409	0.235
Pancreas	2.25	1.64	1.32	0.964	0.776	0.579	0.411	0.235
Skin	2.24	1.63	1.31	0.958	0.772	0.575	0.408	0.234
Spleen	2.24	1.63	1.31	0.958	0.772	0.575	0.409	0.234
Testicle	2.25	1.64	1.31	0.961	0.774	0.577	0.410	0.235
Thyroid gland	2.24	1.64	1.31	0.959	0.773	0.576	0.409	0.234

Table 7.16b Total mass stopping power, S/ρ, for electrons in the energy range 0.6 to 100 MeV (MeV m^2 kg^{-1})

Electron energy, MeV	0.6	1	3	6	10	30	60	100
Blood	0.195	0.184	0.187	0.199	0.213	0.271	0.351	0.458
Bone,cortical	0.174	0.165	0.170	0.185	0.202	0.277	0.385	0.529
Bone,trabecular	0.191	0.180	0.183	0.195	0.210	0.270	0.355	0.468
Bone,red marrow	0.197	0.185	0.188	0.199	0.213	0.267	0.342	0.442
Bone,yellow marrow	0.200	0.188	0.190	0.201	0.214	0.265	0.337	0.431
Brain	0.196	0.185	0.188	0.200	0.214	0.271	0.351	0.458
Breast	0.197	0.186	0.188	0.200	0.213	0.268	0.345	0.446
Cartilage	0.193	0.182	0.185	0.198	0.212	0.271	0.353	0.463
Cell nucleus	0.196	0.185	0.188	0.200	0.214	0.273	0.354	0.463
Eye,lens	0.194	0.183	0.186	0.198	0.211	0.268	0.347	0.452
Fatty tissue	0.200	0.188	0.190	0.201	0.214	0.266	0.338	0.433
Intestine	0.196	0.185	0.188	0.200	0.213	0.271	0.351	0.458
Kidney	0.195	0.184	0.187	0.199	0.213	0.270	0.351	0.457
Liver	0.195	0.184	0.187	0.199	0.213	0.270	0.351	0.457
Lung,inflated	0.196	0.187	0.195	0.208	0.222	0.281	0.363	0.470
Lymph	0.196	0.185	0.188	0.200	0.214	0.272	0.354	0.462
Muscle,cardiac	0.195	0.184	0.187	0.199	0.213	0.271	0.351	0.458
Muscle,skeletal	0.195	0.184	0.187	0.199	0.213	0.270	0.350	0.457
Ovary	0.196	0.185	0.187	0.199	0.213	0.271	0.352	0.459
Pancreas	0.196	0.185	0.188	0.200	0.213	0.270	0.350	0.455
Skin	0.195	0.184	0.186	0.198	0.212	0.269	0.348	0.452
Spleen	0.195	0.184	0.187	0.199	0.213	0.271	0.351	0.458
Testicle	0.196	0.185	0.188	0.200	0.213	0.271	0.352	0.459
Thyroid gland	0.196	0.185	0.187	0.199	0.213	0.271	0.351	0.458

Table 7.17a Mass scattering power, T/ρ, for electrons in the energy range 10 to 300 keV ($radian^2\ m^2\ kg^{-1}$)

Electron energy, keV	10	15	20	30	40	60	100	300
Blood	677	334	202	99.3	60.1	29.8	12.5	2.10
Bone,cortical	886	440	268	132	80.4	40.0	16.9	2.87
Bone,trabecular	707	350	212	104	63.2	31.3	13.1	2.22
Bone,red marrow	635	313	189	92.8	56.2	27.8	11.6	1.96
Bone,yellow marrow	599	295	178	87.4	52.8	26.1	10.9	1.84
Brain	672	332	201	98.7	59.7	29.6	12.4	2.09
Breast	643	317	192	94.1	56.9	28.2	11.8	1.99
Cartilage	691	341	206	102	61.5	30.5	12.8	2.15
Cell nucleus	686	338	205	101	60.9	30.2	12.6	2.13
Eye,lens	663	327	198	97.2	58.8	29.1	12.2	2.06
Fatty tissue	606	298	180	88.4	53.4	26.4	11.1	1.86
Intestine	673	332	201	98.8	59.8	29.6	12.4	2.09
Kidney	674	332	201	98.8	59.8	29.6	12.4	2.09
Liver	674	333	201	98.9	59.8	29.6	12.4	2.09
Lung,inflated	678	334	202	99.4	60.2	29.8	12.5	2.11
Lymph	684	337	204	100	60.7	30.1	12.6	2.13
Muscle,cardiac	675	333	201	99.1	60.0	29.7	12.4	2.10
Muscle,skeletal	673	332	201	98.7	59.7	29.6	12.4	2.09
Ovary	678	335	202	99.5	60.2	29.8	12.5	2.11
Pancreas	667	329	199	97.8	59.2	29.3	12.3	2.07
Skin	663	327	198	97.2	58.8	29.1	12.2	2.06
Spleen	676	334	202	99.2	60.1	29.8	12.5	2.10
Testicle	677	334	202	99.3	60.1	29.8	12.5	2.10
Thyroid gland	675	333	201	99.1	60.0	29.7	12.4	2.10

Table 7.17b **Mass scattering power, T/ρ, for electrons in the energy range 0.6 to 100 MeV (radian^2 m^2 kg^{-1}x10^{-3})**

Electron energy, MeV	0.6	1	3	6	10	30	60	100
Blood	732	340	61.5	19.3	7.92	1.09	0.297	0.108
Bone,cortical	1000	468	85.0	26.8	11.0	1.51	0.408	0.148
Bone,trabecular	774	360	65.2	20.5	8.41	1.15	0.315	0.114
Bone,red marrow	681	316	57.1	17.9	7.34	1.01	0.277	0.101
Bone,yellow marrow	638	296	53.4	16.7	6.86	0.939	0.259	0.094
Brain	727	338	61.0	19.1	7.86	1.08	0.295	0.107
Breast	691	321	57.9	18.2	7.45	1.02	0.281	0.102
Cartilage	750	349	63.0	19.8	8.12	1.11	0.305	0.111
Cell nucleus	743	345	62.4	19.6	8.04	1.10	0.302	0.110
Eye,lens	715	332	60.0	18.8	7.73	1.06	0.291	0.106
Fatty tissue	646	300	54.0	16.9	6.95	0.951	0.262	0.095
Intestine	727	338	61.1	19.2	7.86	1.08	0.296	0.107
Kidney	728	338	61.1	19.2	7.87	1.08	0.296	0.108
Liver	728	339	61.2	19.2	7.88	1.08	0.296	0.108
Lung,inflated	733	341	61.5	19.3	7.92	1.09	0.298	0.108
Lymph	740	344	62.1	19.5	8.00	1.10	0.301	0.109
Muscle,cardiac	730	339	61.3	19.2	7.89	1.08	0.297	0.108
Muscle,skeletal	727	338	61.0	19.1	7.86	1.08	0.295	0.107
Ovary	733	341	61.6	19.3	7.93	1.09	0.298	0.108
Pancreas	720	335	60.5	19.0	7.78	1.07	0.293	0.106
Skin	715	332	60.0	18.8	7.73	1.06	0.291	0.106
Spleen	731	340	61.4	19.3	7.91	1.08	0.297	0.108
Testicle	732	340	61.4	19.3	7.91	1.08	0.297	0.108
Thyroid gland	730	339	61.3	19.2	7.90	1.08	0.297	0.108

Table 7.18 Calculated collision and radiative stopping powers, $(S/\rho)_{col}$ and $(S/\rho)_{rad}$ (MeV cm^2 g^{-1}), csda ranges (g cm^2) and radiation yields for electrons and positrons; skeletal muscle*

Energy MeV	Electrons Stopping power $(S/\rho)_{col}$	Electrons Stopping power $(S/\rho)_{rad}$	Electrons csda range	Electrons Radiation yield	Positrons Stopping power $(S/\rho)_{col}$	Positrons Stopping power $(S/\rho)_{rad}$	Positrons csda range	Positrons Radiation yield
0.01	22.31	3.84×10^{-3}	2.54×10^{-4}	9.37×10^{-5}	24.56	2.18×10^{-3}	2.28×10^{-4}	4.24×10^{-5}
0.02	13.03	3.90×10^{-3}	8.66×10^{-4}	1.66×10^{-4}	14.11	2.60×10^{-3}	7.89×10^{-4}	8.95×10^{-5}
0.03	9.55	3.92×10^{-3}	1.78×10^{-3}	2.29×10^{-4}	10.24	2.84×10^{-3}	1.63×10^{-3}	1.37×10^{-4}
0.05	6.53	3.97×10^{-3}	4.37×10^{-3}	3.42×10^{-4}	6.92	3.17×10^{-3}	4.07×10^{-3}	2.30×10^{-4}
0.1	4.07	4.17×10^{-3}	1.45×10^{-2}	5.82×10^{-4}	4.23	3.64×10^{-3}	1.37×10^{-2}	4.47×10^{-4}
0.2	2.76	4.73×10^{-3}	4.54×10^{-2}	9.80×10^{-4}	2.81	4.36×10^{-3}	4.39×10^{-2}	8.31×10^{-4}
0.3	2.33	5.44×10^{-3}	8.51×10^{-2}	1.33×10^{-3}	2.34	5.11×10^{-3}	8.33×10^{-2}	1.18×10^{-3}
0.5	2.01	7.16×10^{-3}	0.179	1.97×10^{-3}	1.99	6.85×10^{-3}	0.177	1.83×10^{-3}
1.0	1.83	1.26×10^{-2}	0.442	3.57×10^{-3}	1.79	1.23×10^{-2}	0.445	3.46×10^{-3}
2.0	1.80	2.64×10^{-2}	0.990	7.07×10^{-3}	1.76	2.60×10^{-2}	1.01	7.05×10^{-3}
3.0	1.82	4.24×10^{-2}	1.53	1.09×10^{-2}	1.77	4.21×10^{-2}	1.56	1.10×10^{-2}
5.0	1.87	7.81×10^{-2}	2.58	1.91×10^{-2}	1.82	7.79×10^{-2}	2.64	1.94×10^{-2}
10.0	1.95	0.179	5.03	4.07×10^{-2}	1.89	0.179	5.16	4.16×10^{-2}
20.0	2.02	0.403	9.43	8.34×10^{-2}	1.97	0.403	9.66	8.54×10^{-2}
30.0	2.07	0.640	13.3	0.123	2.01	0.640	13.6	0.126
50.0	2.12	1.13	20.1	0.192	2.06	1.13	20.5	0.196
100.0	2.18	2.40	33.0	0.319	2.12	2.40	33.6	0.324
200.0	2.24	5.01	50.2	0.469	2.18	5.01	51.0	0.475
300.0	2.27	7.66	61.9	0.558	2.22	7.66	62.8	0.563
500.0	2.32	13.0	78.0	0.661	2.26	13.0	79.0	0.666
1000.0	2.37	26.5	101.4	0.776	2.32	26.5	102.4	0.779

Source: ICRU (1984).

* Composition from ICRP (1975); mean excitation energy I = 75.3 eV; ρ = 1040 kg m^{-3}.

Total stopping power $S/\rho = (S/\rho)_{col} + (S/\rho)_{rad}$. Values of S/ρ for electrons are also given in Table 7.16.

Table 7.19 Calculated collision and radiative stopping powers, $(S/\rho)_{col}$ and $(S/\rho)_{rad}$ (MeV cm^2 g^{-1}), csda ranges (g cm^2) and radiation yields for electrons and positrons; adipose tissue*

Energy MeV	Electrons: Stopping power $(S/\rho)_{col}$	Electrons: Stopping power $(S/\rho)_{rad}$	Electrons: csda range	Electrons: Radiation yield	Positrons: Stopping power $(S/\rho)_{col}$	Positrons: Stopping power $(S/\rho)_{rad}$	Positrons: csda range	Positrons: Radiation yield
0.01	23.47	3.17×10^{-3}	2.41×10^{-4}	7.40×10^{-5}	25.75	2.02×10^{-3}	2.16×10^{-4}	3.78×10^{-5}
0.02	13.65	3.21×10^{-3}	8.24×10^{-4}	1.30×10^{-4}	14.75	2.36×10^{-3}	7.52×10^{-4}	7.86×10^{-5}
0.03	9.98	3.23×10^{-3}	1.69×10^{-3}	1.80×10^{-4}	10.69	2.57×10^{-3}	1.56×10^{-3}	1.19×10^{-4}
0.05	6.82	3.28×10^{-3}	4.18×10^{-3}	2.69×10^{-4}	7.21	2.81×10^{-3}	3.89×10^{-3}	1.98×10^{-4}
0.1	4.24	3.45×10^{-3}	1.39×10^{-2}	4.61×10^{-4}	4.40	3.18×10^{-3}	1.32×10^{-2}	3.79×10^{-4}
0.2	2.87	3.94×10^{-3}	4.36×10^{-2}	7.81×10^{-4}	2.92	3.78×10^{-3}	4.22×10^{-2}	6.98×10^{-4}
0.3	2.42	4.55×10^{-3}	8.19×10^{-2}	1.06×10^{-3}	2.43	4.42×10^{-3}	8.02×10^{-2}	9.85×10^{-4}
0.5	2.08	6.02×10^{-3}	0.172	1.59×10^{-3}	2.06	5.91×10^{-3}	0.171	1.53×10^{-3}
1.0	1.88	1.07×10^{-2}	0.428	2.92×10^{-3}	1.84	1.06×10^{-2}	0.430	2.89×10^{-3}
2.0	1.85	2.25×10^{-2}	0.962	5.84×10^{-3}	1.80	2.24×10^{-2}	0.978	5.90×10^{-3}
3.0	1.87	3.63×10^{-2}	1.49	9.06×10^{-3}	1.82	3.62×10^{-2}	1.52	9.21×10^{-3}
5.0	1.92	6.70×10^{-2}	2.52	1.60×10^{-2}	1.87	6.70×10^{-2}	2.58	1.63×10^{-2}
10.0	2.00	0.154	4.93	3.43×10^{-2}	1.94	0.154	5.06	3.52×10^{-2}
20.0	2.07	0.349	9.31	7.14×10^{-2}	2.02	0.349	9.54	7.31×10^{-2}
30.0	2.11	0.554	13.2	0.107	2.06	0.554	13.6	0.109
50.0	2.16	0.982	20.1	0.169	2.11	0.982	20.6	0.172
100.0	2.22	2.09	33.7	0.288	2.17	2.09	34.3	0.293
200.0	2.28	4.37	52.2	0.435	2.23	4.37	53.1	0.440
300.0	2.32	6.68	65.1	0.525	2.26	6.68	66.0	0.530
500.0	2.36	11.4	82.9	0.631	2.31	11.4	84.0	0.636
1000.0	2.42	23.1	109.2	0.753	2.37	23.1	110.4	0.756

Source: ICRU (1984).

* Composition from ICRP (1975); mean excitation energy I = 63.2 eV; ρ = 920 kg m^{-3}.

Total stopping power $S/\rho = (S/\rho)_{col} + (S/\rho)_{rad}$. Values of S/ρ for electrons are also given in Table 7.16.

Table 7.20 Calculated collision and radiative stopping powers, $(S/\rho)_{col}$ and $(S/\rho)_{rad}$ (MeV cm^2 g^{-1}), csda ranges (g cm^2) and radiation yields for electrons and positrons; cortical bone*

Energy	Electrons				Positrons			
	Stopping power		csda	Radiation	Stopping power		csda	Radiation
MeV	$(S/\rho)_{col}$	$(S/\rho)_{rad}$	range	yield	$(S/\rho)_{col}$	$(S/\rho)_{rad}$	range	yield
0.01	19.71	5.46x10^{-3}	2.91x10^{-4}	1.47x10^{-4}	21.85	2.39x10^{-3}	2.58x10^{-4}	5.10x10^{-5}
0.02	11.61	5.78x10^{-3}	9.80x10^{-4}	2.68x10^{-4}	12.64	2.98x10^{-3}	8.86x10^{-4}	1.12x10^{-4}
0.03	8.55	5.91x10^{-3}	2.00x10^{-3}	3.78x10^{-4}	9.21	3.34x10^{-3}	1.83x10^{-3}	1.74x10^{-4}
0.05	5.87	6.05x10^{-3}	4.88x10^{-3}	5.72x10^{-4}	6.24	3.82x10^{-3}	4.53x10^{-3}	3.00x10^{-4}
0.1	3.68	6.36x10^{-3}	1.61x10^{-2}	9.81x10^{-4}	3.83	4.56x10^{-3}	1.52x10^{-2}	6.03x10^{-4}
0.2	2.51	7.14x10^{-3}	5.02x10^{-2}	1.64x10^{-3}	2.55	5.67x10^{-3}	4.84x10^{-2}	1.16x10^{-3}
0.3	2.12	8.13x10^{-3}	9.39x10^{-2}	2.21x10^{-3}	2.13	6.78x10^{-3}	9.17x10^{-2}	1.67x10^{-3}
0.5	1.83	1.06x10^{-2}	0.196	3.24x10^{-3}	1.81	9.22x10^{-3}	0.195	2.65x10^{-3}
1.0	1.66	1.82x10^{-2}	0.486	5.76x10^{-3}	1.62	1.67x10^{-2}	0.490	5.12x10^{-3}
2.0	1.64	3.76x10^{-2}	1.09	1.11x10^{-2}	1.60	3.54x10^{-2}	1.10	1.05x10^{-2}
3.0	1.67	5.98x10^{-2}	1.67	1.69x10^{-2}	1.62	5.73x10^{-2}	1.71	1.62x10^{-2}
5.0	1.72	0.109	2.80	2.90x10^{-2}	1.67	0.106	2.87	2.85x10^{-2}
10.0	1.80	0.248	5.37	5.98x10^{-2}	1.75	0.245	5.52	6.01x10^{-2}
20.0	1.87	0.553	9.85	0.118	1.82	0.551	10.1	0.120
30.0	1.92	0.874	13.7	0.169	1.86	0.873	14.0	0.172
50.0	1.96	1.54	20.1	0.254	1.91	1.54	20.5	0.258
100.0	2.02	3.25	31.6	0.396	1.97	3.25	32.2	0.402
200.0	2.08	6.76	46.1	0.550	2.03	6.76	46.8	0.555
300.0	2.11	10.3	55.6	0.633	2.06	10.3	56.4	0.638
500.0	2.15	17.5	68.3	0.726	2.10	17.5	69.1	0.730
1000.0	2.21	35.5	86.4	0.824	2.16	35.5	87.2	0.827

Source: ICRU (1984).

* Composition from ICRP (1975); mean excitation energy I = 106.4 eV; ρ = 1850 kg m^{-3}.

Total stopping power $S/\rho = (S/\rho)_{col} + (S/\rho)_{rad}$. Values of S/ρ for electrons are also given in Table 7.16.

Table 7.21 Calculated collision and radiative stopping powers, $(S/\rho)_{col}$ and $(S/\rho)_{rad}$ (MeV cm^2 g^{-1}), csda ranges (g cm^2) and radiation yields for electrons and positrons; water*

Energy	Electrons				Positrons			
	Stopping power		csda	Radiation	Stopping power		csda	Radiation
MeV	$(S/\rho)_{col}$	$(S/\rho)_{rad}$	range	yield	$(S/\rho)_{col}$	$(S/\rho)_{rad}$	range	yield
0.01	22.56	3.90×10^{-3}	2.52×10^{-4}	9.41×10^{-5}	24.83	2.20×10^{-3}	2.25×10^{-4}	4.23×10^{-5}
0.02	13.17	3.96×10^{-3}	8.57×10^{-4}	1.66×10^{-4}	14.27	2.63×10^{-3}	7.80×10^{-4}	8.93×10^{-5}
0.03	9.65	3.98×10^{-3}	1.76×10^{-3}	2.30×10^{-4}	10.36	2.87×10^{-3}	1.62×10^{-3}	1.37×10^{-4}
0.05	6.60	4.03×10^{-3}	4.32×10^{-3}	3.44×10^{-4}	7.00	3.21×10^{-3}	4.02×10^{-3}	2.29×10^{-4}
0.1	4.12	4.23×10^{-3}	1.43×10^{-2}	5.84×10^{-4}	4.27	3.69×10^{-3}	1.36×10^{-2}	4.47×10^{-4}
0.2	2.79	4.80×10^{-3}	4.49×10^{-2}	9.83×10^{-4}	2.84	4.42×10^{-3}	4.34×10^{-2}	8.33×10^{-4}
0.3	2.36	5.51×10^{-3}	8.42×10^{-2}	1.33×10^{-3}	2.36	5.18×10^{-3}	8.24×10^{-2}	1.18×10^{-3}
0.5	2.03	7.26×10^{-3}	0.177	1.98×10^{-3}	2.01	6.94×10^{-3}	0.175	1.83×10^{-3}
1.0	1.85	1.28×10^{-2}	0.437	3.58×10^{-3}	1.81	1.25×10^{-2}	0.440	3.46×10^{-3}
2.0	1.82	2.68×10^{-2}	0.979	7.09×10^{-3}	1.78	2.64×10^{-2}	0.995	7.05×10^{-3}
3.0	1.85	4.30×10^{-2}	1.51	1.09×10^{-2}	1.80	4.27×10^{-2}	1.55	1.10×10^{-2}
5.0	1.89	7.92×10^{-2}	2.55	1.91×10^{-2}	1.84	7.90×10^{-2}	2.61	1.95×10^{-2}
10.0	1.97	0.181	4.98	4.07×10^{-2}	1.91	0.181	5.10	4.17×10^{-2}
20.0	2.05	0.409	9.32	8.36×10^{-2}	1.99	0.409	9.55	8.56×10^{-2}
30.0	2.09	0.649	13.2	0.123	2.03	0.649	13.5	0.126
50.0	2.14	1.15	19.8	0.192	2.08	1.15	20.3	0.196
100.0	2.20	2.43	32.6	0.319	2.15	2.43	33.2	0.324
200.0	2.26	5.08	49.6	0.470	2.21	5.08	50.4	0.475
300.0	2.30	7.76	61.2	0.558	2.24	7.76	62.0	0.564
500.0	2.34	13.2	77.1	0.661	2.29	13.2	78.0	0.666
1000.0	2.40	26.8	100.2	0.776	2.35	26.8	101.2	0.779

Source: ICRU (1984).

* Mean excitation energy I = 75 eV; ρ = 1000 kg m^{-3}.

Total stopping power $S/\rho = (S/\rho)_{col} + (S/\rho)_{rad}$. Values of S/ρ for electrons are also given in Table 7.15.

Table 7.22 Total mass stopping power, S/ρ, for electrons in tissue substitute materials (MeV m^2 kg^{-1})

Energy,MeV	0.01	0.03	0.1	0.3	1	3	10	100
For muscle and similar tissues								
A150	2.30	0.979	0.416	0.238	0.185	0.186	0.211	0.426
Acrylic	2.22	0.946	0.403	0.231	0.180	0.182	0.207	0.429
Alderson muscle	2.24	0.957	0.407	0.233	0.182	0.185	0.209	0.433
Lincolnshire bolus	2.12	0.908	0.388	0.223	0.176	0.179	0.205	0.443
Mix D	2.39	1.02	0.431	0.246	0.191	0.193	0.217	0.432
Polystyrene	2.24	0.955	0.406	0.232	0.181	0.183	0.206	0.413
Temex	2.29	0.975	0.414	0.237	0.184	0.186	0.210	0.423
For fatty tissues								
AP6	2.24	0.954	0.406	0.232	0.182	0.184	0.209	0.426
Ethoxyethanol	2.31	0.985	0.419	0.239	0.187	0.189	0.214	0.436
For lipid								
Paraffin wax	2.47	1.05	0.443	0.252	0.195	0.196	0.220	0.422
Polyethylene	2.45	1.04	0.440	0.251	0.194	0.195	0.219	0.422
For lung								
Alderson lung	2.16	0.922	0.393	0.225	0.180	0.185	0.211	0.431
LN 10/75	2.20	0.942	0.402	0.230	0.184	0.190	0.217	0.451
For cortical bone								
B110	1.95	0.844	0.363	0.210	0.165	0.170	0.201	0.513
Plaster of Paris	1.85	0.809	0.351	0.203	0.164	0.172	0.207	0.566
SB5	1.93	0.839	0.362	0.209	0.165	0.170	0.201	0.518

Table 7.23 Mass scattering power, T/ρ, for electrons in tissue substitute materials (MeV m^2 kg^{-1})

Energy,MeV	0.01	0.03	0.1	0.3	1	3	10	100
For muscle and similar tissues								
A150	595	86.8	10.9	1.83	0.295	0.0531	6.83×10^{-3}	9.37×10^{-5}
Acrylic	615	89.8	11.2	1.89	0.305	0.0551	7.08×10^{-3}	9.72×10^{-5}
Alderson muscle	616	90.1	11.3	1.90	0.307	0.0553	7.12×10^{-3}	9.76×10^{-5}
Lincolnshire bolus	659	96.6	12.1	2.05	0.331	0.0597	7.69×10^{-3}	1.05×10^{-4}
Mix D	594	86.6	10.8	1.82	0.294	0.0529	6.80×10^{-3}	9.34×10^{-5}
Polystyrene	573	83.3	10.4	1.75	0.281	0.0507	6.51×10^{-3}	8.96×10^{-5}
Temex	585	85.3	10.7	1.79	0.289	0.0521	6.70×10^{-3}	9.20×10^{-5}
For fatty tissues								
AP6	601	87.7	11.0	1.84	0.298	0.0537	6.90×10^{-3}	9.47×10^{-5}
Ethoxyethanol	615	89.7	11.2	1.89	0.305	0.0549	7.06×10^{-3}	9.69×10^{-5}
For lipid								
Paraffin wax	561	81.4	10.1	1.70	0.274	0.0493	6.33×10^{-3}	8.72×10^{-5}
Polyethylene	562	81.6	10.2	1.70	0.274	0.0494	6.34×10^{-3}	8.74×10^{-5}
For lung								
Alderson lung	605	88.4	11.1	1.86	0.300	0.0542	6.97×10^{-3}	9.56×10^{-5}
LN 10/75	642	94.1	11.8	1.99	0.321	0.0580	7.47×10^{-3}	1.02×10^{-4}
For cortical bone								
B110	847	126	16.1	2.74	0.446	0.0810	1.05×10^{-2}	1.41×10^{-4}
Plaster of Paris	965	145	18.5	3.15	0.515	0.0936	1.21×10^{-2}	1.63×10^{-4}
SB5	858	128	16.3	2.77	0.452	0.0822	1.06×10^{-2}	1.43×10^{-4}

7.3 Heavy charged particles

7.3.1 *Terminology and definitions*

Elastic scattering and bremsstrahlung production are generally negligible by comparison with the collision process for heavy charged particles at energies up to 500 MeV (Evans, 1955). Therefore the total mass stopping power, S/ρ, for particles such as protons, alpha particles and pions is essentially equal to the collision mass stopping power:

$$\frac{S}{\rho} = \frac{1}{\rho}\frac{dE}{dl} \approx \left[\frac{S}{\rho}\right]_{col} \tag{7.14}$$

After a negative pion has lost its kinetic energy by ionisation it may cause fragmentation of the nucleus. Molecular structure plays a considerable role and trace concentrations of elements can produce disproportionately large effects (Jackson and O'Leary, 1984).

7.3.2 *Calculated values for proton stopping power for tissue*

Calculated values for proton stopping powers in a range of tissues are given in Table 7.24, from ICRU (1989). The uncertainties are about 1–2%.

7.3.2.1 Tissue substitute materials

Calculated values for proton stopping power in some tissue substitute materials in the energy range 1–300 MeV are given in Table 7.25.

Table 7.24a Proton stopping powers in the range 1 to 20 MeV (MeV m^2 kg^{-1})

Proton energy,MeV	1	2	3	4	6	8	10	20
Blood	25.8	15.7	11.6	9.31	6.79	5.41	4.52	2.58
Bone,cortical	21.3	13.2	9.83	7.94	5.83	4.67	3.92	2.26
Bone,trabecular	25.0	15.3	11.3	9.08	6.62	5.27	4.41	2.52
Bone,red marrow	26.5	16.1	11.9	9.51	6.92	5.51	4.60	2.62
Bone,yellow marrow	27.3	16.6	12.2	9.77	7.10	5.64	4.71	2.68
Brain	26.0	15.8	11.7	9.39	6.84	5.45	4.56	2.60
Breast	26.4	16.0	11.8	9.49	6.91	5.50	4.59	2.62
Cartilage	25.3	15.5	11.4	9.18	6.70	5.33	4.46	2.55
Cell nucleus	25.8	15.7	11.6	9.33	6.80	5.42	4.53	2.59
Eye,lens	25.7	15.7	11.6	9.29	6.77	5.39	4.51	2.57
Fatty tissue	27.2	16.5	12.2	9.73	7.07	5.62	4.70	2.67
Intestine	26.0	15.8	11.7	9.37	6.83	5.44	4.55	2.60
Kidney	25.9	15.7	11.6	9.33	6.81	5.42	4.53	2.59
Liver	25.8	15.7	11.6	9.32	6.80	5.41	4.53	2.58
Lung	25.8	15.7	11.6	9.32	6.80	5.41	4.53	2.58
Lymph	25.9	15.8	11.7	9.35	6.82	5.43	4.54	2.59
Muscle,cardiac	25.8	15.7	11.6	9.33	6.80	5.42	4.53	2.59
Muscle,skeletal	25.8	15.7	11.6	9.33	6.80	5.41	4.53	2.58
Ovary	25.9	15.8	11.6	9.34	6.81	5.42	4.54	2.59
Pancreas	26.1	15.9	11.7	9.39	6.85	5.45	4.56	2.60
Skin	25.9	15.8	11.6	9.34	6.81	5.42	4.53	2.58
Spleen	25.8	15.7	11.6	9.32	6.80	5.41	4.53	2.58
Testicle	25.9	15.8	11.7	9.36	6.82	5.43	4.54	2.59
Thyroid gland	25.9	15.8	11.6	9.34	6.81	5.42	4.53	2.59

Table 7.24b Proton stopping powers in the range 30 to 500 MeV (MeV m^2 kg^{-1})

Proton energy,MeV	30	40	60	80	100	200	300	500
Blood	1.86	1.47	1.07	0.854	0.722	0.445	0.349	0.272
Bone,cortical	1.63	1.30	0.944	0.757	0.641	0.397	0.312	0.243
Bone,trabecular	1.82	1.44	1.04	0.835	0.706	0.435	0.341	0.266
Bone,red marrow	1.89	1.49	1.08	0.866	0.731	0.450	0.353	0.275
Bone,yellow marrow	1.93	1.53	1.10	0.882	0.745	0.459	0.359	0.280
Brain	1.87	1.48	1.08	0.860	0.727	0.448	0.351	0.273
Breast	1.88	1.49	1.08	0.865	0.731	0.450	0.353	0.275
Cartilage	1.84	1.46	1.06	0.845	0.714	0.440	0.345	0.269
Cell nucleus	1.86	1.48	1.07	0.856	0.724	0.446	0.350	0.272
Eye,lens	1.85	1.47	1.06	0.851	0.719	0.443	0.347	0.271
Fatty tissue	1.92	1.52	1.10	0.880	0.743	0.458	0.358	0.279
Intestine	1.87	1.48	1.07	0.859	0.726	0.447	0.350	0.273
Kidney	1.86	1.48	1.07	0.856	0.723	0.446	0.349	0.272
Liver	1.86	1.47	1.07	0.855	0.722	0.445	0.349	0.272
Lung	1.86	1.47	1.07	0.855	0.722	0.445	0.349	0.272
Lymph	1.87	1.48	1.07	0.858	0.725	0.447	0.350	0.273
Muscle,cardiac	1.86	1.48	1.07	0.855	0.723	0.446	0.349	0.272
Muscle,skeletal	1.86	1.48	1.07	0.855	0.723	0.445	0.349	0.272
Ovary	1.86	1.48	1.07	0.857	0.724	0.446	0.350	0.273
Pancreas	1.87	1.48	1.08	0.860	0.727	0.448	0.351	0.273
Skin	1.86	1.48	1.07	0.855	0.722	0.445	0.349	0.272
Spleen	1.86	1.48	1.07	0.855	0.723	0.445	0.349	0.272
Testicle	1.87	1.48	1.07	0.858	0.725	0.447	0.350	0.273
Thyroid gland	1.86	1.48	1.07	0.856	0.724	0.446	0.350	0.272

Table 7.25 Proton stopping powers in tissue substitute materials (MeV m^2 kg^{-1})

Proton energy,MeV	1	3	10	30	100	300
Water	26.0	11.7	4.56	1.87	0.728	0.352
For muscle and similar tissues						
A150	26.9	12.0	4.64	1.90	0.735	0.354
Acrylic	25.6	11.5	4.48	1.84	0.713	0.344
Alderson muscle	26.0	11.7	4.53	1.86	0.720	0.347
Lincolnshire bolus	24.1	11.0	4.29	1.77	0.687	0.332
Mix D	28.2	12.5	4.83	1.97	0.762	0.367
Polystyrene	26.1	11.7	4.53	1.85	0.717	0.346
Temex	26.8	12.0	4.63	1.89	0.732	0.353
for fatty tissues						
AP6	26.0	11.7	4.52	1.85	0.718	0.346
Ethoxyethanol	27.0	12.1	4.67	1.91	0.740	0.357
For lipid						
Paraffin wax	29.3	13.0	4.98	2.02	0.782	0.376
Polyethylene	29.1	12.9	4.94	2.01	0.776	0.373
For lung						
Alderson lung	24.8	11.2	4.37	1.79	0.695	0.336
LN 10/75	25.4	11.5	4.46	1.83	0.711	0.343
For cortical bone						
B110	21.8	9.99	3.96	1.64	0.644	0.313
Plaster of Paris	20.1	9.37	3.77	1.58	0.623	0.304
SB5	21.4	9.88	3.93	1.64	0.641	0.311

7.4 Neutrons

7.4.1 *Terminology and definitions*

Unlike photons and charged particles, both of which interact with electrons, neutrons interact with atomic nuclei. In the energy interval 25 meV to 100 MeV, there are five ways in which neutrons may interact with body tissues, namely, neutron capture, elastic scattering, inelastic scattering, nuclear reactions and spallation (Auxier *et al*, 1968, ICRU, 1977). For simplicity the total neutron cross–section, σ, may be represented as

$$\sigma = \sigma_a + \sigma_s \tag{7.15}$$

where σ_a is the absorption cross–section (resulting from neutron capture and spallation) and σ_s is the scatter cross–section resulting from elastic and inelastic scattering, and nuclear reactions. Neutron capture dominates for thermal neutrons at energies around 25 meV. Between 10 keV and 18 MeV elastic scattering by hydrogen is a major contributor to σ, with thesholds for inelastic scattering and nuclear reactions being reached at 0.5 MeV and 5 MeV respectively.

The definition for mass attenuation coefficient for photons (Equation 7.1) may also be applied to neutrons, and similarly the attenuation coefficient may by separated into absorption and scatter components. The total linear attenuation coefficient for neutrons is sometimes given the symbol $\sum$ and is called the macroscopic cross–section.

It is convenient to quantify the interaction of neutrons and tissue using the **kerma**, K. The kerma is defined as dE_{tr}/dm, where dE_{tr} is the sum of the initial kinetic energies of all the charged ionising particles liberated by uncharged ionising particles in a material of mass m. The unit is joule per kilogram (J kg^{-1}), and has the special name gray (Gy). For neutrons (and other uncharged ionising radiation) of energy E, the relationship between energy fluence, Ψ, and kerma, K, may be written as

$$K = \Psi\left[\frac{\mu_{tr}}{\rho}\right] = \Phi\left[E\left[\frac{\mu_{tr}}{\rho}\right]\right] \tag{7.16}$$

where μ_{tr} is the mass energy transfer coefficient (Equation 7.7) Φ is the particle fluence and $[E(\mu_{tr}/\rho)]$ is termed the **kerma factor**. The kerma factor has units gray per metre^2 (Gy m^{-2}).

7.4.2 *Calculated values of neutron kerma factors for tissue*

Calculated neutron kerma factors for a range of tissues are given in Table 7.26 (ICRU, 1989). Below 5 MeV the uncertainties are about 1% for

Table 7.26a Neutron kerma factors in the energy range 0.1 to 360 eV (Gy m^2 x 10^{-17})

Neutron energy,eV	0.11	0.36	1.1	3.6	11	36	110	360
Blood	1.29	0.711	0.407	0.228	0.140	0.108	0.153	0.397
Bone,cortical	1.63	0.912	0.532	0.307	0.188	0.126	0.114	0.180
Bone,trabecular	1.10	0.610	0.353	0.202	0.126	0.0995	0.137	0.342
Bone,red marrow	1.32	0.730	0.418	0.235	0.144	0.112	0.159	0.409
Bone,yellow marrow	0.293	0.162	0.0938	0.0554	0.0421	0.0583	0.137	0.426
Brain	0.868	0.480	0.276	0.156	0.0985	0.0869	0.145	0.407
Breast	1.17	0.644	0.369	0.208	0.128	0.103	0.155	0.407
Cartilage	0.867	0.480	0.275	0.156	0.0980	0.0838	0.134	0.369
Cell nucleus	1.24	0.685	0.393	0.221	0.136	0.108	0.156	0.404
Eye,lens	2.19	1.21	0.693	0.387	0.230	0.157	0.176	0.389
Fatty tissue	0.293	0.162	0.0937	0.0553	0.0420	0.0579	0.136	0.422
Intestine	0.864	0.478	0.274	0.155	0.0980	0.0862	0.144	0.401
Kidney	1.17	0.646	0.370	0.208	0.128	0.102	0.151	0.397
Liver	1.17	0.646	0.370	0.208	0.128	0.102	0.150	0.393
Lung	1.21	0.669	0.383	0.215	0.132	0.104	0.152	0.399
Lymph	0.453	0.251	0.144	0.0828	0.0568	0.0638	0.133	0.404
Muscle,cardiac	1.25	0.690	0.395	0.222	0.136	0.106	0.153	0.400
Muscle,skeletal	1.32	0.729	0.417	0.234	0.143	0.110	0.155	0.394
Ovary	0.941	0.520	0.298	0.168	0.106	0.0902	0.146	0.399
Pancreas	0.865	0.478	0.274	0.155	0.0980	0.0863	0.144	0.402
Skin	1.63	0.899	0.515	0.288	0.174	0.127	0.162	0.397
Spleen	1.24	0.688	0.394	0.221	0.136	0.106	0.153	0.398
Testicle	0.789	0.436	0.250	0.142	0.0906	0.0821	0.142	0.400
Thyroid gland	0.940	0.520	0.298	0.168	0.105	0.0897	0.144	0.396

Table 7.26b Neutron kerma factors in the energy range 1 to 3500 keV (Gy m^2 x 10^{-17})

Neutron energy,keV	1.1	3.6	11	36	105	350	1050	3500
Blood	1.13	3.63	10.5	30.2	67.1	137	248	414
Bone,cortical	0.431	1.29	3.69	10.6	23.7	50.4	95.6	177
Bone,trabecular	0.966	3.07	8.88	25.6	56.9	116	206	368
Bone,red marrow	1.18	3.76	10.9	31.4	69.7	142	250	440
Bone,yellow marrow	1.28	4.12	12.0	34.5	76.6	155	269	484
Brain	1.18	3.80	11.0	31.7	70.3	143	258	432
Breast	1.18	3.79	11.0	31.6	70.2	143	254	439
Cartilage	1.07	3.42	9.89	28.5	63.2	130	235	391
Cell nucleus	1.18	3.76	10.9	31.3	69.5	142	256	426
Eye,lens	1.08	3.44	9.94	28.6	63.6	130	235	399
Fatty tissue	1.27	4.08	11.8	34.1	75.8	153	267	478
Intestine	1.17	3.76	10.9	31.4	69.6	142	257	427
Kidney	1.14	3.66	10.6	30.5	67.8	138	250	418
Liver	1.13	3.63	10.5	30.3	67.1	137	248	415
Lung	1.14	3.66	10.6	30.5	67.7	138	251	417
Lymph	1.19	3.82	11.1	31.9	70.7	144	262	429
Muscle,cardiac	1.15	3.66	10.6	30.5	67.7	138	250	418
Muscle,skeletal	1.14	3.63	10.5	30.3	67.2	137	248	415
Ovary	1.16	3.72	10.8	31.1	68.9	141	255	422
Pancreas	1.17	3.77	10.9	31.4	69.7	142	256	430
Skin	1.12	3.57	10.3	29.8	66.1	135	243	412
Spleen	1.14	3.66	10.6	30.5	67.7	138	250	417
Testicle	1.17	3.76	10.9	31.4	69.6	142	257	426
Thyroid gland	1.15	3.69	10.7	30.8	68.4	140	253	420

Table 7.27 Neutron kerma factors for tissue substitute materials (Gy m^2 x 10^{-17})

Neutron energy,eV	0.11	1.1	$1.1x10^1$	$1.1x10^2$	$1.1x10^3$	$1.1x10^4$	$1.1x10^5$	$1.1x10^6$
Water	0.0229	0.0084	0.0146	0.123	1.22	11.4	73.1	272
For muscle and similar tissues								
A150	1.35	0.430	0.148	0.158	1.15	10.6	68.2	236
Acrylic	0.0164	0.0061	0.0107	0.0909	0.901	8.44	54.5	198
Alderson muscle	1.65	0.521	0.172	0.150	1.01	9.27	59.6	212
Lincolnshire bolus	0.0124	0.0045	0.0079	0.0675	0.669	6.28	40.7	156
Mix D	0.0278	0.0101	0.0176	0.150	1.48	13.9	88.8	304
Temex	0.0605	0.0214	0.0187	0.112	1.08	10.1	65.2	226
For fatty tissues								
AP6	0.932	0.295	0.103	0.124	0.957	8.88	57.4	203
For lung								
Alderson lung	0.772	0.244	0.0838	0.0907	0.669	6.21	40.4	148
LN 10/75	0.667	0.211	0.0760	0.116	0.949	8.83	57.0	201
For cortical bone								
B110	1.25	0.417	0.155	0.109	0.468	4.04	27.5	94.5
Plaster of Paris	0.0668	0.0543	0.0538	0.0771	0.313	2.52	16.5	72.7
SB5	0.420	0.153	0.0702	0.0715	0.341	2.93	19.2	80.3

hydrogen, and 5% for carbon, nitrogen and oxygen. Above 5 MeV the uncertainty for C, N and O increases to 10–25% (Awschalom *et al*, 1983). Values for tissue substitutes are included in Table 7.27.

7.4.2.1 Tissue substitute materials

Values of neutron kerma factors in several tissue substitutes are included in Table 7.27.

7.5 Naturally occurring radionuclides

The majority of body tissue constituents show no radioactivity, unless they are externally excited, or if a radionuclide is administered. However, two naturally occurring radionuclides do reside in body tissues. Potassium-40 has a relative abundance of about 0.01% naturally, and exists within the potassium in the body tissues. Carbon-14 may also exist within the body tissues due to the natural turnover of carbon by the body, even though its half-life is relatively short. Details of the radioactivity of these two radionuclides are given in Table 7.28.

Table 7.28 Naturally occurring radioisotopes in tissue

Nuclide	Half-life years	Abundance	Emissions
Carbon-14	5730		β^- 0.156 MeV
Potassium-40	1.26×10^9	0.0118%	89% β^- 1.314 MeV 11% K capture 0.001% β^+ 0.483 MeV 11% γ 1.46 MeV

References

Alderson S.W., Lanzl L.H., Rollins M. and Spira J., 1962, An instrumented phantom system for analog computation of treatment plans, Am J Roentgenol, 87, 185–195.

Auxier J.A., Snyder W.S. and Jones T.D., 1968, Neutron interactions and pentration in tissue. In *Radiation Dosimetry*, Vol. 1 (2nd edition), F.H. Attix and W.C. Roesch (eds), Academic Press, New York, pp.275–316.

Awschalom M. and Attix F.H., 1980, A-150 plastic-equivalent gas, Phys Med Biol, 25, 567-576.

Awschalom M., Rosenberg I. and Mravca A., 1983, Kermas for various substances averaged over the energy spectra of fast neutron therapy beams: A study in uncertainties, Med Phys, 10, 395-409.

Berger M.J. and Hubbell J.H., 1987, *XCOM: Photon cross sections on a personal computer*, Report No. NBSIR 87-3597, US Government Printing Office, Washington, D.C.

Bethe H.A. and Ashkin J., 1953, Passage of radiations through matter. In *Experimental Nuclear Physics*, Vol. I, E. Segrè (ed.), Wiley, New York, p.166.

Bichsel H, 1968, Charged particle interactions. In *Radiation Dosimetry*, Vol. 1 (2nd edition), F.H. Attix and W.C. Roesch (eds), Academic Press, New York, pp.157-228.

Brahme A., 1971, Multiple scattering of relativistic electrons in air, TRITA-EPP 71-22, Royal Inst. Technology, Stockholm, Sweden.

Bydder G.M. and Kreel L., 1979, The temperature dependence of computed tomography attenuation values, J Comput Assist Tomog, 3, 506-510.

Bydder G.M. and Kreel L., 1980, Attenuation values of fluid collections within the abdomen, J Comput Assist Tomogr, 4, 145-150.

Cho Z.H., Tsai C.M. and Wilson G., 1975, Study of contrast and modulation mechanisms in x-ray/photon transverse axial transmission tomography, Phys Med Biol, 20, 879-889.

Evans R.D., 1955, *The Atomic Nucleus*, McGraw-Hill, New York.

Frigerio N.A., 1962, Neutron penetration during neutron capture therapy, Phys Med Biol, 6, 541-549.

Goodman L.J., 1969, A modified tissue equivalent liquid, Health Phys, 16, 763-764.

Hermann K-P, Geworski L., Muth M. and Harder D., 1986, Muscle- and fat-equivalent polyethylene-based phantom materials for x-ray dosimetry at tube voltages below 100 kV, Phys Med Biol, 31, 1041-1046.

Hubbell J.H., 1969, *Photon Cross Sections, Attenuation Coefficients and Energy Absorption Coefficients from 10 keV to 100 GeV*, Report No. NSRDS-NBS 29, US Government Printing Office, Washington, D.C.

Hubbell J.H., 1982, Photon mass attenuation and energy-absorption coefficients from 1 keV to 20 MeV, Intl J Appl Radiat Isot, 33, 1269-1290.

ICRP, 1975, *Report of the Task Group on Reference Man*, ICRP Publication 23, International Commission on Radiological Protection, Pergamon Press, Oxford.

ICRU, 1977, *Neutron Dosimetry for Biology and Medicine,* ICRU Report 26, International Commission on Radiation Units and Measurements, Bethesda, MD, USA.

ICRU, 1984, *Stopping Powers for Electrons and Positrons*, ICRU Report 37,International Commission on Radiation Units and Measurements, Bethesda, MD, USA.

ICRU, 1989, *Tissue Substitutes in Radiation Dosimetry and Measurement,* ICRU Report 44, International Commission on Radiation Units and Measurements, Bethesda, MD, USA.

Jackson D.F. and O'Leary K., 1984, Tests of tissue equivalence for negative pions, Phys Med Biol, 29, 257–259.
Johns P.C. and Yaffe M.J., 1987, X–ray characterisation of normal and neoplastic breast tissues, Phys Med Biol, 32, 675–695.
Jones D.E.A. and Raine H.C., 1949, letter, Br J Radiol, 22, 549–550.
Joyet G., Baudraz A and Joyet M.–L., 1974, Determination of the electronic density and the average atomic number of tissues in man by γ–ray attenuation, Experientia, 30, 1338–1341.
Kim Y.S., 1974a, Human tissues: chemical composition and photon dosimetry data, Radiat Res, 57, 38–45.
Kim Y.S., 1974b, Human tissues: chemical composition and photon dosimetry data. A correction, Radiat Res, 60, 361–362.
Lindsay D.D. and Stern B.E., 1953, A new tissue–like material for use as a bolus, Radiology, 60, 355–361.
Mategrano V.C., Petasnick J., Clark J., Bin A.C. and Weinstein R., 1977, Attenuation values in computed tomography of the abdomen, Radiology, 125, 135–140.
McCullough E.C., 1975, Photon attenuation in computed tomography, Med Phys, 2, 307–320.
New P.F.J. and Aronow S., 1976, Attenuation measurements of whole blood and blood fractions in computed tomography, Radiology, 121, 635–640.
Rossi H.H. and Failla G., 1956, Tissue–equivalent ionisation chambers, Nucleonics, 14, 32–37.
Parthasaradhi K., Rao B.M. and Prasad S.G., 1989, Effective atomic numbers of biological materials in the energy region 1 to 50 MeV for photons, electrons, and He ions, Med Phys, 16, 653–654.
Phelps M.E., Hoffman E.J. and Ter–Pogossian M.M, 1975, Attenuation coefficients of various body tissues, fluids, and lesions at photon energies of 18 to 136 keV, Radiology, 117, 573–583.
Pullan B.R., Fawcitt R.A. and Isherwood I., 1978, Tissue characterisation by an analysis of the distribution of attenuation values in computed tomography scans: a preliminary report, J Comput Assist Tomogr, 2, 49–54.
Rao P.S. and Gregg E.C., 1975, Attenuation of monoenergetic gamma rays in tissues, Am J Roentgenology, 123, 631–637.
Rossi B.B, 1952, *High Energy Particles*, Prentice Hall, Englewood Cliffs, NJ.
Rossi H.H. and Failla G., 1956, Tissue–equivalent ionisation chambers, Nucleonics, 14, 32–37.
Rutherford R.A., Pullan B.R. and Isherwood I., 1976, Measurement of effective atomic number and electron density using an EMI scanner, Neuroradiology, 11, 15–21.
Smathers J.B., Otte V.A., Smith A.R., Almond P.R., Attix F.H., Spokas J.J., Quam W.M. and Goodman L.J., 1977, Composition of A–150 tissue–equivalent plastic, Med Phys, 4, 74–77.
Spiers F.W., 1946, Effective atomic number and energy absorption in tissues, Br J Radiol, 19, 52–63.
Spokas J.J. and White D.R., 1982, A conducting plastic simulating cortical bone, Phys Med Biol, 27, 115–121.
Stacy A.J., Bevan A.R. and Dickens C.W., 1961, A new phantom material employing depolymerised natural rubber, Br J Radiol, 34, 510–515.

Storm E. and Israel H.I., 1970, Photon cross sections from 1 keV to 100 MeV for elements Z = 1 to Z = 100, Nucl Data Tables, A7, 565.

Thirumala Rao B.V., Raju M.L.N., Narasimham K.L., Parthasaradhi K. and Mallikarjuna Rao B., 1985, Interaction of low-energy photons with biological materials and the effective atomic number, Med Phys, 12, 745-748.

Uehling E.A., 1954, Penetration of heavy charged particles in matter, An Rev Nucl Sci, 4, 315.

Walter B., 1926, Über die besten Formeln zur Berechnung der Absorption der Röntgenstrahlen in einem beliebigen Stoff, Fortschr Geb Röntgenstr, 35, 929.

Weber J. and van den Berge D.J., 1969, The effective atomic number and the calculation of the composition of phantom materials, Br J Radiol, 42, 378-383.

White D.R., 1976, Tissue substitute materials, 4th Int Conf on Med Phys, Ottawa, July 1976, cited in ICRU (1989).

White D.R., 1977a, Analysis of the Z-dependence of photon and electron interactions, Phys Med Biol, 22, 219-228.

White D.R., 1977b, The formulation of tissue substitute materials using basic interaction data, Phys Med Biol, 22, 889-899.

White D.R., Martin R.J. and Darlison R., 1977, Epoxy resin-based tissue substitutes, Br J Radiol, 50, 814-821.

White D.R. Peaple L.H.J. and Crosby T.J., 1980, Measured attenuation coefficients at low photon energies (9.88-59.32 keV) for 44 materials and tissues, Radiat Res, 84, 239-252.

White D.R., Constantinou C. and Martin R.J., 1986, Foamed epoxy resin-based lung substitutes, Br J Radiol, 59, 787-790.

Yang N.C., Leichner P.K. and Hawkins W.G., 1987, Effective atomic numbers for low-energy total photon interactions in human tissues, Med Phys, 14, 759-766.

Zatz L.M., 1976, The effect of the kVp level on EMI values, Radiology, 119, 683-688.

Chapter 8

Nuclear Magnetism of Tissue

The main body of this chapter is concerned with the magnetic response of the atomic nuclei in tissue. In particular the magnetic relaxation of hydrogen nuclei (protons) in tissue is quantified in the context of the use of nuclear magnetic resonance methods for imaging body tissues. Nuclear magnetic relaxation times for hydrogen protons are especially discussed, together with some comments about proton density and magnetic susceptibility. Finally a brief overview is added of the magnetic relaxation of some other nuclides of biological interest.

8.1 Hydrogen proton magnetic relaxation

8.1.1 *Terminology and definitions*

Atomic nuclei containing an odd number of neutrons protons or both possess a magnetic moment. When placed in a magnetic field of strength B, any substance containing such nuclei will be magnetised. The ratio between the intensity of magnetisation and the field strength is the **magnetic susceptibility** of the substance. The **volume susceptibility**, κ, is given by $J/\mu_0 B$ where J is the magnetic polarisation (intensity of magnetisation) and μ_0 ($4\pi x10^{-7}$ H m^{-1}) is the permeability of free space. The SI unit is metre^{-3}, or, for practical use cm^{-3}. The **mass susceptibility**, χ, is κ/ρ, where ρ is the density. The **molar susceptibility**, χ_M is $\chi.M/4\pi$, where M is the molecular weight.

When exposed to a magnetic field the nuclear magnetic moments precess around the field direction. The angular frequency of this precession, ω, the **Larmor frequency**, is linearly dependent on the local field strength:

$$\omega = \gamma B \qquad (8.1)$$

where γ is the **gyromagnetic ratio** for the particular nucleus, defined as the ratio of its magnetic moment to its angular momentum or spin. The unit for

γ is second^{-1} tesla^{-1} (s^{-1} T^{-1}). It is common to give the nuclear magnetic properties in terms of $\gamma/2\pi$, with unit hertz tesla^{-1} (Hz T^{-1}). For hydrogen nuclei (protons), $\gamma/2\pi \simeq 42.6$ MHz T^{-1}.

The local intensity of magnetisation, J, may be modified as a result of the resonant absorption by nuclei of electromagnetic energy at the Larmor frequency. Following this alteration (which may result in the complete reversal of the magnetic moment) the magnetisation relaxes to its original state. For simple homogenous media the relaxation may be characterised by two exponental decay times. The first is the **longitudinal relaxation time**, T_1, otherwise known as the **spin-lattice relaxation time.** T_1 characterises the recovery of the magnetisation along the field direction, z. The second time is the **transverse relaxation time**, T_2, otherwise known as the **spin-spin relaxation time.** T_2 characterises the recovery of magnetisation from a direction perpendicular to the magnetic field (x or y axes).

Relaxation rates are the reciprocals of the relaxation times: the **longitudinal relaxation rate** is $1/T_1$ and the **transverse relaxation rate** is $1/T_2$.

The longitudinal relaxation rate for hydrogen in water is largely controlled by the modulation of intramolecular dipolar interactions by the Brownian rotation of water molecules. In this relatively simple case the relaxation rate is given by

$$\frac{1}{T_1} = \left[\frac{2I(I + 1)}{5}\right]\left[\frac{\gamma^4\hbar^2}{r^6}\right]\left[\frac{t}{1 + (\omega t)^2} + \frac{t}{1 + 4(\omega t)^2}\right] \qquad (8.2)$$

where r is the distance between protons in the water molecule and $\hbar$ is the Planck constant ($1.055\text{x}10^{-34}$ J s). The correlation time, t, is the characteristic time for the for the correlation function between the local magnetic field and the applied field to become effectively zero. The nuclear spin, I, for protons is $\frac{1}{2}$, and so the value for the first bracket is 3/10.

In tissue the relaxation for water is commonly described as resulting from the rapid exchange between two water components: a large free water component and a bound water component hydrogen-bonded to macromolecules or hydration layers. In this 'fast exchange two-state' (FETS) model the overall relaxation rate becomes the weighted average of the relaxation rates of the two separate states:

$$\frac{1}{T_1} \sim \frac{b}{T_b} + \frac{1 - b}{T_f + \tau_e} \qquad (8.3)$$

where T_b is the T_1 of the bound fraction, b, T_f that of the free water fraction, 1-b, and τ_e is the (short) residence time in the restricted compartment. Free water undergoes rapid translational and rotational diffusion, with a correlation time $t \simeq 1$ ps. So $\omega t \ll 1$ for frequencies less than 100 MHz, meaning that T_f is constant over this lower frequency range (see Equation 8.2). The frequency dependence of tissue depends primarily, therefore, on T_b.

Such a simple model is inadequate to explain all the observed relaxation behaviour of tissues. For instance, Fullerton *et al* (1982) have proposed an alternative model, the fast proton diffusion (FPD) model, which goes some way to predict experimental behaviour. For further general discussion the reader is referred to Abragam (1978), and for tissue to Taylor and Bore (1981), and Bottomley *et al* (1984).

The transverse relaxation rate $1/T_2$ for water is given by

$$\frac{1}{T_2} = \left[\frac{I(I+1)}{5}\right]\left[\frac{\gamma^4 \hbar^2}{r^6}\right]\left[3\tau + \frac{5t}{1+(\omega t)^2} + \frac{2t}{1+4(\omega t)^2}\right] \qquad (8.4)$$

where the symbols have the same meanings as for Equation 8.2. As is the case with T_1, T_2 can also be represented as the sum of free and bound components in an expression analagous to Equation 8.3. T_2 is approximately equal to T_1 for water, and always less than T_1 in tissue.

The dependence of relaxation time on temperature is given roughly by

$$T_1 \sim \exp(-E/kT) \qquad (8.5)$$

where E is the activation energy for the T_1 process, k is the Boltzmann constant and T the absolute temperature.

8.1.2 *Measurement of relaxation time*

The methods for the measurement of relaxation times are well described in standard references (for instance, Beall *et al*, 1984). The inversion- recovery method is the most commonly used technique for the measurement of T_1 both for *in-vitro* and *in-vivo* studies. A radio-frequency (rf) pulse is used to alter the magnetisation direction by 180°, and after a chosen delay the remaining magnetisation is measured by the application of a second rf pulse. The second pulse rotates the magnetisation 90° so that it is perpendicular to the field direction. This allows the emission of a signal termed the **free induction decay**, and the magnetisation is derived from the amplitude of this signal. A single measurement of this sort can be used provided that a single exponential relaxation is assumed; greater accuracy can be achieved by the use of several measurements with different delay times.

Three methods are available for the measurement of T_2. The Hahn spin-echo method uses a 90°-τ-180° pulse sequence (meaning a 90°-pulse followed by a 180° pulse after a delay τ). This sequence does not yield a true value for T_2 if there is molecular diffusion, as is the case in tissue. The Carr-Purcell and Carr-Purcell-Meiboon-Gill (CPMG) sequences both extend the Hahn method to include a series of 180° pulses to invert sequentially the magnetisation perpendicular to the magnetic field. Comparison of T_2 values measured *in-situ* using both the Hahn spin-echo and CPMG methods have demonstrated the underestimates resulting from spin-echo measurements in tissue (Luyten *et al*, 1987).

Other problems attend the application of these methods to the measurement of relaxation times when using magnetic resonance imaging systems. Lerski *et al* (1988) have reported a multicentre study in which many aspects of NMR scanner performance were reviewed, including both precision and accuracy of T_1 and T_2 estimations. Only about two-thirds of the reported measurements achieved a better than 10% coefficient of variation (range/mean), T_2 being slightly poorer than T_1. Whilst the accuracy of some scanners was very good, particularly for shorter relaxation times, it was generally degraded for test samples with longer T_1. In addition, tissue movement may also be a source of error, particularly for measurements on the abdominal contents. Ehman *et al* (1984) have demonstrated that for both liver and spleen a considerably greater spread of measurements was observed under normal body scanning than when respiratory gating was used or when measurements were made on a static phantom. Whilst relaxation times measured *in-vivo* are generally to be preferred in principle, some care is still needed in judging the absolute accuracy of these measurements.

8.1.3 *Historical background*

Early work by Odeblad and co-workers (for instance, Odeblad and Lindstrom, 1955) on the measurement of relaxation times of tissues was brought into focus by the observation that magnetic resonance could provide a tool with which to distinguish between malignant and non-malignant tissue (Damadian, 1970; Damadian *et al*, 1974). Many reports in the early 1970s emphasised this perspective, reporting relaxation times on tissue samples *in-vitro* and mostly at room temperature (for instance, Cottam *et al*, 1972; Frey *et al*, 1972; Hollis *et al*, 1973; Parrish *et al* 1974). It was soon appreciated that, unlike pure water, the spin-lattice relaxation time in tissue is dispersive, varying with the applied magnetic field (Coles, 1976). The availability of magnetic resonance imaging systems from the late 1970s has resulted in a large number of publications giving values of relaxation times from tissues *in- vivo*, mostly human. Extensive reviews of relaxation times for normal tissues from the then available literature were carried out by Bottomley *et al* (1984) and Beall *et al* (1984). The volume by Beall *et al* also contains much information on pathological tissues, and further pathological tissue results were also extensively reviewed by Bottomley *et al* (1987).

8.1.4 *Values of* ^1H *longitudinal relaxation time,* T_1*, for tissue*

Tables 8.1, 8.2 and 8.3 give experimental longitudinal relaxation times, T_1, for normal mammalian tissues covering the frequency range 0.01 to 400 MHz. Entries for tissues are organised alpabetically and by frequency, irrespective of animal species or temperature. Measurements from fetal or neonatal tissues are included with the tissue type, but at the end of the frequency-ordered list. Tissues described as 'fat', 'fatty tissue' and 'adipose tissue' are included together under fatty tissue. Heart tissue is entered under muscle:cardiac. Blood and its constituents are included separately in Table 8.2.

Some selection from the large quantity of available data has occurred, whilst retaining details of the frequency dependency of T_1. The experimental studies from which the relaxation times have been drawn are predominantly

Table 8.1 ^1H proton spin–lattice relaxation time, T_1, for tissue

Tissue	Freq MHz	Temp °C	T_1 ms	Reference
Adrenal gland, rabbit	2.5	20	277±13	Ling *et al* 1980
–,dog	5.1	room	253	Wolf & Conard 1983
–,human	8.5	*in-vivo*	860±510	Rupp *et al* 1983
–,–	15	*in-vivo*	394±209	Moon *et al* 1983a
–,rabbit	24	23	448±13	Ling *et al* 1980
–,human	24.3	26	585	Cottam *et al* 1972
–,–	100	26	608±20	Damadian *et al* 1974
Artery,rabbit	2.5	20	167±17	Ling *et al* 1980
–,dog	5.1	room	212	Wolf & Conard 1983
–,human	8.5	*in-vivo*	860±510	Rupp *et al* 1983
–,rabbit	24	23	364±30	Ling *et al* 1980
Amniotic fluid, human	10^{-4}	37	2200	Béné *et al* 1982
–,–	10^{-3}	37	2600	–
–,–	0.01	37	2800	–
–,–	0.1	37	3000	–
–,–	1	37	3300	–
–,–	10	37	3600	–
Blood	see Table 8.2			
Bile,human	1.7	*in-vivo*	360–400	Smith 1983
–,rabbit	2.5	20	888±388	Ling *et al* 1980
–,dog	5.1	room	383	Wolf & Conard 1983
–,human	8.5	*in-vivo*	890±140	Rupp *et al* 1983
–,rabbit	24	23	1078±44	Ling *et al* 1980
Bladder,human	100	26	891±61	Damadian *et al* 1974
Bone,human	1.7	*in-vivo*	190–220	Smith *et al* 1982
–,–, vertebral body	15	*in-vivo*	505±75	Nyman *et al* 1986
–,–	24	room	125	Mallard *et al* 1979
–,–,	100	26	554±27	Damadian *et al* 1974
Bone marrow, rabbit	4.3	*in-vivo*	306±4*	Gore *et al* 1983
–,human, long bone	8.5	*in-vivo*	280±40	Rupp *et al* 1983
–,–,vertebral	8.5	*in-vivo*	380±50	–
–,–,–	15	*in-vivo*	420±112	Moon *et al* 1983b
–,–	15	*in-vivo*	502±211	Ehman *et al* 1985
–,–,1–10yr	15	*in-vivo*	652±33	Dooms *et al* 1985
–,–,21–40yr	15	*in-vivo*	500±129	–
–,–,51+yr	15	*in-vivo*	427±118	–
–,–	25	room	803±100	Ranade *et al* 1977
–,–	100	26	554±27	Damadian *et al* 1974

cont.

Table 8.1 cont. T_1 **for tissue**

Tissue	f,MHz	T,°C	T_1,ms	Reference
Brain,rat	0.01	37	80	Fischer *et al* 1989a
–,–	0.1	37	130	–
–,–	1	37	270	–
–,–	10	37	500	–
–,human†	20	40	630±18	Grodd & Schmitt 1983
–,rat	60	room	866±42	Kiricuta & Simplăceanu 1975
–,human	100	26	998±16	Damadian *et al* 1974
–,–,unmyelinated				
36 weeks	6.4	*in–vivo*	1000–1340	Johnson *et al* 1983
42 weeks	6.4	*in–vivo*	970–1120	–
6 months	6.4	*in–vivo*	630–780	–
–,–,myelinated				
20 months	6.4	*in–vivo*	350–530	–
9 years	6.4	*in–vivo*	310–440	–
–,rat,fetal	60	room	1361	Kiricuta & Simplăceanu 1975
Brain:grey matter,				
human	0.01	37	120	Fischer *et al* 1989b
–,–	0.1	37	184	–
–,–	1	37	330	–
–,–	1.7	*in–vivo*	275–300	Smith 1983
–,	3.4	*in–vivo*	465±36	Bell *et al* 1987
–,–	6.4	*in–vivo*	375–525	Bydder *et al* 1982
–,–	12	*in–vivo*	625±50	Just & Thelen 1988
–,–	15	*in–vivo*	702±68	Kjos *et al* 1985
–,cow	21.5	40	940	Kamman *et al* 1988
–,human	50	37	1000	Fischer *et al* 1989b
–,–	200	37	1500	–
–,–,neonate	1.7	*in–vivo*	410	Smith 1983
Brain:white matter,				
human	0.01	37	88	Fischer *et al* 1989b
–,–	0.1	37	118	–
–,–	1	37	180	–
–,–	1.7	*in–vivo*	225–250	Smith 1983
–,–	6.5	*in–vivo*	220–350	Bydder *et al* 1982
–,–	12	*in–vivo*	401±38	Just & Thelen 1988
–,–	15	*in–vivo*	419±34	Kjos *et al* 1985
–,cow	21.5	40	507	Kamman *et al* 1988
–,human	50	37	590	Fischer *et al* 1989b
–,–	200	37	1200	–
–,–,neonate	1.7	*in–vivo*	390–430	Smith 1983
Brain:cerebellum,				
rabbit	2.5	20	326±11	Ling *et al* 1980
–,human	12.8	*in–vivo*	585	Wehrli *et al* 1983
–,–,grey	15	*in–vivo*	652±46	Kjos *et al* 1985
–,–,white	15	*in–vivo*	456±38	–
–,rabbit	24	23	570±24	Ling *et al* 1980

cont.

Table 8.1 cont. T_1 **for tissue**

Tissue	f,MHz	T,°C	T_1,ms	Reference
Brain:corpus callosum, human	12.8	*in-vivo*	380	Wehrli *et al* 1983
Brain:pons,human	12.8	*in-vivo*	445	-
Breast,human	0.02	37	120	Koenig & Brown 1984
-,-	0.1	37	140	-
-,-	1.0	37	165	-
-,-	1.8	*in-vivo*	126±46	Ross *et al* 1982
-,-	10	37	240	Koenig & Brown 1984
-,-	22.5	room	447±136	Goldsmith *et al* 1978b
-,-	25	23	420±198	Kasturi *et al* 1976
-,-	30.3	25	682±32	Medina *et al* 1975
-,-	50	37	305	Koenig & Brown 1984
-,-	60	24	907±23	Bovée *et al* 1978
-,-	100	26	367±79	Damadian *et al* 1974
-,-,lactating	100	26	978	-
CSF,human	1.7	*in-vivo*	800-1000	Smith 1983
-,-	6.25	*in-vivo*	4360±600	Hopkins *et al* 1986
-,-	6.5	*in-vivo*	900->2000	Bydder *et al* 1982
-,-	15	*in-vivo*	2720±406	Kjos *et al* 1985
-,-	25.4	*in-vivo*	4220±280	Hopkins *et al* 1986
-,-	60.1	*in-vivo*	4310±520	-
Cervix,human	22.5	room	510±83	Fruchter *et al* 1978
-,-	25	23	825±286	Kasturi *et al* 1976
-,-	100	26	827±26	Damadian *et al* 1974
Colon,human	0.02	37	100	Koenig & Brown 1984
-,-	0.1	37	140	-
-,-	1.0	37	288	-
-,rabbit	2.5	20	227±20	Ling *et al* 1980
-,human	10	37	660,610	Koenig & Brown 1984
-,-	22.5	room	330±129	Koutcher *et al* 1978
-,rabbit	24	23	466±29	Ling *et al* 1980
-,human	40	37	1150,1050	Koenig & Brown 1984
-,-	100	26	641±43	Damadian *et al* 1974
Embryo,rat	60	room	1422±7	Kiricuta & Simplăceanu 1975
Endometrium, human	22.5	room	801	Fruchter *et al* 1978
Epididymis, rabbit	2.5	20	309±6	Ling *et al* 1980
-,-	24	23	625±5	-
Eye,human, vitreous	15	*in-vivo*	2800±426	Kjos *et al* 1985
-,-,aqueous	21		3000	Huggert & Odeblad 1959
-,-,cornea	21		300-400	-
-,-,optic nerve	21		100	-
-,-,retina	21		1000	-
-,-,sclera	21		100	-

cont.

Table 8.1 cont. T_1 **for tissue**

Tissue	f,MHz	T,°C	T_1,ms	Reference
Eye:lens,human,				
nucleus	15	*in-vivo*	600±41	Kjos *et al* 1985
–,–,cortex	21		200–400	Huggert & Odeblad 1959
–,rabbit	25	36	490	Neville *et al* 1974
–,–,cortex	25	36	551	–
–,–,nucleus	25	36	268	–
–,mouse	60	27	352±5	Bakker & Vriend 1984
Fallopian tube,				
human	22.5	room	546±22	Fruchter *et al* 1978
–,human	100	26	1247	Damadian *et al* 1974
Fat tissue,rat	0.01	37	145	Fischer *et al* 1989a
–,–	0.1	37	160	–
–,–	1	37	185	–
–,human	1.7	*in-vivo*	130–160	Smith 1983
–,rabbit,perirenal	4.3	*in-vivo*	183±6*	Gore *et al* 1983
–,dog	5.1	room	75	Wolf & Conard 1983
–,human	8.5	*in-vivo*	240±20	Rupp *et al* 1983
–,rat	10	37	230	Fischer *et al* 1989a
–,human	12	*in-vivo*	209±17	Just & Thelen 1988
–,rat	15	*in-vivo*	305±38	Moon *et al* 1983c
–,human	15	*in-vivo*	266±45	Nyman *et al* 1986
–,–†	20	40	192±26	Grodd & Schmitt 1983
–,–	30	room	144±6	Inch *et al* 1974
–,rat	50	37	350	Fischer *et al* 1989a
–,human	60	24	144,200	Bovée *et al* 1978
–,mouse	60	27	252±2	Bakker & Vreind 1984
–,human	100	26	279±8	Damadian *et al* 1974
Fetus,mouse	30	room	990±34	Inch *et al* 1974
Gall bladder				
rabbit	4.3	*in-vivo*	290±12	Gore *et al* 1983
Hair,mouse	60	27	26±1	Bakker & Vriend 1984
Intestine,human	22.5	room	416±103	Goldsmith *et al* 1978a
–,–	24.3	23	584	Eggleston *et al* 1975
–,mouse	30	25	366±19	Frey *et al* 1972
–,human	100	26	641±80	Damadian *et al* 1974
Kidney,rat	0.01	30	50	Koenig *et al* 1984
–,–	0.1	30	70	–
–,–	1	30	165	–
–,human	1.7	*in-vivo*	300–340	Smith *et al* 1982
–,–	8.5	*in-vivo*	670±60	Rupp *et al* 1983
–,rat	15	*in-vivo*	418–733	Brasch *et al* 1983
–,human†	20	40	765±26	Grodd & Schmitt 1983
–,–	100	26	862±33	Damadian *et al* 1974
–,human fetal	25	23	993	Kasturi *et al* 1976

cont.

Table 8.1 cont. T_1 **for tissue**

Tissue	f,MHz	T,°C	T_1,ms	Reference
Kidney:cortex				
rabbit	2.5	20	206±27	Ling *et al* 1980
–,–	4.3	*in–vivo*	230±5	Gore *et al* 1983
–,dog	5.1	room	312	Wolf & Conard 1983
–,human	15	*in–vivo*	590±171	Ehman *et al* 1985
–,rabbit	24	23	406±41	Ling *et al* 1980
–,–	60	*in vivo*	685±155	Kundel *et al* 1986
–,rat	85.5	22	839±36	Dockery *et al* 1989
–,–	300	22	1035±137	–
Kidney:medulla				
rabbit	2.5	20	426±61	Ling et al 1980
–,–	4.3	*in–vivo*	520±10	Gore *et al* 1983
–,dog	5.1	room	604	Wolf & Conard 1983
–,human	15	*in–vivo*	696±188	Ehman *et al* 1985
–,rabbit	24	23	801±35	Ling *et al* 1980
–,–	60	*in vivo*	1245–1453	Kundel *et al* 1986
–,rat	85.5	22	1516±72	Dockery *et al* 1989
–,–	300	22	1593±153	–
Liver,rabbit	0.01	35	37	Koenig *et al* 1984
–,–	0.1	35	46	–
–,–	1	35	103	–
–,human	1.7	*in–vivo*	140–170	Smith *et al* 1981,1982
–,rabbit	2.5	20	141±16	Ling *et al* 1980
–,–	3	35	150	Koenig *et al* 1984
–,–	4.3	*in–vivo*	215±6*	Gore *et al* 1983
–,human	6.5	*in–vivo*	210–270	Doyle *et al* 1982
–,mouse	8	*in–vivo*	228±26	Barroilhet & Moran 1975
–,human	8.5	*in–vivo*	380±20	Rupp *et al* 1983
–,rat	10.7	37	226	Mathur–De Vré *et al* 1988
–,human	15	*in–vivo*	443±76	Nyman *et al* 1986
–,–	15	*in–vivo*	377±76	Ehman *et al* 1985
–,human	20	40	397±11	Grodd & Schmitt 1983
–,pig	20	40	182±7	–
–,rat	20	40	205±43	–
–,–	20	37	349±15	Barthwal *et al* 1986
–,human†	24	room	339±42	Mallard *et al* 1979
–,rat	24.5	37	251	Mathur–De Vré *et al* 1988
–,mouse	30	25	386±13	Frey *et al* 1972
–,human	30	room	298±23	Inch *et al* 1974
–,rat	54.7	37	435	Mathur–De Vré *et al* 1988
–,human	100	26	570±29	Damadian *et al* 1974
–,human fetal	25	23	721	Kasturi *et al* 1976
Lung,rabbit	2.5	20	283±19	Ling *et al* 1980
–,rabbit	4.3	*in–vivo*	268,423	Gore *et al* 1983
–,dog	5.1	room	374	Wolf & Conard 1983
–,human†	20	40	756±17	Grodd & Schmitt 1983

cont.

Table 8.1 cont. T_1 for tissue

Tissue	f,MHz	T,°C	T_1,ms	Reference
Lung,human	22.5	room	505±63	Koutcher *et al* 1978
–,rabbit	24	23	624±37	Ling *et al* 1980
–,mouse	30	25	641±9	Frey *et al* 1972
–,human	100	26	788±63	Damadian *et al* 1974
–,human fetal	25	23	1051	Kasturi *et al* 1976
Lymph node, human	20	40	725	Grodd & Schmitt 1983
–,–	100	26	720±76	Damadian *et al* 1974
–,rat	0.01	30	58	Koenig *et al* 1984
–,–	0.1	30	87	–
–,–	1	30	183	–
–,human	1.7	*in-vivo*	240–260	Smith 1983
–,rabbit	2.5	20	243±4	Ling *et al* 1980
–,dog	5.1	room	333	Wolf & Conard 1983
–,–	15	*in-vivo*	650±87	Higgins *et al* 1983
–,human†	20	40	644±12	Grodd & Schmitt 1983
–,–	24.3	26	873±118	Cottam *et al* 1972
–,mouse	30	25	650±6	Frey *et al* 1972
–,rat	60	room	873±27	Kiricuta & Simplăceanu 1975
–,human	100	26	906±46	Damadian *et al* 1974
–,rat,fetus	60	room	936	Kiricuta & Simplăceanu 1975
Muscle:skeletal, rat	0.01	37	45	Fischer *et al* 1989a
–,–	0.1	37	70	–
–,–	1	37	160	–
–,human	1.7	*in-vivo*	120–140	Smith *et al* 1982
–,rabbit	4.3	*in-vivo*	259±7	Gore *et al* 1983
–,mouse	5	*in-vivo*	343±17	Kroeker *et al* 1985
–,human	8.5	*in-vivo*	400±40	Rupp *et al* 1983
–,rat	10.7	37	511	Mathur-De Vré *et al* 1988
–,human	15	*in-vivo*	514±138	Ehman *et al* 1985
–,rat	15	*in-vivo*	744±98	Moon *et al* 1983c
–,human†	20	40	629±11	Grodd & Schmitt 1983
–,mouse	20	*in-vivo*	610±30	Kroeker *et al* 1985
–,rat	22.5	37	580	Mathur-De Vré *et al* 1988
–,human	25	23	643±187	Kasturi *et al* 1976
–,–	43.5	37	650–800	Borghi *et al* 1983
–,rat	54.7	37	880	Mathur-De Vré *et al* 1988
–,mouse	57	*in-vivo*	960±50	Kroeker *et al* 1985
–,human	100	26	1023±29	Damadian *et al* 1974
–,rat	200	27	1450±165	LeBlanc *et al* 1986
Myometrium, human	22.5	room	553±42	Fruchter *et al* 1978
Nerve,dog	5.1	room	111	Wolf & Conard 1983
–,human	100	26	557±158	Damadian *et al* 1974

cont.

Table 8.1 cont. T_1 **for tissue**

Tissue	f,MHz	T,°C	T_1,ms	Reference
Oesophagus, rabbit	2.5	20	250±14	Ling *et al* 1980
–,–	24	23	534±42	–
–,human	25	23	765±180	Kasturi *et al* 1976
–,–	100	26	804±108	Damadian *et al* 1974
Ovary,rabbit	2.5	20	310	Ling *et al* 1980
–,dog	5.1	room	380	Wolf & Conard 1983
–,human	22.5	room	504±88	Fruchter *et al* 1978
–,rabbit	24	23	573	Ling *et al* 1980
–,human	100	26	989±47	Damadian *et al* 1974
Pancreas,human	1.7	*in-vivo*	180–200	Smith *et al* 1982
–,dog	5.1	room	231	Wolf & Conard 1983
–,human	8.5	*in-vivo*	290±20	Rupp *et al* 1983
–,–	15	*in-vivo*	463±177	Ehman *et al* 1985
–,–	20	40	572	Grodd & Schmitt 1983
–,–	24.3	26	320±15	Cottam *et al* 1972
–,–	100	26	605±36	Damadian *et al* 1974
Parotid gland, rabbit	2.5	20	234±38	Ling *et al* 1980
–,human	8.5	*in-vivo*	350	Rupp *et al* 1983
–,rabbit	24	23	340±51	Ling *et al* 1980
Peritoneum, human	100	26	476	Damadian *et al* 1974
Penis,human	25	23	617±218	Kasturi *et al* 1976
Pineal gland, human	100	26	1291	Damadian *et al* 1974
Pituitary,dog	5.1	room	345	Wolf & Conard 1983
–,human	12	*in-vivo*	583±40	Just & Thelen 1988
Placenta,human	22.5	room	668±161	Fruchter *et al* 1978
Prostate gland human	1.7	*in-vivo*	200	Hutchinson & Smith 1983
–,–	20	40	808	Grodd & Schmitt 1983
–,–	24.3	26	767	Cottam *et al* 1972
–,–	100	26	803±14	Damadian *et al* 1974
Rectum,rabbit	2.5	20	245±6	Ling *et al* 1980
–,–	24	23	492±56	–
–,human	25	23	762±209	Kasturi *et al* 1976
Salivary gland, mouse	60	27	365±4	Bakker & Vriend 1984
Skin,rabbit	2.4	20	198±13	Ling *et al* 1980
–,–	4.3	*in-vivo*	220±10	Gore *et al* 1983
–,–	24	23	328±28	Ling *et al* 1980
–,human	25	23	362,357	Kasturi *et al* 1976
–,mouse	30	25	390±39	Frey *et al* 1972
–,human	100	26	616±19	Damadian *et al* 1974

cont.

Table 8.1 cont. T_1 for tissue

Tissue	f,MHz	T,°C	T_1,ms	Reference
Spinal chord, human	1.7	*in-vivo*	230-280	Hutchinson & Smith 1983
-,rabbit	2.4	20	325±19	Ling *et al* 1980
-,-	24	23	464±22	-
Spleen,rabbit	0.01	35	77	Koenig *et al* 1984
-,-	0.1	35	120	-
-,-	1	35	235	-
-,human	1.7	*in-vivo*	250-290	Smith *et al* 1981,1982
-,rabbit	3	35	330	Koenig *et al* 1984
-,dog	5.1	room	344	Wolf & Conard 1983
-,human	6.5	*in-vivo*	440-580	Doyle *et al* 1982
-,-	8.5	*in-vivo*	420±50	Rupp *et al* 1983
-,rabbit	10	35	550	Koenig *et al* 1984
Spleen,human	15	*in-vivo*	915±162	Nyman *et al* 1986
-,-	15	*in-vivo*	646±154	Ehman *et al* 1985
-,human	20	40	760±26	Grodd & Schmitt 1983
-,pig	20	40	507±7	-
-,rat	20	40	431±58	-
-,dog	24	*in-vivo*	902±77	Johnson *et al* 1985
-,-	25	25	658±215	Shah *et al* 1982
-,rabbit	50	35	825	Koenig *et al* 1984
-,human	100	26	701±45	Damadian *et al* 1974
Stomach,rabbit	2.5	20	227±20	Ling *et al* 1980
-,-	24	23	468±94	-
-,human	25	23	841±229	Kasturi *et al* 1976
-,mouse	30	25	294±23	Frey *et al* 1972
-,human	100	26	765±75	Damadian *et al* 1974
Tendon, several animals	10.7	room	164±12	Fullerton *et al* 1985
-,human	15	*in-vivo*	864±206	Moon *et al* 1983b
Testis,rabbit	2.5	20	463±47	Ling *et al* 1980
-,dog	5.1	room	410	Wolf & Conard 1983
-,human	20	40	974	Grodd & Schmitt 1983
-,rabbit	24	23	855±110	Ling *et al* 1980
-,human	24.3	23	640-791	Eggleston *et al* 1975
-,mouse	60	27	914±10	Bakker & Vriend 1984
-,human	100	26	1200±48	Damadian *et al* 1974
Thalamus,human	15	*in-vivo*	610±35	Kjos *et al* 1985
Thymus,human	15	*in-vivo*	458±137	Dooms *et al* 1984
-,-,child	15	*in-vivo*	948±186	-
-,-	24.3	26	809	Cottam *et al* 1972
Thyroid gland, human	10.7	37	521 (363-829)	Tennvall *et al* 1987
-,-	15	*in-vivo*	641±186	Dooms *et al* 1984
-,-	20	40	605	Grodd & Schmitt 1983

cont.

Table 8.1 cont. T_1 **for tissue**

Tissue	f,MHz	T,°C	T_1,ms	Reference
–,–	24.3	26	586	Cottam *et al* 1972
–,–	100	26	882±45	Damadian *et al* 1974
Tongue,human	20	40	584	Grodd & Schmitt 1983
Trachea,rabbit	2.5	20	199±17	Ling *et al* 1980
–,–	24	23	276±11	-
–,human	100	26	693	Damadian *et al* 1974
Uterus,rabbit	2.5	20	296	Ling *et al* 1980
–,–	24	23	653	-
–,human	25	23	842	Kasturi *et al* 1976
–,–	100	26	924±38	Damadian *et al* 1974
Ureter,rabbit	2.5	20	172±10	Ling *et al* 1980
–,–	24	23	232±13	-
Urethra,human	24.3	23	503	Eggleton *et al* 1975
Urine,human	0.83	37	1990–3496 (2859)	Alanen *et al* 1987
–,dog	5.1	room	2650	Wolf & Conard 1983
–,human	8.5	*in–vivo*	2200±610	Rupp *et al* 1983
–,rat	15	*in–vivo*	3740±4537	Moon *et al* 1983c
Vertebral disk human	15	*in–vivo*	2623±2030	Moon *et al* 1983b
Vagina,human	22.5	room	609±101	Fruchter *et al* 1978
–,–	100	26	1019	Damadian *et al* 1974
Vein,human	24	room	400–650	Mallard *et al* 1979
–,IVC,rabbit	2.5	20	201±5	Ling *et al* 1980
–,–,–	24	23	316±56	-
Vocal chord, human	25	23	593	Kasturi *et al* 1976
Vulva,human	22.5	room	451	Fruchter *et al* 1978

* τ=300 ms: values at τ=200 and 400 ms also reported.

† T_1 for animal tissue also given.

one of two types. Firstly relaxation times have been measured from a relatively large range of tissue types, but at only one or perhaps two frequencies. From such studies the natural range of relaxation time for one particular tissue may be estimated. The second type of study specifically invesigates dispersion, but on a relatively limited range of tissue samples. When data from these studies have been included, generally only single values are noted. When selection from a wide range of data of the same tissue type was felt to be necessary, for instance for muscle, liver and brain tissue, some general selection principles were applied. Measurements at body temperature were preferred to those at room temperature and those on human tissue were preferred to those from other animals. *In–vivo* studies were preferred to those carried out on tissue samples *in–vitro*. There is some discussion in the succeeding sections concerning species–dependence, temperature–dependence and tissue state to which reference may be made.

Table 8.2 T_1 for blood and constituents

Tissue	Freq MHz	Temp ºC	T_1 ms	Reference
Whole blood,				
rabbit	0.01	35	145	Koenig *et al* 1984
–,human	0.02	25	85,106*	Brooks *et al* 1975
–,rabbit	0.1	35	160	Koenig *et al* 1984
–,human	0.3	25	125,131*	Brooks *et al* 1975
–,rabbit	1.0	35	300	Koenig *et al* 1984
–,human	1.7	*in–vivo*	340–370	Smith *et al* 1981,82
–,rabbit	2.5	20	372±34	Ling *et al* 1980
–,–	5.0	35	670	Koenig *et al* 1984
–,human	6	25	559	Brooks *et al* 1975
–,rabbit	10	35	910	Koenig *et al* 1984
–,human	10.7	37	830	Fullerton *et al* 1982
–,–	19.8	33	900±90	Koivula *et al* 1982
–,–	20	40	893±33	Grodd & Schmitt 1983
–,rabbit	20	35	1100	Koenig *et al* 1984
–,–	24	23	872±43	Ling *et al* 1980
–,human	50	25	925,947*	Brooks *et al* 1975
Blood:red cells,				
human	0.02	25	45	Brooks *et al* 1975
–,–	0.3	25	71	–
–,–	6	25	382	–
–,–	10.7	37	540	Fullerton *et al* 1982
–,–	10.7	37	495,513*	Thompson *et al* 1975
–,–	19.8	33	590±60	Koivula *et al* 1982
–,–	24.3	37	621,662*	Thompson *et al* 1975
Blood:leukocytes,				
human	24	room	715±23	Ekstrand *et al* 1977
Blood:plasma,				
human	0.02	25	269,288*	Brooks *et al* 1975
–,–	0.3	25	407,392*	–
–,–	6	25	1098,1171*	–
–,–	10.7	37	1450	Fullerton *et al* 1982
–,–	19.8	33	1410±80	Koivula *et al* 1982
–,–	20	20	1473±86	Schuhmacher *et al* 1987
–,–	24	room	1260±60	Ekstrand *et al* 1977
Blood:serum,				
–,rabbit	2.5	20	820±12	Ling *et al* 1980
–,dog	5.1	room	860	Wolf & Conard 1983
–,human	10.7	7	906±25	Raeymaekers *et al* 1988
–,–	20	23	1374±60	De Certaines *et al* 1981
–,–	20	40	2030±27	Grodd & Schmitt 1983
–,–	24	room	1230±170	Ekstrand *et al* 1977
–,rabbit	24	23	1590±113	Ling *et al* 1980

* Oxygenated, deoxygenated.

8.1.4.1 Dispersion of T_1

Unlike water, the longitudinal relaxation time T_1 of protons in the macromolecular solutions and water constituents of tissue depends upon the precessional frequency, and hence on field strength. The general shape of the dispersion curve is shown in Figure 8.1, and detailed measurements of T_1 dispersion are included for many tissues in Tables 8.1 and 8.2. At high frequencies, T_1 for soft tissue tends towards a constant value, although evidence suggests that the asymptotic limit has not been reached by 100 MHz (see for instance, Fischer *et al*, 1989b). Even at frequencies of several hundred megahertz soft tissue still relaxes at least twice as rapidly as pure water at the same temperature.

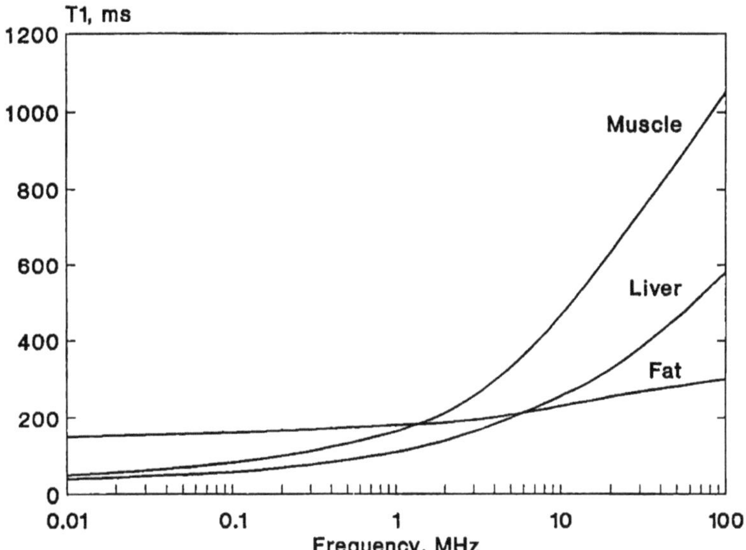

Figure 8.1 The dispersion of longitudinal relaxation time T_1 for typical soft tissues and for fat.

Koenig and co-workers (Koenig *et al*, 1984; Koenig and Brown, 1984) have demonstrated T_1 dispersion for a range of animal and human tissues over the frequency range 0.01 to 60 MHz. With the exception of fat, all other tissues studied showed a similar pattern of relative change of T_1 with frequency. Fat was noticeably different, having the longest T_1 at frequencies below 0.1 MHz, and the shortest T_1 above 10 MHz.

In general the longest values of T_1 are associated with body fluids, in some cases approaching the relaxation time for water. Studies on blood components appear to show a trend towards a constant value for T_1 at low frequencies (Brooks *et al*, 1975; Koenig *et al*, 1984). Extending relaxation measurements in amniotic fluid to very low magnetic fields, in which the precessional frequency was as low as 10 Hz, Béné *et al* (1982) showed a further reduction in T_1 which was attributed to proton exchange in free water. Hopkins *et al* (1986) investigated human cerebrospinal fluid *in-vivo*, and concluded that the relaxation was largely independent of frequency in the range 6–60 MHz and is dominated by the T_1 of free water.

Dispersion of T_1 has been fitted to empirical expressions for predictive purposes. Escayne *et al* (1982) investigated T_1 dispersion in mouse tissue *in-vitro* at room temperature and over the frequency range 6.7 MHz to 90 MHz. The relationship:

$$\frac{1}{T_1} = a\nu^{-\frac{1}{2}} + b \qquad (8.6)$$

was found to provide a good fit to the data, where ν is the Larmor frequency and a and b are tissue-dependent constants.

Table 8.3 Average spin-lattice relaxation time, T_1, for hydrogen in normal tissues, expressed as $T_1 = Af^B$, f in Hz

Tissue	A,ms	B	SD
Brain, grey matter	3.62	0.3082	17%
Brain, white matter	1.52	0.3477	17%
Fatty tissue	1.13	0.1743	28%
Kidney	7.45	0.2488	27%
Liver	0.534	0.3799	22%
Lung	4.07	0.2958	19%
Muscle:cardiac	1.30	0.3618	16%
Muscle:skeletal	0.455	0.4203	18%
Spleen	2.00	0.3321	19%

Source: Bottomley *et al* (1984).

As part of an extensive survey of data from the literature, Bottomley *et al* (1984, 1987) have used the simple expression

$$T_1 = A\nu^B \qquad (8.7)$$

as a fit for T_1 in the frequency range 1–100 MHz. Values for A and B from these surveys for normal tissues are included in Table 8.3. The exponent of frequency varies between 0.17 for fatty tissue and 0.42 for skeletal muscle, being typically about 0.33. It should be remembered, however, that the data set on which the analysis was based included normal tissue from many animal species, from *in-vitro* measurements as well as *in-vivo* measurements, and measurements made at both room and body temperatures. Nyman *et al* (1986) have noted that the predicted values for T_1 are consistently lower than those reported from measurements *in-vivo* at 15 MHz. In addition it is clear that Equation 8.7 has potential validity only above 1 MHz, and does not predict, for instance, the low frequency behaviour presented in Figure 8.1.

Assuming that the dispersion of spin-lattice relaxation time, T_1, can be principally assigned to the reduced mobility of 'bound' water on the surfaces of macromolecular structures, Fullerton *et al* (1984) derived an empirical expression based largely on available data from the literature. This expression,

$$T_1 = 1.83f + 25.0 \tag{8.8}$$

describes the relaxation time of the 'hydration water' in tissue. It was assumed that the remainder of tissue water is 'free' and behaves non-dispersively, as pure water.

Careful investigation of the dispersion of T_1 in tissues at reduced temperatures has demonstrated minor peaks on an otherwise monotonically smooth function. Koenig *et al* (1984), showed that at about 3 MHz, rat myocardium showed two small peaks in the variation of $1/T_1$ with frequency. The measurements were made at 7°C. These peaks were also observed by Winter and Kimmich (1982) in pig muscle at 0°C who identified them as originating from the cross-relaxation of protons by the ^{14}N nuclei of protein when the proton energy becomes equal to a quadripolar transitions of ^{14}N.

8.1.4.2 Temperature dependence of T_1

Proton spin-lattice relaxation time in tissue varies with temperature (Figure 8.2), a fact which may be of considerable importance when relating the measured relaxation times *in-vitro* at room temperature to those *in-vivo*. Table 8.4 shows values of T_1 temperature coefficients for some soft tissues. Tissues have an approximately linear increase in T_1 with temperature of about 1% $°C^{-1}$ in the range 0-40°C. There is evidence that the temperature coefficient decreases at temperatures above about 40°C, and then increases again, going through a minimum value at 45°C and 60°C (Lewa & Majewska, 1980; Kamman *et al*, 1988). This high temperature reduction in T_1 is irreversible and was explained on the basis of protein denaturation. The temperature coefficient for soft tissues has also been shown to increase with frequency (Fung *et al*, 1975; Fung, 1977).

At temperatures below 0°C, T_1 for tissue initially drops, reaching a minimum between 0°C and -40°C (Mathur-De Vré *et al*, 1983; Fung *et al*, 1975). For a particular tissue the temperature at which the minimum occurs depends upon the frequency. Rustgi *et al* (1978) investigated the relaxation of frozen mouse tissue, spleen muscle and kidney at 33.8 MHz, reporting minimum T_1 values of about 60 ms. T_1 increases with a further decrease in temperature, reaching relaxation times comparable with those at room temperature at between -150°C and -200°C.

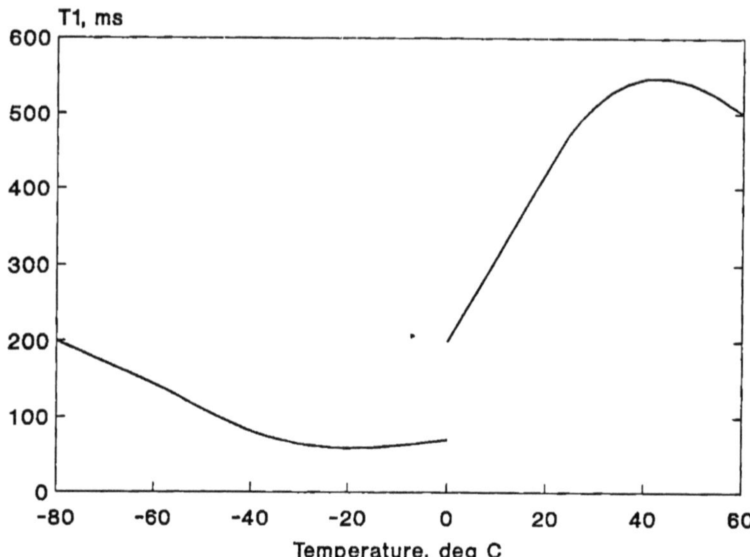

Figure 8.2 Generalised temperature variation of T_1 for a typical soft tissue.

8.1.5 *Values of ^1H transverse relaxation time, T_2, for tissue*

Tables 8.5 and 8.6 give measured values of the transverse relaxation time, T_2, for normal tissues: Table 8.6 for blood and constituents and Table 8.5 for all other normal tissues. The frequency and temperature of the measurement are both noted, although the available evidence suggests minimal dependence on either. As is the case for T_1, some selection of data has occurred based on the criteria outlined above. In particular, average values from Bottomley *et al* (1984) have been included for some common tissues and the original source references have been omitted.

8.1.5.1 Dispersion of T_2

Bottomley *et al* (1984) reported that no T_2 frequency dependence could be observed in their compilation of data. Some evidence has since been published which suggests that, at least for some tissues, there may be a

Table 8.4 Temperature coefficients for T_1 for tissue

Tissue	Freq MHz	Temp °C	Coefficient ms/°C	%/°C	Reference
Blood,rabbit	0.01	5–35	2	1.9	Koenig *et al* 1984
–,–	1	5–35	2.5	1.0	–
–,human		25–37		1.6	Brooks *et al* 1975
Blood:red cells, human	10.7	11–37	8	2	Thompson *et al* 1975
–,–	24.3	11–37	7	1.4	–
Brain:white matter, cow	21.5	0–40	4	1.0	Kamman *et al* 1988
Brain:grey matter, cow	21.5	0–40	12	1.7	–
Heart,rat & cow	25.6	15–40	5.5	0.9	Lewa & Majewska 1980
Liver, mouse	4.5	0–37	3	1.6	Fung *et al* 1975
–,rat	10.7	5–37	1.2	0.6	Mathur–De Vré *et al* 1988
–,–	20	0–37	1.5	0.7	Barthwal *et al* 1986
–,–	22.5	5–37	0.8	0.3	Mathur–De Vré *et al* 1988
–,mouse	60	0–37	9	2.2	Fung *et al* 1975
Lung,rat & cow	25.6	15–40	6	0.9	Lewa & Majewska 1980
Muscle, mouse	1.0	0–37	1.0	0.7	Fung 1977
–,rat	10.7	5–37	5	1.2	Mathur–De Vré *et al* 1988
–,rat & cow	25.6	15–40	6	1.0	Lewa & Majewska 1980
–,rat	54.7	5–37	11	1.6	Mathur–De Vré *et al* 1988
–,mouse	60	0–37	20	2.5	Fung 1977
Spleen,rat & cow	25.6	15–40	4	0.8	Lewa & Majewska 1980

small but finite reduction in T_2 with increased frequency. For instance Kroeker *et al* (1985) report a reduction of T_2 for mouse muscle measured *in–vivo* from 39±3 ms at 10 MHz to 31±2 ms at 57 ms. A comparable reduction was noticed by Johnson *et al* (1985) in dog fat *in–vivo*, although no similar reduction was observed in muscle.

8.1.5.2 Temperature dependence of T_2

The temperature dependence of T_2 for soft tissue is very much lower than that of T_1. Some estimates of temperature coefficients are given in Table 8.7. In the temperature range 0–40°C both small positive and small negative temperature coefficients have been reported. There is no evidence that the rate of change of T_2 with temperature, $\Delta T_2/\Delta T$, is dependent on frequency.

Table 8.5 ^{1}H proton spin–spin relaxation time, T_2, for tissues

Tissue	Freq MHz	Temp °C	T_2 ms	Reference
Adrenal gland,				
dog	5.1	room	79	Wolf & Conard 1983
–,human	15	*in–vivo*	73±49	Moon *et al* 1983a
Amniotic fluid, human	100	37	2300	Béné *et al* 1982
Aorta,human	8.5	*in–vivo*	90±50	Rupp *et al* 1983
Artery,dog	5.1	room	47	Wolf & Conard 1983
Bile,dog	5.1	room	339	–
–,human	8.5	*in–vivo*	80±20	Rupp *et al* 1983
Blood	see Table 8.6			
Bone,human	15	*in–vivo*	55	Heller *et al* 1983
–,–, vertebral body	15	*in–vivo*	66±5	Nyman *et al* 1986
Bone marrow,				
survey	all	all	59±24	Bottomley *et al* 1984
–,human	15	*in–vivo*	49±9	Ehman *et al* 1985
–,–,1–10yr	15	*in–vivo*	61±7	Dooms *et al* 1985
–,–,21–40yr	15	*in–vivo*	57±10	–
–,–,51+yr	15	*in–vivo*	54±12	–
Brain,survey	all	all	76±21	Bottomley *et al* 1984
–,rat	10	37	85±5	Hayman *et al* 1988
–,–	200	37	50±3	–
–,–,fetal	60	20	149	Kiricuta & Simplăceanu 1975
Brain:grey matter,				
survey	all	all	101±13	Bottomley *et al* 1984
–,human	6	*in–vivo*	90–105	Rinck *et al* 1985
–,–	12	*in–vivo*	143±17	Just & Thelen 1988
–,–	15	*in–vivo*	60±4	Kjos *et al* 1985
–,cow	21.4	40	69	Kamman *et al* 1988
Brain:white matter,				
survey	all	all	92±22	Bottomley *et al* 1984
–,human	6	*in–vivo*	90–113	Rinck *et al* 1985
–,–	12	*in–vivo*	122±13	Just & Thelen 1988
–,–	15	*in–vivo*	53,57±4	Kjos *et al* 1985
–,cow	21.5	40	107	Kamman *et al* 1988
Brain:cerebellum				
human	6	*in–vivo*	107–154	Rinck *et al* 1985
–,–,grey	15	*in–vivo*	60±4	Kjos *et al* 1985
–,–,white	15	*in–vivo*	57±4	–
Brain:pons	6	*in–vivo*	100–134	Rinck *et al* 1985
Brain:corpus callosum	6	*in–vivo*	106–134	–
Breast,survey	all	all	49±16	Bottomley *et al* 1984
CSF,human	6.2	*in–vivo*	1760±430	Hopkins *et al* 1986
–,–	25.4	*in–vivo*	2190±380	–

cont.

Table 8.5 cont. T_2 **for tissue**

Tissue	f,MHz	T,°C	T_2,ms	Reference
Cervix,human	22.5	room	41±6	Fruchter *et al* 1978
Embryo,rat	60	room	200±4	Kiricuta & Simplăceanu 1975
Endometrium, human	22.5	room	95	Fruchter *et al* 1978
Eye:vitreous, human	15	*in-vivo*	160±91	Kjos *et al* 1985
Eye:lens,human	15	*in-vivo*	57±3	-
-,rabbit	25	36	29	Neville *et al* 1974
-(cortex),-	25	36	40	-
-(nucl.),-	25	36	7.7	-
Fallopian tube, human	22.5	room	73±23	-
Fatty tissue, survey	all	all	84±36	Bottomley *et al* 1984
-,human	6	*in-vivo*	145-179	Rinck *et al* 1985
-,-	12	*in-vivo*	135±16	Just & Thelen 1988
-,-	15	*in-vivo*	57±3	Nyman *et al* 1986
Intestine, mouse	15	25	35	Frey *et al* 1972
-,human(colon)	22.5	room	58±10	Goldsmith *et al* 1978a
Kidney,survey	all	all	58±24	Bottomley *et al* 1984
Kidney:cortex,dog	5.1	room	69	Wolf & Conard 1983
-,human	15	*in-vivo*	76±77	Ehman *et al* 1985
-,rat	85.5	22	38±2	Dockery *et al* 1989
-,-	300	22	18±2	-
Kidney:medulla, dog	5.1	room	171	Wolf & Conard 1983
-,human	15	*in-vivo*	81±24	Ehman *et al* 1985
-,rat	85.5	22	75±8	Dockery et al 1989
-,-	300	22	35±6	-
Liver,survey	all	all	43±14	Bottomley *et al* 1984
-,human	15	*in-vivo*	51±6	Nyman *et al* 1986
-,-	15	*in-vivo*	45±8	Ehman *et al* 1985
-,rat	20	37	63±4	Barthwal *et al* 1986
-,-	22.5	37	31	Mathur-De Vré *et al* 1988
-,-,fetal	60	20	43	Kiricuta & Simplăceanu 1975
Lung,survey	all	all	79±29	Bottomley *et al* 1984
Lymph node human	20	40	91	Grodd & Schmitt 1983
Muscle:cardiac, survey	all	all	57±16	Bottomley *et al* 1984
-,rat,fetal	60	20	69	Kiricuta & Simplăceanu 1975
Muscle:skeletal, survey	all	all	47±13	Bottomley *et al* 1984
-,human	6	*in-vivo*	47±3	LeBlanc *et al* 1986
-,mouse	10	*in-vivo*	39±3	Kroeker *et al* 1985
-,human	15	*in-vivo*	32±7	Ehman *et al* 1985

cont.

Table 8.5 cont. T_2 **for tissue**

Tissue	f,MHz	T,°C	T_2,ms	Reference
Muscle,rat	22.4	37	32	Mathur-De Vré *et al* 1988
–,mouse	57	*in-vivo*	31±2	Kroeker *et al* 1985
–,rat	200	27	36±3	LeBlanc *et al* 1986
Myometrium, human	22.5	room	66±10	Fruchter *et al* 1978
Nerve,dog	5.1	room	58	Wolf & Conard 1983
Ovary,dog	5.1	room	126	-
–,human	22.5	room	59±13	Fruchter *et al* 1978
Pancreas,dog	5.1	room	43	Wolf & Conard 1983
–,human	8.5	*in-vivo*	60±40	Rupp *et al* 1983
–,–	15	*in-vivo*	49±12	Ehman *et al* 1985
–,–	20	40	189	Grodd & Schmitt 1983
Parotid gland, human	8.5	*in-vivo*	30	Rupp *et al* 1983
Pituitary gland, dog	5.1	room	114	Wolf & Conard 1983
–,human	12	*in-vivo*	129±13	Just & Thelen 1988
Placenta	22.5	room	114±28	Fruchter *et al* 1978
Prostate,human	20	40	98	Grodd & Schmitt 1983
Skin,mouse	15	25	35	Frey *et al* 1972
Spinal medulla, human	6	*in-vivo*	160–264	Rinck *et al* 1985
Spleen,survey	all	all	62±27	Bottomley *et al* 1984
–,human	15	*in-vivo*	87±18	Nyman *et al* 1986
–,–	15	*in-vivo*	75±21	Ehman *et al* 1985
Stomach,mouse	15	25	35	Frey *et al* 1972
Tendon,cow	10.7	room	83(long)*	Fullerton *et al* 1985
–,human	15	*in-vivo*	107±16	Moon *et al* 1983b
Testis,dog	5.1	room	111	Wolf & Conard 1983
–,human	20	40	153	Grodd & Schmitt 1983
Thalamus,human	15	*in-vivo*	55±2	Kjos *et al* 1985
Thymus,human	15	*in-vivo*	71±11	Dooms *et al* 1984
Thyroid gland, human	10.7	37	72(45–114)	Tennvall *et al* 1987
–,–	15	*in-vivo*	45±4	Dooms *et al* 1984
–,–	20	40	102	Grodd & Schmitt 1983
Tongue,human	20	40	61	-
Urine,dog	5.1	room	2600	Wolf & Conard 1983
–,human	8.5	*in-vivo*	570±230	Rupp *et al* 1983
–,rat	15	*in-vivo*	159	Moon *et al* 1983c
Vagina,human	22.5	room	60±14	Fruchter *et al* 1978
Vulva,human	22.5	room	34	-
Vertebral disk, human	15	*in-vivo*	48±7	Moon *et al* 1983b

* Angle dependent.

Table 8.6 Transverse relaxation time, T_2, for blood and its constituents

Tissue	Freq MHz	Temp °C	T_2 ms	Reference
Whole blood,				
–,rat,43%Hct	10	37	316±10	Hayman *et al* 1988
–,human	20	40	362±6	Grodd & Schmitt 1983
–,–	20	20	253±15	Schuhmacher *et al* 1987
–,rat,oxy	80	37	182±2	Thulborn *et al* 1982
–,–,79% O_2	200	37	60±9	Hayman *et al* 1988
–,–,oxy	270	37	92±1	Thulborn *et al* 1982
–,–,–	469	37	65±1	–
–,–,deoxy	80	37	72±.5	–
–,–,43% O_2	200	37	24±2	Hayman *et al* 1988
–,–,deoxy	270	37	15±2	Thulborn *et al* 1982
–,–,–	469	37	9±2	–
Red cells,rat	10	37	140±2	Hayman *et al* 1988
–,human	10.7	37	230,180*	Thompson *et al* 1975
–,–	10.7	37	154	Fullerton *et al* 1982
–,–	24.3	37	272,235	Thompson *et al* 1975
–,rat,79%O_2	200	37	42±1	Hayman *et al* 1988
Plasma,human	10.7	37	282	Fullerton *et al* 1982
–,–	20	23	582±47	De Certaines *et al* 1981
–,–	20	20	646±34	Schuhmacher *et al* 1987
Serum,dog	5.1	room	526	Wolf & Conard 1983
–,human	20	23	627±47	De Certaines *et al* 1981
–,–	20	40	903±39	Grodd & Schmitt 1983

* Oxygenated, deoxygenated.

Table 8.7 Temperature coefficients for T_2 for tissue

Tissue	Freq MHz	Temp °C	Coefficient ms/°C	%/°C	Reference
Blood,human	80–470	20–37		<0.2	Thulborn *et al* 1982
Blood:red cells, human	10.7, 24.3	11–37	4	2	Thompson *et al* 1975
Brain:white matter, cow	21.5	0–40	–0.6	–0.7	Kamman *et al* 1988
Brain:grey matter, cow	21.5	0–40	0.2	0.2	–
Liver,rat	10.7	5–40	0.2	0.7	Mathur-De Vré *et al* 1988
–,–	20	0–37	0.5	1.2	Barthwal *et al* 1986
–,–	22.5	5–37	0.0	0.0	Mathur-De Vré *et al* 1988
Muscle,rat	22.5	5–37	–0.3	–0.9	–
–,–	54.7	5–37	–0.3	–0.7	–

8.1.6 *Other factors affecting relaxation times*

8.1.6.1 Multiple relaxation times

The observed magnetic relaxation of tissue is not properly characterised by a single exponential decay (Barroilhet & Moran, 1974; Bakker & Vriend, 1984; Barthwal *et al*, 1986; Brix *et al*, 1990). Fullerton *et al* (1982), in a description of the FDP model, point out that the transverse relaxation is always multi-phasic in character with at least crystalline (T_2~10μs), hydration (T_2~10ms) and free water (T_2~100ms) components. The free water component can also subdivide into extracellular and intracellular components. Similarly the longitudinal relaxation time is multi-exponential (Equation 8.3). Gore *et al* (1983) demonstrated the effect of altering the data aquisition time on both T_1 and T_2 measurement *in-vivo*, showing an apparent lengthening of both with aquisition time. The lengthening was greater for kidney and bone marrow and least for muscle and fat, and was associated with the inadequate measurement of the long component in a multicomponent decay.

The multi-exponential character of the transverse relaxation time T_2 has been studied both *in-vitro* and *in-vivo* for a number of tissues at 20 MHz by Brix *et al* (1990). It was concluded that whilst a single exponential decay adequately described the relaxation in liver, kidney and brain grey matter, fat and brain white matter required two exponentials and muscle required three. These conclusions assume that fast decay during the first 2 ms is ignored. The longer decay times were derived from only between 81% and 85% of the tissue ^1H content for most soft tissues. The remaining fraction of tissue protons do not contribute to tissue relaxation measurements because of their associated very short relaxation times. A higher percentage contributed from fat (88%) and a lower fraction (70%) from brain white matter.

The relaxation of protons in tissue has been spectrally resolved by Luyten *et al* (1987), separating the contributions to both T_1 and T_2 from H_2O, $-CH_2-$, $-CH_3$ and $-CH=$ in subcutaneous fat and muscle, and in bone marrow, *in-situ* at a field strength of 1.5 T (69 MHz). For water in fat and muscle T_1 and T_2 were 963±77 ms and 44±3 ms respectively, whilst for $-CH_2-$ the relaxation times were 252±7 ms and 131±7 ms respectively. Under more conventional measurement techniques these components would remain unresolved.

Investigating tooth enamel, Funduk *et al* (1984) separated the transverse relaxation into components from solid-like interstitial water (T_2~14μs), enamel apatite (T_2~61μs), and semi-liquid water (T_2~240μs). The very short T_2 is caused by the strong sample heterogeneity causing the relaxation to shorten by about an order of magnitude.

The relaxation times included in the tables assume a single exponential relaxation. Details given by some authors of multiple relaxations, and of long and short components, have not been included. However, as pointed out by Koenig *et al* (1984), the sum of two exponentials that differ in time constant by a factor of three can be represented by a single exponential with mean squared uncertainty of less than ±3%. The single observed time constant is the appropriately weighted average of the two.

8.1.6.2 Changes following death

The question as to the relevance of measurements on *in-vitro* tissue samples to *in-vivo* studies, and the manner by which tissue samples should best be handled following excision, has been widely discussed (Bottomley *et al*, 1984; Beall *et al*, 1984; Small *et al*, 1983; Fischer *et al*, 1989a). Several authors report that for many tissues both T_1 and T_2 remain stable for several hours following excision, even when stored at body temperature. For rat liver and spleen tissue, Thickman *et al* (1983) reported T_1 to be stable for at least 6 hours, and T_2 stable for 48 hours at room temperature. A 10% reduction in relaxation rate ($1/T_1$) for periods up to 6 hours after excision, at 37°C, for rat liver tissue at frequencies below 1 MHz was noted by Fischer *et al* (1989a). Above about 4 MHz, slight increases in $1/T_1$ were observed. For longer storage times T_1 became progressively longer at all frequencies. This general lengthening of T_1 with long-term storage was also reported by Grodd and Schmitt (1983), for a range of human tissues stored at 40°C for 20 hours. An overall slight reduction in T_1 was reported for most tissues after 6 days storage at 0°C. The same authors reported a general increase in T_2 after long-term storage at 40°C, with the smallest alterations observed for muscle, heart and brain tissue.

According to Fischer *et al* (1989), quick deep freezing is a satisfactory storage method for samples of brain tissue which does not affect T_1 nor T_2 significantly. However, the method is not applicable to liver or muscle tissues which show a characteristic flattening in the shape of the relaxation profile. A comparable difference between the responses of brain and liver tissue to freezing was also noted by Mathur-De Vré *et al* (1983). Beall (1982) also warns of the possible effects of cooling and freezing, citing decreases of up to 18% following 24 hours refrigeration for some breast tissues.

Some authors have also considered the separate effects on the components in a multi-exponential relaxation process resulting from storage, but give conflicting evidence. Small *et al* (1983), report an approximately 10% reduction in both short and long components of T_2 of breast tissue 24 hours after excision. Barroilhet and Morran (1974) noticed a reduction in T_1 long component by up to 25% at 3.5 hours following death (mouse liver); the short component reduced slightly and then increased.

8.1.6.3 Tissue fixation

There is a general decrease in T_1 and increase in T_2 as a result of tissue fixation in formalin, the change depending on the strength of the fixing fluid (Grodd and Schmitt, 1983). Alterations of up to 50% were observed for a range of human tissues. Thickman *et al* (1983) reported that T_1 measurements on fixed tissues (heart and spleen) would only be representative if made within 30 minutes after fixation, and that no T_2 measurements on fixed tissues would be representative of tissue in an unfixed state.

Table 8.8 Mean T_1 and T_2 relaxation times for pathological tissues
$T_1 = Af^B$; f in Hz

Tissue	T_1			T_2
	A,ms	B	SD,%	ms
Bone,misc. tumours	124	0.09811	78	106±60
Brain,misc. tumours	9.18	0.2649	36	121±63
–,glioma	273	0.06981	35	*
–,oedema	7.85	0.2745	23	113±73
Breast,misc.tumours	0.846	0.3923	28	80±35
–,carcinoma	0.698	0.3999	25	94±48
–,fibroadenoma	0.271	0.4669	29	*
–,fibrocystic displasia	3.41	0.2822	62	80±41
Kidney,misc. tumours	107	0.1189	37	*
–,occlusion	*	*	*	63±20
Liver,misc. tumours	82.3	0.1334	26	84±31
–,hepatoma	4.34	0.3068	16	84±26
Lung,misc. tumours	0.677	0.3954	51	68±45
Lymph,misc. tumours	*	*	*	65±16
Muscle,misc.tumours	2.76	0.3323	32	87±40
–,carcinoma	4.52	0.3029	16	*
–,fibrosarcoma	142	0.1092	15	*
–,rhabdomyosarcoma	3.88	0.3178	27	*
–,oedema	0.380	0.4603	26	*

Note: Only sample sizes of 10 or more have been included; smaller samples are marked with an asterisk *.
Source: Bottomley *et al* (1987).

8.1.6.4 Tissue pathology

No attempt has been made to assemble an extensive tabulation of T_1 for pathological tissues. Instead, values for A and B (Equation 8.7) for some pathological tissues are drawn from an extensive survey by Bottomley *et al* (1987) and included in Table 8.8. Only values based on at least ten data points have been included.

8.1.6.5 Animal species

In their wide review of relaxation times of tissue, Bottomley *et al* (1984) concluded that there was insufficient evidence to separate T_1 or T_2 values from different mammalian species, and pooled all the available data for any particular tissue when applying their curve-fitting routine. Such a global conclusion is probably simplistic. Grodd and Schmitt (1983) compared T_1 and T_2 at 20 MHz for eight human soft tissues with those from equivalent pig and rat tissues. For fat, muscle (both cardiac and skeletal) and brain, the spin–lattice relaxation time differed between species by no more than about one standard deviation. Conversely for liver, spleen and kidney, and for rat lung, animal tissue spin–lattice relaxation times were shorter than those for equivalent human tissue. Shorter spin–spin relaxation times were also reported for liver and lung from both animals, for pig spleen and for rat kidney in comparison with human tissue. Mallard *et al* (1979) and Koenig *et al* (1984) suggest that liver tissue is more likely to show a species difference than spleen tissue. In an early study, Cottam *et al* (1972) reported mouse kidney tissue at 24.3 MHz to have a substantially lower T_1 than equivalent human kidney tissue. The explanation for such differences is obscure, molecular composition and structure being probably as important as water content as causative factors.

8.1.6.6 Age and sex

Several authors have recorded the considerably raised relaxation times associated with fetal and neonatal tissues in comparison with those for equivalent adult tissue. The lengthened relaxation time is generally assumed to result from an increased proportion of free water in these immature tissues. Selected values included in Tables 8.1 and 8.5 are from Kiricuta and Simplăceanu (1975) for rat fetal brain, from Smith (1983) for human neonatal tissue *in–vivo*, and from Kasturi *et al* (1976) for several human fetal tissues.

The alteration in both T_1 and T_2 in adult bone marrow with age has been studied by Dooms *et al* (1985). A slight increase in relaxation time with age was observed. Such ageing may also be a contributing factor to the observed spread of relaxation times from nominally normal tissues.

No differences in relaxation times associated with sex have been reported.

8.1.6.7 Tissue composition

Water content is recognised as being an important factor in determining relaxation times (Ling *et al*, 1980; Kasturi *et al*, 1976; Kiricuta and Simplăceanu, 1975). However, attempts to relate T_1 or T_2 uniquely to tissue water content have been unsuccessful and it is clear that other factors such as macromolecular composition and structure are equally important in determining relaxation in tissue. At present no clear explanation emerges and the reader is referred to reviews, for instance Bottomley *et al* (1984, 1987) and Beall *et al* (1984), for further discussion.

8.1.6.8 *Anisotropy*

The spin–spin relaxation time T_2 for hydrated tendon shows considerable anisotropy, being shortest when the magnetic field is aligned with the tendon. Fullerton *et al* (1985) demonstrated that the well established anisotropy of poorly hydrated tendon may also be observed at higher levels of hydration, and hence may affect the relaxation signal obtained *in–vivo*. The highest values of T_2 were observed when the angle θ between the magnetic field and the tendon was 55° and 125°, or when $3 \cos^2\theta - 1 = 0$.

8.1.6.9 Tissue variability

Natural variability of relaxation times within a particular organ is well documented for some tissues. The region–to–region variation within the brain has been investigated both *in–vivo* at particular frequencies (eg Wehrli *et al*, 1983; Kjos *et al*, 1985), and over a wide range of frequencies for particular tissues (Fischer *et al*, 1989b). T_1 for grey matter is consistently shorter than that for white matter at all frequencies. Similar differences within the kidney have been reported (Ling *et al*, 1980; Ehman *et al*, 1985; Dockery *et al*, 1989), the renal medulla T_1 being consistently longer than that for the renal cortex. The variation of T_1 with position across the kidney appears to be well correlated with tissue water content. Whilst some evidence exists for parallel differences in T_2, the larger spread in the measured T_2 values serves to obscure them.

Polak *et al* (1988) have reported a significant difference in relaxation time between different skeletal muscles (soleus and gastrocnemius) from rats and rabbits, even though the water content was the same. This was explained on the basis of the differing proportions of intracellular to extracellular fluid space in the two muscles. In other tissues, liver and spleen for instance, there is a normal variability both between organs from different animals, and within the parenchyma of a single organ. This variation may be noted in the standard deviations given in the tables of relaxation times. Nyman *et al* (1986) noted that the spread of relaxation times for spleen is consistently greater than for other tissues. It is at present not clear how much of the reported spread of relaxation times originates from errors associated with the measurement technique used, especially *in–vivo*.

8.1.7 *Relaxation times in blood*

Values of T_1 and T_2 for blood and its components are given in Tables 8.2 and 8.6. The relaxation of protons in blood has been studied in order to obtain diagnostic information in sickle cell disease (Thompson *et al*, 1975) and in malignancy (Ekstrand *et al*, 1977; de Certaines *et al*, 1981; Koivula *et al*, 1982; Schuhmacher *et al*, 1987). In addition the dispersion of T_1 and T_2 in blood has been measured in order to evaluate theories of relaxation mechanisms. Relaxation rates $1/T_1$ and $1/T_2$ both increase linearly with haematocrit over the full range from pure plasma to pure erythrocytes at both 10 MHz and 200 MHz (Fullerton *et al*, 1982; Hayman *et al*, 1988). Relaxation is dominated by the concentration of haemoglobin. Hyman reported only a modest increase in T_2 when the blood was lysed, at any haematocrit. Similarly, Brooks *et al* (1975) noted that values of T_2 for erythrocytes agree with those from haemoglobin solutions except at very low frequencies when membrane and other effects may be important. In addition Koenig and Brown (1984) noticed only small increases in T_1 between fresh rabbit blood and congealed blood after 55 hours storage, in spite of a substantial alteration in its physical state.

The longitudinal relaxation of normal human serum has been examined by Raeymaekers *et al* (1988) who give an empirical expression relating serum T_1 to total protein:

$$\frac{1}{T_1} = 0.0067s + 0.601 \qquad (8.9)$$

where s is the serum protein concentration in g l^{-1} and the relaxation rate $1/T_1$ is in s^{-1}. The measurements were at 10.7 MHz and 7°C, and both temperature and dispersion corrections should be applied for other conditions.

The effect of oxygenation on relaxation has also been noted by several authors. The difference in T_1 and T_2 between oxygenated and deoxygenated blood is relatively small at low frequencies. However, as demonstrated by Thulborn *et al* (1982) and Hayman *et al* (1988) differences in T_2 become increasingly marked at frequencies above 100 MHz, T_2 increasing with the level of oxygenation.

Brooks *et al* (1975) also noted that increasing pH causes T_1 to decrease because of its effect on the shape of the protein molecule. The magnitude of this effect is on the order of 25% per pH unit at low frequencies.

8.1.8 *Proton relaxation in substances other than tissue*

8.1.8.1 Water

Measurements of proton spin–lattice relaxation time for pure oxygen-free water are given in Table 8.9, covering a temperature range 0 to 100°C. The values are from Krynicki (1966), and differ from those reported by Hindman

Table 8.9 Spin–lattice relaxation time, T_1, for water hydrogen.

Temperature °C	T_1 s
0	1.73
10	2.39
20	3.15
30	4.03
40	5.00
50	6.03
60	7.20
80	9.80
100	12.75

Source: Krynicki (1966).

et al (1973) by only a few per cent at the lowest temperatures. At body temperature $T_1 \simeq 4.7$ s.

Over most of the frequency range of interest, T_1 for protons in water is constant and non-dispersive. However, Graf *et al* (1980) have reported dispersion between 0.1 and 1 kHz, resulting in a reduction in T_1 of about 25% at very low frequencies.

It is generally accepted that, for pure water, $T_2 \simeq T_1$.

8.1.8.2 Tissue substitute materials

A variety of gel mixtures have been suggested as suitable materials to act as tissue substitutes, particularly in the evaluation of magnetic resonance imaging performance. Several include paramagnetic ions in order to modify the relaxation times of the basic gel. Mathur-De Vré *et al* (1985) have reported T_1 and T_2 measurements on a water-based agar gel with and without doping by $MnCl_2$. The properties of more complex gels including variable amounts of agar, gelatin, water and glycerol, together with stabilising agents have been reported by Blechinger *et al* (1988). An alternative is the polysaccharide gel, agarose. This has been used successfully by Walker *et al* (1989) as the basis for a range of tissue substitute materials, showing satisfactory T_1 dispersion, by including gadolinium chloride chelated to EDTA. Mano *et al* (1986) have used a polyvinyl alcohol gel, noting possible problems with long-term

Table 8.10 Relaxation times of some tissue substitute materials

Material	Freq MHz	T_1 ms	T_2 ms	Reference
Agar gel 1.5%	100	2640	143	Mathur-De Vré *et al*
Agar gel 3%	100	2802	55	1985
Agar gel 4.5%	100	2760	25	-
Agar gel 3%, 0.2mM $MnCl_2$	100	380	30	-
-, 0.1mM $MnCl_2$	100	650	39	-
-, 0.05mM $MnCl_2$	100	1070	48	-
Agar/gelatin, 25% glycerol	10	541	75	Blechinger *et al* 1988
-,-	40	871	79	-
Agar/geletin, 45% glycerol	10	285	69	-
-,-	40	403	88	-
Agarose 0.6%, 0.32mM Gd-EDTA	4.2	214	104	Walker *et al* 1989
-,-	46	421	137	-
Agarose 1.5%, 0.14mM Gd-EDTA	8	509	72	-
-,-	61	791	76	-
Agarose 1.5%, 0.53mM Gd-EDTA	4.2	133	53	-
-,-	20	234	61	-
Agarose 2.5%, 0.21mM Gd-EDTA	20	493	45	-
-,-	46	572	45	-

stability. De Luca *et al* (1987) have discussed the use of a polyacrylamide gel doped with sodium chloride. Selected relaxation times and formulations from some of these reports are collected in Table 8.10.

8.2 Hydrogen proton density

Calculated values giving estimates of proton density for some tissues are included in the section on density in Chapter 5, Table 5.1. As noted above, proton density estimated from magnetic resonance measurements may derive from only about 80% of the ^{1}H present. This is because some magnetic relaxation occurs with very short relaxation times (Brix *et al*, 1990).

8.3 Magnetic susceptibility

The volume susceptibility, κ, of water at 20°C is -9.060×10^{-6} cm^{-3}. The molar susceptibility of water is -12.97×10^{-6} cm^3 mol^{-1}. Values of magnetic susceptibility for solutions of NaCl are -9.05 cm^{-3} and -10.02 cm^{-3} for 50 mM and 5 M solutions respectively (Chu *et al*, 1990). Susceptibility of normal tissue is commonly considered to be the same as that of water.

8.4 Magnetic relaxation of other nuclei; ^{39}K, ^{23}Na, ^{17}O, ^{31}P

Any nucleus with non-zero spin may exhibit magnetic resonance. Some of those nuclei of biological interest are listed in Table 8.11 giving their gyromagnetic ratio, nuclear spin, natural abundance, and NMR sensitivity relative to ^{1}H. Relaxation times for some of these nuclei have been investigated in tissue. In particular, phosphorous-31 has been investigated widely, particularly when studying the chemical shifts of tissue metabolites.

Generally relaxation times for other nuclei are shorter than those associated with ^{1}H. Damadian and Cope (1974) reported longitudinal relaxation times for potassium-39 from a variety of rat tissues at 10 MHz to be in the range 6–11 ms.

Sodium-23 has been more widely studied. Goldsmith and Damadian (1975) reported ^{23}Na relaxation in normal rat tissue at 10 MHz and 25°C. The shortest T_1 was for liver (6.5±0.5 ms) and the longest that from testis (17.8±0.7 ms). Values from other authors are broadly in agreement, with similar values of T_2 (Beall *et al*, 1984). ^{23}Na relaxation in blood and plasma at 95 MHz, and at 21°C has been reported (Shinar and Navon, 1984, 1986). Human plasma T_1 was 37.3±1.0 ms, and T_2 24.5±0.8 ms. Slightly longer relaxation times were recorded for serum. Human red blood cells gave values

Table 8.11 Gyromagnetic ratio, $\gamma/2\pi$, and other NMR properties for selected nuclei

Nucleus	$\gamma/2\pi$ MHz/T	Natural % abundance	Spin	Relative sensitivity
^{1}H	42.58	99.98	1/2	1.000
^{13}C	10.71	1.11	1/2	0.016
^{14}N	3.08	99.64	1	0.001
^{17}O	5.77	0.037	5/2	0.029
^{23}Na	11.26	100	3/2	0.093
^{31}P	17.24	100	1/2	0.066
^{35}Cl	4.17	75.53	3/2	0.047
^{39}K	1.99	93.10	3/2	0.0005

of T_1 in the range 23–28 ms, and two components for T_2, 6.0–7.3 ms and 15.2–18.7 ms. Relaxation of ^{23}Na in CSF was almost identical to that measured from 0.15 M NaCl solutions (T_1=T_2=54 ms). For CSF, the measured relaxation times were T_1=54.0±0.9 ms, and T_2=52.8±0.3 ms. There was no measurable frequency dependence of relaxation of intracellular sodium over the range 25–95 MHz, and both T_1 and T_2 decreased with temperature.

Oxygen-17 relaxation in human blood, and in mouse and frog muscle is about 1–2ms for both T_1 and T_2 (Beall *et al*, 1984).

The T_1 relaxation times of phosphorous-31, studied by Zaner and Damadian (1975) at 100 MHz on a variety of rat tissues, lay in the range from 1.03 ms (for kidney) to 2.33±0.14 ms (for liver). ^{31}P relaxation has been reported as being independent of field strength for liver, and to decrease linearly with increasing field strength for other rat tissues, *in-vivo* (Evelhoch *et al*, 1985). At 4.7 T, T_1 varied between 2.0 ms and 4.5 ms for leg muscle, 1.5 ms to 5.6 ms for brain tissue, and 0.2 ms to 1.3 ms for liver, depending on the metabolite.

References

Abragam A., 1978, *The Principles of Nuclear Magnetism*, Oxford University Press, London.

Alanen A., Nummi P., Kormano M. and Irjala K., 1987, Proton T_1 relaxation time of normal and abnormal urine, Acta Radiol, 28, 601–602.

Bakker C.J.G. and Vriend J., 1984, Multi-exponential water proton spin-lattice relaxation in biological tissues and its implications for quantitative NMR imaging, Phys Med Biol, 29, 509–518.

Barthwal R., Höhn-Berlage M. and Gersonde K., 1986, *In vitro* proton T_1 and T_2 studies on rat liver: analysis of multiexponential relaxation processes, Magn Reson Med, 3, 863–875.
Barroilhet L.E. and Moran P.R., 1975, Nuclear magnetic resonance (NMR) relaxation spectroscopy in tissues, Med Phys, 2, 191–194.
Beall P.T., 1982, Practical methods for biological NMR sample handling, Magn Reson Imaging, 1, 165–181.
Beall P.T., Amtey S.R. and Kasturi S.R., Eds, 1984, *NMR Data Handbook for Biomedical Applications*, Pergamon Press, Oxford.
Bell B.A., Kean D.M., MacDonald H.L., *et al*, 1987, Brain water measured by magnetic resonance imaging, Lancet i, 66–69.
Béné G.J., Borcard B., Graf V., *et al*, 1982, Proton NMR-relaxation dispersion in meconium solutions and healthy amniotic fluid: possible applications to medical diagnosis, Z Naturforsch, 37, 394–398.
Blechinger J.C., Madsen E.L. and Frank G.R., 1988, Tissue-mimicking gelatin-agar gels for use in magnetic resonance imaging phantoms, Med Phys, 15, 629–636.
Borghi L., Savoldi F., Scelsi R. and Villa M., 1983, Nuclear magnetic resonance response of protons in normal and pathological human muscles, Exp Neurol, 81, 89–96.
Bottomley P.A., Foster T.H., Argersinger R.E and Pfeifer L.M., 1984, A review of normal tissue hydrogen NMR relaxation times and relaxation mechanisms from 1–100 MHz: dependence on tissue type, NMR frequency, temperature, species, exision, and age, Med Phys, 11, 425–448.
Bottomley P.A., Hardy C.J., Argersinger R.E. and Allen-Moore G., 1987, A review of ^1H nuclear magnetic resonance relaxation in pathology: are T_1 and T_2 diagnostic? Med Phys, 14, 1–37.
Bovée W.M.M.J, Getreuer K.W., Smidt J. and Lindeman J., 1978, Nuclear magnetic resonance and detection of human breast tumors, J Natl Cancer Inst, 61, 53–55.
Brasch R.C., London D.A., Wesbey G.E., *et al*, 1983, Nuclear magnetic resonance study of a paramagnetic nitrxide contrast agent for enhancement of renal structures in experimental animals, Radiology, 147, 773–779.
Brix G., Schad L.R. and Lorenz W.J., 1990, Evaluation of proton density by magnetic resonance imaging: phantom experiments and analysis of multiple component proton transverse relaxation, Phys Med Biol, 35, 53–66.
Brooks R.A., Battocletti J.H., Sances A., *et al*, 1975, Nuclear magnetic resonance in blood, IEEE Trans Biomed Eng, BME-22, 12–18.
Bydder G.M., Steiner R.E., Young I.R., *et al*, 1982, Clinical NMR imaging of the brain: 140 cases, Am J Roentgenol, 139, 215–236.
Chu S.C.-K., Xu Y., Balschi J.A. and Springer C.S., 1990, Bulk magnetic susceptibility shifts in NMR studies of compartmentalized samples: use of paramagnetic agents, Magn Reson Imaging, 13, 239–262.
Coles B.A., 1976, Dual-frequency proton spin relaxation measurements on tissues from normal and tumor-bearing mice, J Natl Cancer Inst, 57, 389–393.

Cottam G.L., Vasek A. and Lusted D., 1972, Water proton relaxation rates in various tissues, Res Commun Chem Pathol Pharmacol, 4, 495–502.

Damadian R., 1970, Tumor detection by nuclear magnetic resonance, Science, 171, 1151–1153.

Damadian R. and Cope F.W., 1974, NMR in cancer. V. Electronic diagnosis of cancer by potassium (^{39}K) nuclear magnetic resonance: Spin signatures and T_1 beat patterns, Physiol Chem Phys, 6, 309–322.

Damadian R., Zaner K., Hor D. and DiMaio T., 1974, Human tumors detected by nuclear magnetic resonance, Proc Nat Acad Sci USA, 71, 1471–1473.

De Certaines J., Bernard A.M., Benoist L., *et al*, 1981, Nuclear magnetic resonance study of cancer: systemic effect on the proton relaxation times (T_1 and T_2) of human serum, Cancer Detect Prevent, 4, 267–271.

De Luca F., Maraviglia B. and Mercurio A., 1987, Biological tissue simulation and standard testing material for MRI, Magn Reson Med, 4, 189–192.

Dockery S.E., Suddarth S.A. and Johnson G.A., 1989, Relaxation measurements at 300 MHz using MR microscopy, Magn Reson Med, 11, 182–192.

Dooms G.C., Hricak H., Crooks L.E. and Higgins C.B., 1984, Magnetic resonance imaging of the lymph nodes: comparison with CT, Radiology, 153, 719–728.

Dooms G.C., Fisher M.R., Hricak H., *et al*, 1985, Bone marrow imaging: magnetic resonance studies related to age and sex, Radiology, 155, 429–432.

Doyle F.H., Pennock J.M., Banks L.M., *et al*, 1982, Nuclear magnetic resonance imaging of the liver: initial experience, Am J Roentgenol, 138, 193–200.

Eggleston J.C., Saryan L.A. and Hollis D.P., 1975, Nuclear magnetic resonance investigations of human neoplastic and abnormal nonneoplastic tissues, 1975, Cancer Res, 35, 1326–1332.

Ehman R.L., McNamara M.T., Pallack M., Hricak H. and Higgins C.B., 1984, Magnetic resonance imaging with respiratory gating: techniques and advantages, Am J Roentgenol, 143, 1175–1182.

Ehman R.L., Kjos B.O., Hricak H., *et al*, 1985, Relative intensity of abdominal organs in MR images, J Comput Assist Tomog, 9, 315–319.

Ekstrand K.E., Dixon R.L., Raben M. and Ferree C.R., 1977, Proton NMR relaxation times in the peripheral blood of cancer patients, Phys Med Biol, 22, 925–931.

Escayne J.M., Canet D. and Robert J., 1982, Frequency dependence of water proton longitudinal nuclear magnetic relaxation times in mouse tissues at 20°C, Biochim Biophys Acta, 721, 305–311.

Evelhoch J.L., Ewy C.S., Siegfried B.A., *et al*, 1985, ^{31}P spin-lattice relaxation times and resonant linewidths of rat tissue *in vivo*: dependence upon the static magnetic field strength, Magn Reson Med, 2, 410–417.

Fischer H.W., Van Haverbeke Y., Rinck P.A., Schmitz-Feuerhake I. and Muller R.N., 1989a, The effect of aging and storage conditions on excised tissues as monitored by longitudinal relaxation dispersion profiles, Magn Reson Med, 9, 315–324.

Fischer H.W., Van Haverbeke Y., Schmitz-Feuerhake I. and Muller R.N., 1989b, The uncommon longitudinal relaxation dispersion of human brain white matter, Magn Reson Med, 9, 441-446.

Frey H.E., Knispel R.R., Kruuv J., *et al*, 1972, Proton spin-lattice relaxation studies of nonmalignant tissues of tumorous mice, J Natl Cancer Inst, 49, 903-906.

Fruchter R.G., Goldsmith M., Boyce J.G., *et al*, 1978, Nuclear magnetic resonance properties of gynecological tissues, Gynecol Oncol, 6, 243-255.

Fullerton G.D., Potter J.L. and Dornbluth N.C., 1982, NMR relaxation of protons in tissues and other macromolecular water solutions, Magn Reson Imaging, 1, 209-228.

Fullerton G.D., Cameron I.L. and Ord V.A., 1984, Frequency dependence of magnetic resonance spin-lattice relaxation of protons in biological materials, Radiology, 151, 135-138.

Fullerton G.D., Cameron I.L. and Ord V.A., 1985, Orientation of tendons in the magnetic field and its effect in T2 relaxation times, Radiology, 155, 433-435.

Funduk N., Kydon D.W., Schreiner L.J., *et al*, 1984, Composition and relaxation of the proton magnetization of human enamel and its contribution to the tooth NMR image, Magn Reson Med, 1, 66-75.

Fung B.M., 1977, Proton and deuteron relaxation of muscle over wide ranges of resonance frequencies, Biophys J, 18, 235-239.

Fung B.M., Durham D.L. and Wassil D.A., 1975, The state of water in biological systems as studied by proton and deuterium relaxation, Biochim Biophys Acta, 399, 191-202.

Goldsmith M. and Damadian R., 1975, NMR in cancer VII. Sodium-23 magnetic resonance of normal and cancerous tissues, Physiol Chem Phys, 7, 263-269.

Goldsmith M., Koutcher J. and Damadian R., 1978a, NMR in cancer XI. Application of the NMR malignancy index to human gastro-intestinal tumors, Cancer, 41, 183-191.

Goldsmith M., Koutcher J.A. and Damadian R., 1978b, NMR in cancer, XIII: Application of the NMR malignancy index to human mammary tumours, Br J Cancer, 38, 547-554.

Gore J.C., Doyle F.H. and Pennock J.M., 1983, Relaxation rate enhancement observed *in vivo* by NMR imaging. In *Nuclear Magnetic Resonance Imaging*, C.L. Partain, A.E. James, F.D. Rollo and R.R. Price (eds), Saunders, Philadelphia, pp.94-106.

Graf V., Noack F. and Béné G.J., 1980, Proton spin T_1 relaxation dispersion in liquid H_2O by slow proton-exchange, J Chem Phys, 72, 861-863.

Grodd von W. and Schmitt W.G.H., 1983, Protonenrelaxationsverhalten menschlicher und tierischer Gewebe in vitro, Änderungen bei Autolyse und Fixierung, Fortschr Röntgenstr, 139, 233-240.

Hayman L.A., Ford J.J., Taber K.H., *et al* 1988, T2 effect of hemoglobin concentration: assessment with *in vitro* MR spectroscopy, Radiology 168, 489-491.

Higgins C.B., Herfkens R., Lipton M.J., *et al*, 1983, Nuclear magnetic resonance imaging of acute myocardial infarction in dogs: alterations in magnetic relaxation times, Am J Cardiol, 52, 189-195.

Hindman J.C., Svirmickas A. and Wood M., 1973, Relaxation processes in water. A study of the proton spin-lattice relaxation time, J Chem Phys, 59, 1517-1522.

Hollis D.P., Economou J.S., Parks L.C., *et al*, 1973, Nuclear magnetic resonance studies of several experimental and human malignant tumors, Cancer Res, 33, 2156-2160.

Hopkins A.L., Yeung H.N. and Bratton C.B., 1986, Multiple field strength *in vivo* T_1 and T_2 for cerebrospinal fluid protons, Magn Reson Med, 3, 303-311.

Hutchinson J.M.S. and Smith F.W., 1983, NMR clinical results: Aberdeen. In *Nuclear Magnetic Resonance Imaging*, C.L. Partain, A.E. James, F.D. Rollo and R.R. Price (eds), Saunders, Philadelphia, pp.231-249.

Huggert A. and Odeblad E., 1959, Proton magnetic resonance studies of some tissues and fluids of the eye, Acta Radiol, 51, 385-392.

Inch W.R., McCredie J.A., Knispel R.R, Thompson R.T. and Pintar M.M., 1974, Water content and proton spin relaxation time for neoplastic and non-neoplastic tissues from mice and humans, J Natl Cancer Inst, 52, 353-356.

Johnson G.A., Herfkens R.J. and Brown M.A., 1985, Tissue relaxation time: *in vivo* field dependence, Radiology, 156, 805-810.

Johnson M.A., Pennock J.M., Bydder G.M., *et al*, 1983, Clinical NMR imaging of the brain in children: normal and neurologic disease, Am J Roentgenol, 141, 1005-1018.

Just M. and Thelen M., 1988, Tissue characterization with T1, T2, and proton density values: results in 160 patients with brain tumors, Radiology, 169, 779-785.

Kamman R.L., Go K.G., Brouwer W. and Berendsen H.J.C., 1988, Nuclear magnetic resonance relaxation in experimental brain edema: effects of water concentration, protein concentration, and temperature, Magn Reson Med, 6, 265-274.

Kasturi S.R., Ranade S.S. and Shah S.S., 1976, Tissue hydration of malignant and uninvolved human tissues and its relevance to proton spin-lattice relaxation mechanism, Proc Indian Acad Sci, 84 B, 60-74.

Kiricuta I-C. and Simplăceanu V., 1975, Tissue water content and nuclear magnetic resonance in normal and tumor tissues, Cancer Res, 35, 1164-1167.

Kjos B.O., Ehman R.L., Brant-Zawadzki M., *et al*, 1985, Reproducibility of relaxation times and spin density calculated from routine MR imaging sequences: clinical study of the CNS, Am J Roentgenol, 144, 1165-1170.

Koenig S.H. and Brown R.D., 1984, Determinants of proton relaxation rates in tissue, Magn Reson Med, 1, 437-449.

Koenig S.H., Brown R.D., Adams D., Emerson D. and Harrison C.G., 1984, Magnetic field dependence of $1/T_1$ of protons in tissue, Invest Radiol, 19, 76-81.

Koivula A., Suominen K., Timonen T. and Kiviniitty K., 1982, The spin-lattice relaxation time in the blood of healthy subjects and patients with malignant blood disease, Phys Med Biol, 27, 937-947.

Koutcher J.A., Goldsmith M. and Damadian R., 1978, NMR in cancer X. A malignancy index to descriminate normal from cancerous tissue, Cancer, 41, 174–182.

Kroeker R.M., McVeigh E.R., Hardy P., Bronskill M.J. and Henkelman R.M., 1985, *In vivo* measurements of NMR relaxation times, Magn Reson Med, 2, 1–13.

Krynicki K., 1966, Proton spin–lattice relaxation in pure water between 0°C and 100°C, Physica, 32, 167–178.

Kundel H.L., Schlakman B., Joseph P.M., Fishman J.E. and Summers R., 1986, Water content and NMR relaxation time gradients in the rabbit kidney, Invest Radiol, 21, 12–17.

LeBlanc A., Evans H., Schonfeld E., *et al*, 1986, Relaxation times of normal and atrophied muscle, Med Phys, 13, 514–517.

Lerski R.A., McRobbie D.W., Staughan K., *et al*, 1988, V. Multi–center trial with protocols and prototype test objects for the assessment on MRI equipment, Magn Reson Imaging, 6, 201–214.

Lewa C.J. and Majewska Z, 1980, Temperature relationships of proton spin–lattice relaxation time T_1 in biological tissues, Bull Cancer, 67, 525–530.

Ling C.R., Foster M.A. and Hutchinson J.M.S., 1980, Comparison of NMR water proton T_1 relaxation times of rabbit tissues at 24 MHz and 2.5 MHz, Phys Med Biol, 25, 748–751.

Luyten P.R., Anderson C.M. and den Hollander J.A., 1987, ^{1}H NMR relaxation measurements of human tissues *in situ* by spatially resolved spectroscopy, Mag Reson Med, 4, 431–440.

Mathur–De Vré R., 1984, Biomedical implications of the relaxation behaviour of water related to NMR imaging, Br J Radiol, 57, 955–976.

Mathur–De Vré R., Grimée R. and Rosa M.P., 1983, Experimental protocol for tissue discrimination *in vitro* by n.m.r., Bioscience Rep, 3, 599–608.

Mathur–De Vré R., Grimee R., Parmentier F. and Binet J., 1985, The use of agar gel as a basic reference material for calibrating relaxation times and imaging parameters, Magn Reson Med, 2, 176–179.

Mathur–De Vré R., Binet J., Bovée W.M.M.J. and Foster M.A., 1988, III. Multi–center trial with an *in vitro* NMR protocol, Magn Reson Imaging, 6, 185–194.

Mallard J., Hutchinson J.M.S., Edelstein W., Ling R. and Foster M., 1979, Imaging by nuclear magnetic resonance and its bio–medical implications, J Biomed Eng, 1, 153–160.

Mano I., Goshima H., Nambu M. and Iio M., 1986, New polyvinyl alcohol gel material for MRI phantoms, Magn Reson Med, 3, 921–926.

Mansfield P. and Morris P.G., 1982, *NMR Imaging in Biomedicine*, Academic Press, New York, pp.10–32.

Medina D., Hazlewood C.F., Cleveland G.G, *et al*, 1975, Nuclear magnetic resonance studies on human breast dysplasias and neoplasms, J Natl Cancer Inst, 54, 813–818.

Moon K.L., Hricak H., Crooks L.E., *et al*, 1983a, Nuclear magnetic resonance imaging of the adrenal gland: a preliminary report, Radiology, 147, 155–160.

Moon K.L., Genant H.K., Helms C.A., *et al*, 1983b, Musculoskeletal applications of nuclear magnetic resonance, Radiology, 147, 161–171.

Moon K.L., Davis P.L., Kaufman L., *et al*, 1983c, Nuclear magnetic resonance imaging of a fibrosarcoma tumor implanted in the rat, Radiology, 148, 177–181.

Neville M.C., Paterson C.A., Rae J.L. and Woessner D.E., 1974, Nuclear magnetic resonance studies and water "ordering" in the crystalline lens, Science, 185, 1072–1074.

Nyman R., Ericsson A., Hemmingsson A., *et al*, 1986, T_1, T_2, and relative proton density at 0.35 T for spleen, liver, adipose tissue, and vertebral body: normal values, Magn Reson Med, 3, 901–910.

Odeblad E. and Lindstrom G., 1955, Some preliminary observations on the PMR in biological samples, Acta Radiol, 43, 469–476.

Parrish R.G., Kurland R.J., Janese W.W. and Bakay L., 1974, Proton relaxation rates of water in brain and brain tumors, Science, 183, 438–439.

Polak J.F., Jolensz F.A. and Adams D.F., 1988, NMR of skeletal muscle Differences in relaxation parameters related to extracellular/intracellular fluid spaces, Invest Radiol, 23, 107–112.

Raeymaekers H.H., Borghys D. and Eisendrath H., 1988, Determinants of water proton T_1 in blood serum, Magn Reson Med, 6, 212–216.

Ranade S.S., Shah S., Advani S.H. and Kasturi S.R., 1977, Pulsed nuclear magnetic resonance studies of human bone marrow, Physiol Chem Phys, 9, 297–299.

Rinck P.A., Meindl S., Higer H.P., Bieler E.U. and Pfannenstiel P., 1985, Brain tumors: detection and typing by use of CPMG sequences and *in vivo* T2 measurements, Radiology, 157, 103–106.

Ross R.J., Thompson J.S., Kim K. and Bailey R.A., 1982, Nuclear magnetic resonance imaging and evaluation of human breast tissue: preliminary clinical trials, Radiology, 143, 195–205.

Rupp N., Reiser M. and Stetter E., 1983, The diagnostic value of morphology and relaxation times in NMR–imaging of the body, Eur J Radiol, 3, 68–76.

Rustgi S.N., Peemoeller H., Thompson R.T., Kydon D.W. and Pintar M.M., 1978, A study of molecular dynamics and freezing phase transition in tissues by proton spin relaxation, Biophys J, 22, 439–452.

Schuhmacher J.H., Clorius J.H., Semmler W., *et al*, 1987, NMR relaxation times T_1 and T_2 of water in plasma from patients with lung carcinoma: correlation of T_2 with blood sedimentation rate, Magn Reson Med, 5, 537–547.

Shah S.S., Ranade S.S., Phadke R.S. and Kasturi S.R., 1982, Significance of water proton spin–lattice relaxation times in normal and malignant tissues and their subcellular fractions–I, Magn Reson Imaging, 1, 91–104.

Shinar H. and Navon G., 1984, NMR relaxation studies of intracellular Na^+ in red blood cells, Biophys Chem, 20, 275–283.

Shinar H. and Navon G., 1986, Sodium–23 NMR relaxation times in body fluids, Magn Reson Med, 3, 927–934.

Small W.C., McSweeney M.B., Goldstein J.H., Sewell C.W. and Powell R.W., 1983, Handling of *in vitro* human breast tissue samples: protocol requirements for accurate NMR relaxation measurements, Biochem Biophys Res Commun, 112, 991–999.

Smith F.W., 1983, The value of NMR imaging in pediatric practice: a preliminary report, Pediatr Radiol, 13, 141–147.

Smith F.W., Mallard J.R., Reid A. and Hutchinson J.M.S., 1981, Nuclear magnetic resonance tomographyic imaging in liver disease, Lancet i, 963–966.

Smith F.W., Reid A., Hutchinson J.M.S., *et al*, 1982, Nuclear magnetic resonance imaging of the pancreas, Radiology, 142, 677–680.

Taylor D.G. and Bore C.F., 1981, A review of the magnetic resonance response of biological tissue and its applicability to the diagnosis of cancer by NMR radiology, CT: J Comput Tomog, 5, 122–133.

Tennvall J., Olsson M., Möller T., *et al*, 1987, Thyroid tissue characterization by proton magnetic resonance relaxation time determination, Acta Oncol, 26, 27–32.

Thickman D.I., Kundel H.L. and Wolf G., 1983, Nuclear magnetic resonance characteristics of fresh and fixed tissue: the effect of elapsed time, Radiology, 148, 183–185.

Thompson B.C., Waterman M.R. and Cottam G.L., 1975, Evaluation of the water environments in deoxygenated sickle cells by longitudinal and transverse water proton relaxation rates, Arch Biochem Biophys, 166, 193–200.

Thulborn K.R., Waterton J.C., Matthews P.M. and Radda G.K., 1982, Oxygenation dependence of the tranverse relaxation time of water protons in whole blood at high field, Biochim Biophys Acta, 714, 265–270.

Walker P.M., Balmer C., Ablett S. and Lerski R.A., 1989, A test material for tissue characterisation and system calibration in MRI, Phys Med Biol, 34, 5–22.

Wehrli F.W., MacFall J.R. and Glover G.H., 1983, The dependence of nuclear magnetic resonance (NMR) image contrast on intrinsic and operator-selectable parameters, Proc SPIE, 419, 256–264.

Winter F. and Kimmich R., 1982, NMR field-cycling relaxation spectroscopy of bovine serum albumin, muscle tissue, *micrococcus luteus* and yeast. $^{14}N^{1}H$-quadrupole dips, Biochim Biophys Acta, 719, 292–298.

Wolf G.L. and Conard B., 1983, NMR proton T_1 and T_2 relaxation times from fresh, *in vitro* canine tissues at 5.1 MHz, Physiol Chem Phys Med NMR, 15, 19–22.

Zaner K.S. and Damadian R., 1975, Phosphorus-31 as a nuclear probe for malignant tumors, Science, 189, 729–731.

Chapter 9

Tissue Composition

The variation of the physical properties between tissues and their variability within any particular organ or tissue type may, for many properties, be directly related to the composition of the tissue. The relevant factor may be the biochemical composition, for instance proportions of lipid, protein and water in the tissue, or it may be its composition at an elemental level, or a combination of both. A wide survey of the composition of tissues was reported by the ICRP Task Group on Reference Man (ICRP, 1975), containing a substantial review of the then available compositional data on human tissue, together with much further developmental, physiological and other information. Whilst these data were primarily collected for radiation dosimetry purposes, they have a broader value since the composition of tissue in some way controls all its physical properties.

9.1 Biochemical composition

The composition of normal, adult human tissues in terms of the proportions of water, lipid, protein, ash and carbohydrate have been evaluated and reviewed thoroughly (Kim, 1974; ICRP, 1975). These data have been reassessed by Woodard and White (1986) and White *et al* (1987). Tissue compositions based on these reviews have been compiled in Table 9.1. Woodard and White listed the composition of fifty–six body tissues, and for seven of them (adipose tissue, heart, kidney, liver, mammary gland, skeletal muscle and skin) reviewed sufficient data to allow valid statements to be made about the spread of their composition. The original references have not been cited here; readers are referred to these reviews for details.

9.1.1 *Water content*

Water constitutes the greatest fractional part of most tissues, and for many of the physical properties of tissue the presence and proportion of this water,

Table 9.1 Percentage constituents of adult human tissues

Tissue	Water,%	Ash,%	Lipid,%	Protein,%
Whole body	60	4.8–5.8	19 (5.3)	15–30
Fluids				
Blood,whole	79.0–80.8	1.0	0.65	18–19
Erythrocytes	64	1.1	0.5	34.7
Plasma	92–95	0.95 (94)	0.45–1.26 (0.74)	6.5–7.2
Bile,liver	97	...	0.3	0.3
–,gallbladder	86–88	1.0	0.9–1.9	0.3–0.5
Breast milk	82.5–89.7	0.16–0.27 (0.2)	1.3–8.3 (4.5)	0.73–2 (1.1)
Cerebrospinal fluid	99	0.85–1.7 (1.08)	...	.012–.043 (0.028)
Gastric juice	99.4–99.5	...	trace	0.22–0.34 (0.28)
Saliva	99.4	0.3–0.8 (0.6)	...	0.14–0.64
Semen	89.1–94.4 (91.8)	...	0.17–0.21 (0.19)	3.3–6.9 (4.5)
Sweat	99–99.5	...	...	...
Synovial fluid	96–98.8	0.12–0.48 (0.34)	...	0.4–3.1 (0.72)
Urine	96	0.9		...
Soft tissues				
Adrenal gland	58.1	0.27–0.8 (0.46)	5.1–69 (26)	15.5
Blood vessels				
Arteries	53–78 (70)	0.65–3.1* (1.4)	1.54–1.9*	27–23*
Vein	72.6	0.59	...	24
Brain	76.3–78.5 (77.4)	1.4–2 (1.5)	9–17	8–12
–,grey matter	83–86	1.5	5.3	8–12
–,white matter	68–77	1.4	18	11–12
Eye,aqueous humor	98.1	0.85–0.94		$(5–16)x10^{-3}$
–,cornea	75–80			20
–,lens	64.1	0.07–0.73	1.7–2.3	33.6
–,sclera(animal)	68–75	...	0.62	26–27
–,vitreous humor	99	...	...	$(11–16)x10^{-3}$
Fatty,adipose	[11.4–30.5] (21.2)	0.3	[61.4–87.3] (71.4)	[1.0–7.9] (4.4)
Gall bladder	64,81.5	0.65	...	...
Intestine	77.4–82.2 (79)	0.4–1.3 (0.8)	1.3–9.2 (6.2)	10.9–14.9 (13)

cont.

Table 9.1 cont. Tissue constituents

Tissue	Water,%	Ash,%	Lipid,%	Protein,%
Kidney	[72.3–80.5] (76.6)	0.9	[2.8–6.9] (4.8)	[15.8–19.9] (17.7)
Larynx	68	1.4–5.6(3.0)		...
Liver	[72.8–75.6] (74.5)	1.2	[1.5–7.8] (4.6)	[16.1–19.6] (17.6)
Lungs	71.8–84	0.98–1.3 (1.1)	1.0–1.5	16.4–19.2
Mammary gland	[30.2–72.6] (51.4)	0.4	[5.6–56.2] (30.9)	[13.3–21.5] (17.4)
Muscle,cardiac	[71.0–80.9] (75.9)	0.9	[2.4–10.0] (6.2)	[15.9–18.2] (17.1)
–,skeletal	[70.0–78.6] (74.1)	1.0	[1.6–6.8] (4.2)	[17.9–21.3] (19.8)
Oesophagus	76	0.5–1.1 (0.9)	...	...
Ovary	83	0.8–1.4 (0.97)	1.3–2.3 (1.6)	14
Pancreas	66.7–73.3 (71)	0.7–1.5 (1.2)	2.9–20.4 (8.0)	13.1
Placenta	84.6±.18	1.0±.1	0.11±.02	12±.12
Prostate gland	83.3	0.68–1.52 (1.1)	1.2	15.0
Skin	[58.6–72.1] (65.3)	0.7	[5.2–13.5] (9.4)	[22.0–27.2] (24.6)
Horny layer	~10			
Spinal chord, nerve	63–75 (71)	1.4	2.3–18.5	9
Spleen	72–79 (77)	0.85–3 (1.6)	1–1.8 (1.4)	19–20
Stomach	60–78 (75)	0.8	6.2	17.0
Tendon,ox	63	37–42	1	35–40
Testis	82.7	0.9–1.3 (1.1)	4.5	12.0
Thymus	81,82	0.77	...	...
Thyroid gland	72–78	0.81–1.8 (1.1)	4.4	14±1.6
Tongue	60–72	1	15–24	16–18
Trachea	60	1.2–2.9 (1.6)	...	...
Urinary bladder	65	0.5–1.1(0.8)		...
Uterus	79	0.9–1.2 (1.0)	0.9–2.2 (1.4)	2

cont.

Table 9.1 cont.

Tissue	water,%	ash,%	lipid,%	protein,%
Skeleton and other hard tissues				
Bone,cortical	12–15	55–58	~1	25–26
–,trabecular	23	34	~1	~30
Cartilage	55–85 (72)	2.6–5.6	1.3	11
Red marrow	39.7	0.6	39.7	20
Yellow marrow	15.3	0.06–0.5	80.4	4.0
Hair	4–13	0.23–0.9	2.3	85–91
Nails	0.07–13.7	...	...	...
Teeth	4–14.3 (9.2)	68–80	...	18
Enamel	0.5–6.6 (2.8)	92–94		0.2–0.4
Dentin	4.2–16.7 (11.1)	64		15.5
Pulp	...	21.2	0.9	60–65

Sources: ICRP (1975); ICRU (1989); Woodard and White (1986); Diem and Lentner (1970).
* Age dependent.
The range in square brackets [] represents ±1 SD; otherwise approximate overall ranges are given.
Other constituents over 1%; condroitin sulphate 11%–bone; polysaccharide 2.2%–liver; monosaccharide 5.2%–cortical bone, 2.5%–thyroid; cholic acid 5.4% and mucin 4.1%–bile.

Table 9.2 Percentage water content of animal tissues

	Cow	Dog	Rabbit	Rat
Bone		45.5–57.6	39.2–58.1	34.0
Brain	77.9	74.5–84.6	78–85	78.0–83.2
Kidney	74.9	79.1–81.3 (80.2)	74.0–78.6	74.0–78.6
Liver	69.0	74.6	70–76 (73.0)	70.0–76.0
Lung	80.0	77.1–80.1 (78.6)	80.1–82.0	78.7–82.0
Muscle, cardiac	70.0	76.6–79.9 (78.3)	78.2	75.5–78.4
–,skeletal	70.0	68.0–87.3	77.0–80.0	75.5–78.4
Skin		44.2–82.3	54.0–75	56.7–64.7
Spleen		77.4	78.0	78.0
Testis	86.0	78.5–96.1 (87.3)	68.0–85.0	68.0–85.0

Source: Altman and Ditmer (1972).

whether it lies in the intracellular or extracellular space, and the proportion bound to the protein molecules are the primary factors determining the magnitude of the physical quantity. Water content for human tissue is included in Table 9.1. In order to assist in comparisons between the physical properties of tissues from different animal species, a summary of the water content of some animal tissues is also included (Table 9.2, Altman and Ditmer, 1972). Generally there is a substantial overlap between the normal spread in water content for any particular animal tissue and that for the equivalent human tissue.

The water fraction is higher in fetal tissue than in infant or adult tissue (ICRP, 1975). It decreases from about 95% in early pregnancy to about 70% at birth (Table 9.3). The ICRP report also includes an empirical equation relating total body water (TBW) to fetal weight, W, in kg:

$$\text{TBW}(\%) = 90.394 - 8.1011W + 0.7936W^2 \qquad (9.1)$$

Table 9.3 Total body water for human fetus

Fetal weight,g	% water
0.5	93–95
10	92–94
100	90
200	88
1000	82–84
3450	70–72

Source: ICRP (1975)

For adults the following equations relate total body water in litres to body weight in kilograms and age in years, A:

Male $$\frac{\text{TBW}}{W} \times 100 = 79.45 - 0.24W - 0.15A \qquad (9.2)$$

Female $$\frac{\text{TBW}}{W} \times 100 = 69.81 - 0.26W - 0.12A \qquad (9.3)$$

Extracellular water (ECW) in the fetus reduces from about 75g per 100g fat-free body at 10 weeks to about 46g per 100g at term.

The following relationships are given for ECW in litres for infants, and for children and adults depending on sex:

Infant $$ECW = 0.0239W + 0.325 \quad SE \pm 0.181 \tag{9.4}$$

Children

male $$ECW = 0.227W + 0.916 \quad SD \pm 0.621 \tag{9.5}$$

female $$ECW = 0.211W + 0.989 \quad SD \pm 0.891 \tag{9.6}$$

Adult

male $$ECW = 0.135W + 7.35 \quad 95\% \text{ range } \pm 3.26 \text{ l} \tag{9.7}$$

female $$ECW = 0.135W + 5.27 \quad 95\% \text{ range } \pm 1.39 \text{ l} \tag{9.8}$$

Intracellular water (ICW) in the fetus lies in the range 17% at 10 weeks gestation to 38% at term. For adults, ICW as a percentage of TBW is given by the following equations:

Male $$\frac{ICW}{TBW} \times 100 = 55.3 - 0.07A \tag{9.9}$$

Female $$\frac{ICW}{TBW} \times 100 = 62.3 - 0.16A \tag{9.10}$$

9.1.2 *Fat content and age*

The percentage of body fat increases with age. For males the average percentage fat increases from 11% at 20 years to 31% at 70 years. For females the increase is from 29% at 20 years to 45% at 70 years.

Table 9.4 Average elemental composition of body tissues

Component	Composition,% by mass					
	H	C	N	O	S	P
Carbohydrate,chondroitin sulphate	4.6	36.6	3.0	48.8	7.0	
–,monosaccharide	6.7	40.0		53.3		
–,polysaccharide	6.2	44.5		49.3		
Cholic acid	9.9	70.5		19.6		
Lipid,cerebrocide	9.8	63.5	3.0	23.7		
–,cholesterol	12.0	83.9		4.1		
–,glycerol trioleate	11.8	77.4		10.8		
–,sphingomyelin	11.7	68.9	3.9	11.2		4.3
Mucin	4.8	34.3		60.9		
Protein	6.6	53.4	17.0	22.0	1.0	
Urea	6.7	20.0	46.7	26.6		
Water	11.2			88.8		

Source: Woodard and White (1986).

Table 9.5 Average elemental composition of tissues; percentage by mass.

Tissue	H	C	N	O	Na	Mg	P	S	Cl	K	Ca	Fe
Adrenal gland	10.6	28.4	2.6	57.8	0.1		0.2	0.2	0.1			
Bile,gallbladder	10.8	6.1	0.1	82.2	0.4				0.4			
Bladder,urinary	10.5	9.6	2.6	76.1	0.2		0.2	0.2	0.3	0.3		
Blood,whole	10.2	11.0	3.3	74.5	0.1		0.1	0.2	0.3	0.2		0.1
–,erythrocytes	9.5	19.0	5.9	64.6			0.1	0.3	0.2	0.3		0.1
–,plasma; lymph	10.8	4.1	1.1	83.2	0.3			0.1	0.4			
Blood vessels	9.9	14.7	4.2	69.8	0.2		0.4	0.3		0.1	0.4	
Bone,cortical	3.4	15.5	4.2	43.5	0.1	0.2	10.3	0.3			22.5	
–,–,2–5yrs	4.0	15.7	4.5	45.4		0.2	10.1				20.1	
–,spongiosa	8.5	40.4	2.8	36.7	0.1	0.1	3.4	0.2	0.2	0.1	7.4	0.1
–,cranium	5.0	21.2	4.0	43.5	0.1	0.2	8.1	0.3			17.6	
–,vertebral column	7.1	25.8	3.6	47.2	0.1	0.1	5.1	0.3	0.1	0.1	10.5	
–,femur,total bone	6.3	33.3	2.9	36.2	0.1	0.1	6.6	0.2			14.3	
–,femoral head	7.1	37.9	2.6	34.2	0.1	0.1	5.6	0.2			12.2	
Bone marrow, red	10.5	41.4	3.4	43.9	0.1		0.2	0.2	0.2		0.1	
–, yellow	11.5	64.4	0.7	23.1	0.1			0.1	0.1			
Brain,grey matter	10.7	9.5	1.8	76.7	0.2		0.3	0.2	0.3	0.3		
–,white matter	10.6	19.4	2.5	66.1	0.2		0.4	0.2	0.3	0.3		
Breast	[10.9–10.2]	[50.6–15.8]	[2.3–3.7]	[35.8–69.8]	0.1		0.1	0.2	0.1			
	10.6	33.2	3.0	52.7								
Cartilage	9.6	9.9	2.2	74.4	0.5		2.2	0.9	0.3			
Cerebrospinal fluid	11.1	..	..	88.0	0.5				0.4			
Connective tissue	9.4	20.7	6.2	62.2	0.6			0.6	0.3			
Eyes	10.7	6.9	1.7	80.3			0.1	0.1		0.2		
Eye lens	9.6	19.5	5.7	64.6	0.1		0.1	0.3	0.1			

cont.

Table 9.5 cont.

Tissue	H	C	N	O	Na	Mg	P	Cl	K	Ca	Fe
Fatty (adipose)	[11.2–11.6]	[51.7–68.1]	[1.3–0.2]	[35.5–19.8]	0.1		0.1	0.1			
	11.4	59.8	0.7	27.8							
Heart,blood filled	10.3	12.1	3.2	73.4	0.1	0.1	0.2	0.3	0.2		0.1
Intestine,small	10.6	11.5	2.2	75.1	0.1	0.1	0.1	0.2	0.1		
–,contents	10.0	22.2	2.2	64.4	0.1	0.1	0.3	0.1	0.4	0.1	
Kidney	[10.2–10.4]	[16.0–10.6]	[3.4–2.7]	[69.3–75.2]	0.2	0.2	0.2	0.2	0.1		
	10.3	13.2	3.0	72.4							
Liver	[10.3–10.1]	[15.6–12.6]	[2.7–3.3]	[70.1–72.7]	0.2	0.3	0.3	0.2	0.3		
	10.2	13.9	3.0	71.6							
Lung	10.3	10.5	3.1	74.9	0.2	0.2	0.3	0.3	0.2		
Muscle,cardiac	[10.3–10.4]	[10.3–17.5]	[2.7–3.1]	[68.1–75.6]	0.1	0.2	0.2	0.2	0.3		
	10.4	13.9	2.9	71.8							
Muscle,skeletal	[10.1–10.2]	[17.1–11.2]	[3.6–3.0]	[68.1–74.5]	0.1	0.2	0.3	0.1	0.4		
	10.2	14.3	3.4	71.0							
Ovary	10.5	9.3	2.4	76.8	0.2	0.2	0.2	0.2	0.2		
Pancreas	10.6	16.9	2.2	69.4	0.2	0.2	0.1	0.2	0.2		
Skin	[10.0–10.1]	[25.0–15.8]	[4.6–3.7]	[59.4–69.5]	0.2	0.1	0.2	0.3	0.1		
	10.0	20.4	4.2	64.5							
Spinal chord	10.7	14.5	2.2	71.2	0.2	0.4	0.2	0.3	0.3		
Spleen	10.3	11.3	3.2	74.1	0.1	0.3	0.2	0.2	0.3		
Stomach	10.4	13.9	2.9	72.1	0.1	0.1	0.2	0.1	0.2		
Testis	10.6	9.9	2.0	76.6	0.2	0.1	0.2	0.2	0.2		
Thyroid gland	10.4	11.9	2.4	74.5	0.2	0.1	0.1	0.2	0.1	plus I,0.1%	
Trachea	10.1	13.9	3.3	71.3	0.1	0.4	0.4	0.1	0.4		
Urine	11.0	0.5	1.0	86.2	0.4	0.1		0.6	0.2		

Sources: ICRU (1989), Woodward and White (1986), White *et al* (1987).

9.2 Elemental composition

The percentage elemental composition by mass may be calculated from knowlege of the biochemical composition of tissue (Table 9.1) together with values of the percentage elemental composition of the constituent parts of tissue (Table 9.4). Estimated elemental compositions for a range of tissues are listed in Table 9.5. The values have been drawn from ICRP (1975), updated by Woodard and White (1986), together with some additional composition estimates for several whole bones given by White *et al* (1987). These latter values were selected from data for 24 different bone analyses listed by White *et al*. In addition, White gives the average elemental compositions for seven approximately homogenous groups of tissues including 'all soft tissues', 'soft tissues excluding adipose and yellow marrow' and tissues lying in particular bands of percentage water content.

For 'reference man', the average adult human body, ICRP (1975) lists the total mass and percentage of body weight for oxygen, carbon, hydrogen, nitrogen and 32 other elements (Table 9.6).

Table 9.6 Average elemental composition of total body for 'reference man'

Element	Amount g	Percentage of total body weight
Oxygen	43000	61
Carbon	16000	23
Hydrogen	7000	10
Nitrogen	1800	2.6
Calcium	1000	1.4
Phosphorus	780	1.1
Sulphur	140	0.20
Potassium	140	0.20
Sodium	100	0.14
Chlorine	95	0.12
Magnesium	19	0.027
Silicon	18	0.026
Iron	4.2	$6x10^{-3}$
Fluorine	2.6	$3.7x10^{-3}$
Zinc	2.3	$3.3x10^{-3}$
Rubidium	0.32	$4.6x10^{-4}$
Strontium	0.32	$4.6x10^{-4}$
Bromine	0.20	$2.9x10^{-4}$
Lead	0.12	$1.7x10^{-4}$
Copper	0.072	$1.0x10^{-4}$
Aluminium	0.061	$9x10^{-5}$
Cadmium	0.050	$7x10^{-5}$

cont.

Table 9.6 cont. Elemental composition

Element	Amount,g	% body weight
Boron	<0.048	$<7x10^{-5}$
Barium	0.022	$3x10^{-5}$
Tin	<0.017	$<2x10^{-5}$
Manganese	0.012	$2x10^{-5}$
Iodine	0.013	$2x10^{-5}$
Nickel	0.010	$1x10^{-5}$
Gold	<0.010	$<1x10^{-5}$
Molybdenum	$<9.3x10^{-3}$	$<1x10^{-5}$
Chromium	$<1.8x10^{-3}$	$<3x10^{-6}$
Caesium	$1.5x10^{-3}$	$2x10^{-6}$
Cobalt	$1.5x10^{-3}$	$2x10^{-6}$
Uranium	$9x10^{-5}$	$1x10^{-7}$
Beryllium	$3.6x10^{-5}$	
Radium	$3.1x10^{-11}$	

Source: ICRP (1975).

References

Altman P.L. and Ditmer D.S., 1972, *Biology Data Book*, 2nd edition, Federation of American Societies for Experimental Biology, pp.392–398.

Diem K. and Lentner C. (eds), 1970, *Documenta Geigy Scientific Tables*, 7th edition, Geigy, Macclesfield.

ICRP, 1975, *Report of the Task Group on Reference Man*, ICRP Publication 23, International Commission on Radiological Protection, Pergamon Press, Oxford.

ICRU, 1989, *Tissue Substitutes in Radiation Dosimetry and Measurement*, ICRU Report 44, International Commission on Radiation Units and Measurements, Bethesda, MD, USA.

Kim Y.S., 1974, Human tissues: Chemical composition and photon dosimetry data, Radiat Res, 57, 38–45. A correction, Radiat Res, 60, 361–362.

White D.R., Woodard H.Q. and Hammond S.M., 1987, Average soft-tissue and bone models for use in radiation dosimetry, Br J Radiol, 60, 907–913.

Woodard H.Q. and White D.R., 1986, The composition of body tissues, Br J Radiol, 59, 1209–1219.

Appendix A

Tissue Perfusion Rates

Values for tissue perfusion for adult humans in the resting state have been compiled from a wide literature base by Williams and Leggett (1989). Proposed reference values drawn from this survey are assembled in Table A.1. An approximate range is also given. This is the overall range from all the reported literature expressed as a percentage of the reference value for the particular organ. The perfusion rates are for an average, healthy 35-year-old. Williams and Leggett also discuss the variation of perfusion depending on sex; where clear sex differences occur these differences have been included in the table.

Table A.1 Resting human blood perfusion rates

Organ or tissue	Blood perfusion rate, ml kg^{-1} min^{-1}	range,%
Adrenal gland	2000	300
Brain	560	100
Fatty tissue	28	100
Intestine,small	1000	90
–,large,upper	800	70
–,–,lower	700	65
Kidney,male	4000	60
–,female	3500	60
Liver,male	1000	90
–,female	1100	90
Lung	400	270
Lymph node	500	
Muscle:cardiac,male	800	100
–,female	1000	100
Muscle:skeletal	38	240
Pancreas	600	130
Skeleton	30	210
Skin	120	100
Spleen	1200	115
Stomach and oesophagus	400	220
Thyroid gland	5000	130

Reference: Williams L.R. and Leggett R.W., 1989, Reference values for resting blood flow to organs of man, Clin Phys Physiol Meas, 10, 187–217.

Appendix B

pH

The values of pH for some normal human body fluids and faeces are given in Table B.1.

Table B.1 pH of human body fluids

Fluid	pH mean	range
Blood,arterial	7.424	7.386–7.462*
Blood:red cells	7.209	7.175–7.243*
Blood:plasma,arterial	7.39	7.35–7.43*
–,venous	7.398	7.378–7.418*
Bile,hepatic	7.5	6.2–8.5
–,gall bladder	6.0	5.6–8.0
Breast milk	7.01	6.4–7.6
Cerebrospinal fluid	7.349	7.327–7.371*
Faeces,adult	7.15	5.85–8.45*
–,infant	4.9	4.6–5.2
Gastric juice, male	1.92±1.28	
–,female	2.59±2.08	
–,child	3.27	0.9–7.7
–,newborn	2.52	1.2–7.4
Pancreatic juice		7.5–8.8
Saliva	6.4	5.8–7.1
Semen	7.19	6.9–7.36
Sweat		4.0–6.8
Synovial fluid	7.434	7.31–7.64
Urine	5.75	4.8–7.5

* 95% range.

Reference: Diem E. and Lentner C. (eds), 1975, *Documenta Geigy Scientific Tables,* 7th Edition, Geigy Pharmaceuticals, Macclesfield.

Index

Entries in normal type refer to the page number. Entries in **bold type** refer to the table number.

www.ingramcontent.com/pod-product-compliance
Ingram Content Group UK Ltd.
Pitfield, Milton Keynes, MK11 3LW, UK
UKHW051127260726
13967UKWH00010B/2905